GEOMETRICAL ANALYSIS.

Benjn Hallowell

GEOMETRICAL ANALYSIS,

OR THE

CONSTRUCTION AND SOLUTION OF VARIOUS GEOMETRICAL PROBLEMS

FROM ANALYSIS,

BY

GEOMETRY, ALGEBRA, AND THE DIFFERENTIAL CALCULUS;

ALSO,

THE GEOMETRICAL CONSTRUCTION OF ALGEBRAIC EQUATIONS,

AND

A MODE OF CONSTRUCTING CURVES OF THE HIGHER ORDER BY MEANS OF POINTS.

BY

BENJAMIN HALLOWELL,

FORMERLY PROPRIETOR AND PRINCIPAL OF THE ALEXANDRIA, VA., BOARDING-SCHOOL.

PHILADELPHIA:
J. B. LIPPINCOTT & CO.
1872.

Entered, according to Act of Congress, in the year 1871, by
J. B. LIPPINCOTT & CO.,
In the Office of the Librarian of Congress at Washington.

TO

SWARTHMORE COLLEGE,

including the Youthful Laborers of both sexes, its successive inmates, who are devoting themselves to the pursuit of a knowledge of the True, the Beautiful, and the Good in every Department of Science and Nature,

This Volume,

which is designed to assist in training and strengthening the Intellectual Faculties and thus securing the needed Discipline in the exercise of the fullest freedom in the whole range of human thought, which is their inherent privilege, and so essential to the progress of Truth, and the complete development of humanity,

Is Respectfully Inscribed

BY THE AUTHOR,

with ardent desires that the Blessing of the Good Providence may rest upon the Institution, and the highest hopes for its usefulness be fully realized.

BENJAMIN HALLOWELL.

SANDY SPRING, MD., 8mo. 17th, 1871.

TABLE OF CONTENTS.

INTRODUCTORY NOTE.

In an engagement of nearly forty years' duration in teaching Mathematics, first in the boarding-schools of Fair Hill, Maryland, and Westtown, Pennsylvania, and afterwards in a boarding-school of my own, for boys and young men, at Alexandria, Virginia, I became convinced that the analytic or algebraic method of Descartes, Delambre, and Laplace, while it is a most efficient instrument in the hands of a mathematician, is not so well adapted as the geometric or Greek method, to impart to the student a knowledge of mathematical principles, or to inspire such student with an affection and taste for the science.

The mind of the young is less capable than that of an older person of *abstract thought*, and it needs assistance to the concentration of its ideas, such as is afforded by a mathematical diagram; and with this aid continued for some time, the faculties of perception and conception become cultivated and strengthened, till the mind of the student can readily grasp, without a diagram, a problem of considerable intricacy, and be able to apply to it Descartes' method efficiently.

So entirely has the analytic method taken the place of the geometric in our prominent schools and colleges, that no work on pure Geometrical Analysis has, to my knowledge, ever been published in this country.

Thomas Simpson, F.R.S., and Professor of Mathematics in the Royal Academy of Woolwich, England, in his "Select Exercises for Young Proficients in the Mathematics," which was published in London in 1752, gave a number of geometrical problems, with the method of constructing them.

And, in a treatise on Geometry, by the same valued author, published in London in 1760, the fifth and sixth books are entirely devoted to the Construction and Demonstration of Geometrical Problems; and nearly fifty problems, of great variety, and some of them of much elegance and beauty, are constructed and demonstrated at the end of the volume.

To his Algebra, also, which was published in England about the same time, and reprinted in Philadelphia, by Mathew Carey, in 1809, "from the Eighth London Edition," Thomas Simpson added an "Appendix, containing the Construction of Geometrical Problems, with the Manner of resolving them numerically." This work was of very great value to the mathematical student, and of much service to my revered preceptor, John Gummere, in the preparation of his admirable treatise on Surveying, which was published a few years after Carey's edition of Simpson's Algebra was issued.

But in all these works, including the problems in Gummere's Surveying, the *mode of construction was arbitrarily given*, without the *least clue* to the line of thought that had led to it; and the thoughtful student would naturally inquire, and even *wonder*, in a problem like the seventy-sixth or seventy-seventh of the Appendix to Simpson's Algebra (Problems six and eight in "Algebraic Analysis" in this book), *how the author ever came to think of so complicated a method;* and he would feel discouraged, in view of its complexity, from an apprehension that his own mind would never be endowed with a penetration sufficient to accomplish a result of such intricacy and depth.

Whereas, had the problems been *analyzed*, and the student shown that it was all done by taking one step at a time, and that an easy one,—*gradually* acquiring a knowledge of what we *do not know* by means of what *we do know*, which Doctor Johnson says is the only way it can be done,—he would have felt increased confidence in his own powers, and been prepared to *make an effort himself*—which is a great point—in a similar direction.

In 1821, Professor John Leslie, of Edinburgh College, published, in Edinburgh, a very valuable treatise on "Geometrical Analysis, and Geometry of Curve Lines." The "Geometrical Analysis" had been previously published, annexed to his "Elements of Geometry," and it greatly "conspired to advance the study of Geometry, by reviving the fine models bequeathed by the Greeks." As has been said with great truth of this work of Professor Leslie, "the study of such a digest, appears admirably fitted to improve the intellect, by training it to habits of precision, arrangement, and close investigation. The spirit of Geometrical Analysis may be carried, with the happiest effect, into the domains of Inductive Philosophy, which are to be explored by a similar procedure."

But, about the time of the publication of this work of Professor Leslie, the algebraic method of Descartes and Leibnitz came rapidly into vogue in this country through the popularity of some eminent French writers upon Mathematics, and Professor Leslie's work was never, as far as I know, published in the United States, or, to much extent, placed within the reach of the American student.

The want that I experienced, in my engagement of teaching, of a suitable text-book upon Geometrical Analysis, induced me to make a collection of problems, for the benefit of my students, analyzing, constructing, demonstrating, and giving a method of calculation ;—solving in this way all I could meet with or frame myself, and difficult ones forwarded to me for solution, through a number of years, by former students, and many other persons, known and unknown, so that the number of such problems became considerable, from which those in the present volume have been selected, as best adapted to supply the want I had experienced.

These problems were given to the more advanced mathematical students as "extras," or "*trial questions;*" and in their solution, much ingenuity was frequently manifested, and also gratifying evidence of progress, in the skill with which the student would proceed in the different steps of the analysis; and it was from the solicitations of many of my former students who had become teachers, to publish the problems from which they had derived, as they believed, great benefit, that the preparation of the present work was undertaken.

The *practical teaching* of young persons consists of two parts: —instructing them *how to do* something; and *giving the reason* for doing it in that way. Now, it accords with reason and sound philosophy to do *one thing at a time.* Hence these two parts of teaching, as a general rule, should, with the young, be kept as separate as possible. Youth should first learn, *well,* the *practical* part,—*how to do; then the reason* In acquiring a knowledge of the primary rules of Arithmetic, for instance, to undertake to give a child the reason for every step he takes in the processes of subtraction, multiplication, division, etc. would but confuse him. It would be too complicated. He needs that the ideas he is to acquire shall be placed, with well-defined outline, in the simplest possible light, and one at a time, so that his mind can clearly and concentratedly comprehend the truth to be imparted. And when this is effected, the sparkling eye and animated countenance will attest the pleasurable sensations of the soul, from the conscious acquisition of a truth previously unknown to him.

The *practical part—how to do*—first; *then the why.* We must *know a fact,* before we *care about the reason for it.* And this is the natural mode of the mind's progress in knowledge. Children learn to *use* words, before they learn the *definitions* of them. They form phrases, before they are able to "construe" or "parse" them. And the more nearly a teacher keeps to this natural process, the more successful will he be in developing the minds of his students, and in pleasurably educating them.

Accordingly, whenever practicable, in teaching young persons, objects, models, maps, globes, diagrams, apparatus, specimens, and other means of illustration of the truths designed to be taught, should be employed, as an aid to assist in *concentrating their thoughts*, and imparting a *clear idea* of the subject. In children, the *perceptive* faculties are most prominent and active, and must principally be brought into requisition; and thus the reflective powers will gradually develop and strengthen, when there will be less need of this educational machinery.

Mathematics is an elevated study. Its constant aim is the *discovery* of truth. There is no *new* truth. Every fact in science, every mathematical principle, every property of the triangle or the circle, is *eternal*. No matter when or by whom it was *discovered*, it *pre-existed*, and is the living embodiment of a divine thought. All the mathematician or scientist does in this way is to *discover what was previously unknown*, although eternally existing. When Dr. Herschel and Prof. Leverrier discovered each a "new planet," as it was termed, in Uranus and Neptune, they only saw for the first time what had existed from the period of Creation. The *discovery* was new, not the *object discovered*. So of every truth, principle, or property throughout the whole range of mathematics and all science.

It seems remarkable, and, as I get older, the astonishment does not in the least abate, that the five plane figures formed by the cutting of a cone by a plane in different positions — the Triangle, the Circle, Parabola, Ellipse, and Hyperbola—should possess such a great number and diversity of singular and interesting properties, many of which are embodied in the following pages, and which, *although eternally existing*, have been gradually disclosing themselves to the patient research of mathematicians for over two thousand years, and we are by no means at liberty to suppose that all of them are yet known.

For instance, that in *any* triangle, of *whatever size* or *shape*, the three perpendiculars let fall from each angular point upon the opposite sides (the sides and the perpendiculars being produced if necessary) should all pass *through the same point*. Also, that the three lines *bisecting the angles* respectively, and the three lines drawn from the angular points to the middle point of the opposite sides respectively, should in each case *all pass through* the same point. (See Scholiums 1, 2, 3 to "Theorem" 6, in the following pages.)

Of the same singular character are the properties of the circle and triangle combined, in "Theorems" 11, 12, and 13 of this work, which were original discoveries with the author, although they may have been made by some one before me, whose writings I have not met with.

It is deemed proper, in this Introductory Note, to give some explanation of the arrangement of the work, and of the mode in which it is believed it may be used to the greatest advantage.

Each problem is first analyzed, then constructed, demonstrated, and the method of calculation by Plane Trigonometry clearly indicated. In many cases the limits are shown within which the problem is restricted.

The greatest number of problems are analyzed by pure Geometry. Where they are analyzed by Algebra, or the Differential Calculus, the constructions and demonstrations are by pure Geometry, and the method of calculation is given by Plane Trigonometry, showing the harmonious results by the different modes of investigation.

The references in the text (as III. 8* in the first problem of the circle) are to Davies's edition of Legendre's Geometry,—the III. 8 meaning the 8th Proposition of the Third Book of Legendre, and the * referring to a marginal note or foot-note where the Proposition is given in which the same property is demonstrated in Playfair's Euclid. In this example, the note at the bottom of the page is III. 14, meaning the 14th Proposition of the Third Book of Playfair's Euclid. So of other cases.

The design of the work is principally *as an aid* to professors and teachers, to supply a want which the author had severely experienced. The problems are distributed through the volume without much reference to their intricacy or difficulty of solution, in order that the instructor may have a wide field from which to select extra or trial problems, so as to give different problems to different classes in successive years.

The importance of the student's *constructing* the different problems given him for solution, by means of the Scale and Dividers, with the *greatest possible precision,* can scarcely be too much insisted upon, as a very improving engagement in training to accuracy, and one of the best preparations for draughting and other duties in the Coast

Survey or Civil Engineering; or for the business of a machinist, or any mechanical vocation requiring neatness and precision. If this constructing is done thoughtfully, the eye will soon become practiced in judging of distances and positions, so that drawing by hand can be more rapidly and accurately effected.

The solutions of the problems in relation to surfaces and solids, in the brief article on the Differential and Integral Calculus, and some other portions of that article, if read by a student or a class before entering upon, or while pursuing, that study, it is believed will be of material assistance in mastering the subject satisfactorily.

The same may be said of the article on Analytical Geometry. As an exercise or lesson preparatory to using a standard text-book upon the subject, it will be a great assistance, and all the problems there given should be carefully constructed, and the results measured with the Scale and Dividers.

Or, the teacher or professor may make the explanations, and give these or similar problems, *orally*, to his class. This method possesses great advantages. The students generally understand and retain what they hear from their instructor, better than what they read in their text-books. Besides, their respect and regard for their instructor are increased when the students discover that the contents of the text-book are not the limits of his knowledge of the subject they are studying, and see his willingness thus to render the knowledge he possesses advantageous to them.

In like manner, before entering upon the study of the Properties of the Parabola, in a treatise on Conic Sections, let the three pages on that Curve, in this volume, be read, or equivalent ideas be imparted by the instructor, the student constructing *accurately* the three problems given, or their equivalents proposed by the teacher or professor.

The same course is recommended in regard to the Ellipse and Hyperbola. The student will thus acquire a *practical* acquaintance with the Axes, Ordinates, Abscissas, Asymptotes, Opposite Hyperbolas, etc., which, as these will then be familiar to him, will prove of the greatest advantage in gaining a knowledge of the properties of the Curve, by leaving his mind more free to concentrate its powers upon the particular idea or truth to be acquired.

Also, in regard to the remaining Curves,—the Conchoid, Cissoid, Quadratrix, Spiral, etc.,—the student will find it to his decided advantage to *construct each Curve* by the rules herein given, *accurately* with the Scale and Dividers, previous to entering upon the study of its properties in his text-book.

A Table of Square Roots, carried to two and three places of decimals, of numbers from 1 to 200, is added at the close of the volume, to facilitate the construction of Curves from their Equations, by means of points.

The author does not expect or desire any *pecuniary* return from his work. It is a *labor of love for the youth* of our country, and of *interest* and *sympathy* for those to whom their education may be intrusted, in their most arduous and responsible engagement.

If the work shall prove of some service to the *student*, and lighten the labor of the *instructor*, the highest aim of the author will be reached. The more good it does, and the more service it performs, the more fully will his object be attained.

Should anything in this Introductory Note appear like *dictating* or *prescribing to teachers a particular mode of proceeding*, the author trusts it will be excused, and understood as only a *suggestion* from one who has now completed two more than his "threescore years and ten," and who, although he has not been practically engaged in that vocation for a number of years past, still feels a deep interest in the noble profession of Teaching, and those actively concerned in it, and who regards the preparation of this work, in which he has spent this, his seventy-second birthday, as the closing and crowning labor of his life in that direction.

BENJAMIN HALLOWELL.

SANDY SPRING, MARYLAND, 8mo. 17, 1871.

Postscript.—It seems only proper to add, that my son, HENRY C. HALLOWELL, read the whole work carefully in manuscript, corrected several clerical errors, and, in some instances, where there appeared to be abstruseness, suggested one or two intermediate steps in the process, in order that the idea might be more readily comprehended by the student.

Note.—For the gratification of his former students, who are widely scattered over our country, some of whom have not seen him for many years, a likeness of the author is prefixed to the volume, which is accompanied by his kindest remembrances.

EXPLANATION OF SOME SYMBOLS AND ABBREVIATIONS EMPLOYED.

$\angle$, angle.

$\angle$s, angles.

$\triangle$, triangle.

$\triangle$s, triangles.

$\perp$ or perp., perpendicular to, or at right angles to.

$\parallel$, parallel to.

$\therefore$, therefore.

Hyp., hypothenuse.

$<$ placed between two quantities, implies that the one which comes *before* the symbol is *less* than the one which follows it. Thus, $\angle A < \angle B$ is to be understood and read, "the angle A is *less* than the angle B."

$>$ placed between two quantities denotes that the one which *precedes* is *greater* than the one which *follows* it. Thus, $\angle B > \angle A$, is to be read, "the angle B is *greater* than the angle A." It may assist the memory to observe that the *greater* quantity is always on the side of the *open* part of the symbol, and the *less* quantity on the side of the *closed* part.

A', B', x', y', etc. are read, A prime, B prime, x prime, y prime, etc.

A'', B'', x'', y'', etc. are read, A second, B second, x second, y second, etc.

A''' is read, A third; A^{iv}, A fourth; B^{v}, B fifth, etc.

CASES IN PLANE TRIGONOMETRY AS REFERRED TO IN THE FOLLOWING CALCULATIONS.

Case 1. When the angles and one side are given.

Case 2. When two sides and an angle opposite one of them are given.*

Case 3. When two sides and the included angle are given.*

Case 4. When all the sides are given.

* In Cases 2 and 3, when the angle is a right one, the solution may be effected by the rules for right-angled triangles.

REMARKS TO THE STUDENT.

1. A triangle is said to be *determined* in an analysis or a construction when a sufficient number of parts are known to solve it by any one of the preceding cases of plane trigonometry.

2. References are made to Davies's translation of Legendre's Geometry, the Roman numerals being put for the book, and common figures for the proposition. With these is an asterisk (*) referring to a marginal or foot-note, where the proposition is designated in like manner in which the same property is demonstrated in Playfair's edition of Euclid's Elements of Geometry.

3. In performing the trigonometrical calculation of a constructed problem, the rule manifestly must be, to begin the *calculation* where was begun the *construction;* that is, in *the triangle first formed.* For, if there was sufficient given to *construct* the triangle, there must be sufficient to *calculate* it. Then, with what is found in this triangle, proceed to the next one that was formed, and so on till what is desired is obtained.

4. When a *circle, semicircle,* or *circular arc* is used in *construction,* its *radius,* as a *general* rule, must be used in the *calculation.* An exception to this rule exists when a circular segment is described on a line *merely* to include an angle of a given magnitude.

5. In the analysis of geometrical problems the *ingenuity* and *inventive powers of the student* must be brought into close requisition. No definite rules to meet all cases can be given. We must, however, always draw a figure, *supposing the problem constructed as required,* and then, when practicable, work on those lines which are given *in position* or *length,* or *both,* and continue on until a triangle is obtained which has a sufficient number of parts given to construct it. Then recall, and observe carefully, the process that has been pursued, and construct the figure accordingly.

An attempt will be made in the following problems to render this general advice familiar, and to lead the student on to the analysis of problems of considerable intricacy.

6. It will prove of great benefit to the student, in constructing the problems, to assume *definite quantities* for the parts given, and then work with the scale and dividers with delicate precision, measuring the results accurately from the scale of equal parts, or, if angles, from the scale of chords, and compare these results with those obtained by trigonometrical calculations. This is an admirable preparation for "field work," architectural drafting, machinists, engineering, etc.

7. At the ends of the problems the limits are frequently specified within which the given quantities must be taken; and certain varied conditions of the problem are occasionally referred to, which cannot generally be comprehended to full advantage by the figure that is given. In such cases the student will derive decided benefit, and progress with much greater rapidity and satisfaction by drawing on paper, with pen or pencil, representations of these different conditions, so as clearly to comprehend the idea designed to be conveyed. In this way he will *master* all that is before him as he goes along, and acquire greater courage and power to overcome future difficulties as they arise.

GEOMETRICAL ANALYSIS.

PROBLEM I.—In a plane triangle are given one angle, an adjacent side, and the sum of the other two sides, to determine the triangle.

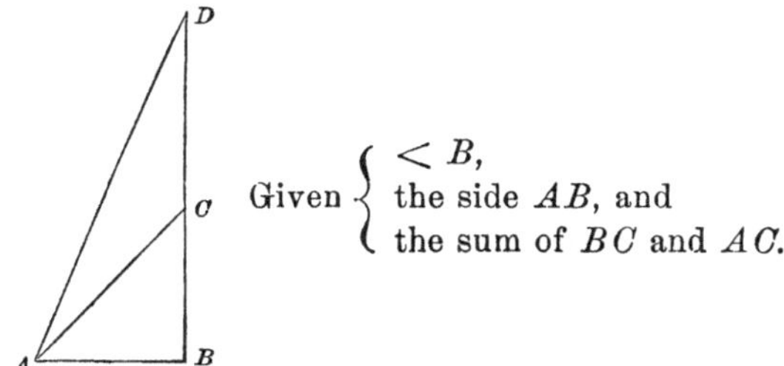

Analysis.—Let ABC represent the required triangle. The first thing to do is to get BC and AC into *one line*, and, if practicable, one whose position is given. Now, the position of AC is not given, but that of BC is, because the angle B is given. Hence, produce BC until the produced part, CD, is equal to CA; then BD becomes known, being equal to the given sum of BC and AC. Join AD, and then the triangle ABD is determined. (Remark 1.) Also, since $CD = CA$, the angle $CAD =$ angle CDA (I. 11*). Whence this

Construction.—Make $AB =$ the given side, the angle $B =$ the given angle, and the line $BD =$ the given sum of BC and AC. Join AD, and make the angle $DAC = < ADC$, then will ABC be the triangle required.

Demonstration.—All we have to prove is that $BC + CA = BD$, the given sum. Since, by construction, $< DAC = < ADC$, we

* I. 5.

have (I. 12*) $CA = CD$. Add each to BC, and we have $BC + CA = BC + CD = BD$. Q. E. D.

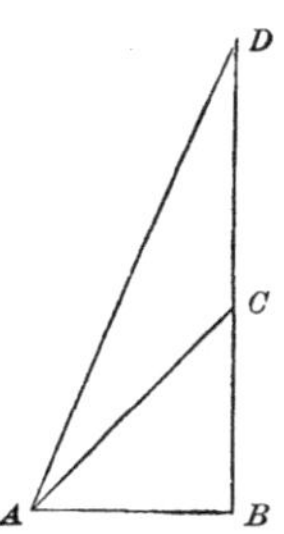

Calculation.—1. In $\triangle$ ABD, by Case 3, find $< ADB$, which is equal to $< DAC$. Then (I. 25, cor. 6†) $< BCA = DAC + ADC = 2\ ADB$.

2. In $\triangle$ ABC, by Case 1, find the sides AC and BC, which together $= BD$.

Limits.—1. The angle B can be *any quantity* between 0 and 180°.

2. The sum of BC and AC may be *any quantity greater than AB.*

Problem II.—In a plane triangle are given one angle, an adjacent side, and the *difference* between the other two sides, to determine the triangle.

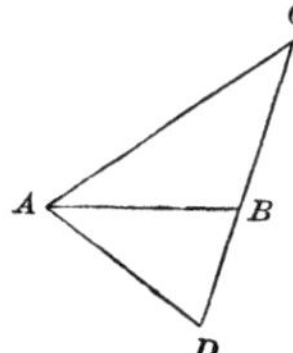

Given $\begin{cases} < B, \\ \text{the side } AB, \text{ and} \\ \text{the } \textit{difference} \text{ between } AC \text{ and } BC. \end{cases}$

Analysis.—Let ABC represent the required triangle. Now, we must first obtain a line equal to $AC - BC$, and whose *position* is known. Since the angle B is given, the line BC is *given in position.* Therefore, produce CB until $CD = CA$; then BD is the given difference. Join AD, and the $\triangle$ ABD is determined. Also, since $CD = CA$, the angle $CAD = CDA$, whence the $\triangle$ ADC, and consequently ABC, is determined. Whence this

Construction.—Make $AB =$ the given side, and the angle $ABC =$ the given angle. Produce CB, making $BD =$ the given difference. Join AD, and make the angle $DAC = < ADC$, and ABC will be the required triangle.

Demonstration.—We have to prove, only, that $AC - BC = BD$, the given difference. Since the angles DAC and ADC are equal by construction, we have (12 I.‡) $AC = CD$. Taking BC from each, we have $AC - BC = CD - BC = BD$, the given difference. Q. E. D.

Calculation.—1. In $\triangle$ ABD, by Case 3, find $< ADB = < DAC$; then $< ACD = 180° - 2\ ADC$.

2. In $\triangle$ ABC, Case 1, find sides AC and BC; then $AC - BC = BD$.

* I. 6. † I. 32. ‡ I. 6.

Limits.—1. The angle B may be *any quantity* from 0 to 180°.

2. The difference between AC and BC may be any quantity *less than* AB.

PROBLEM III.—The *difference* between the diagonal of a square and one of its sides being given, to determine the square.

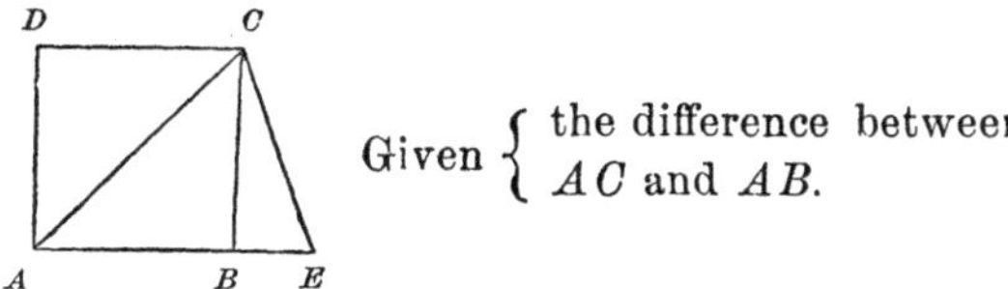

Analysis.—Suppose $ABCD$ to be the required square, having the difference between the diagonal AC and the side AB equal to a given quantity; then produce the side AB until $AE = AC$, and BE will be equal to that given quantity. Join CE. Then in the isosceles triangle AEC, the angle CAE, being half a right angle, is 45°; hence, each of the equal angles AEC and $ACE = \frac{1}{2}(180—45) = 67\frac{1}{2}°$; and the triangles EBC and EAC are determined. Whence this

Construction.—In any line, make $EB =$ the given difference, the angle $BEC = 67\frac{1}{2}°$, and let EC meet a perpendicular erected at B, in C. On BC describe the square $ABCD$, which will be the one required.

Demonstration.—Draw the diagonal AC. We have to prove that $AC - AB = BE$. In $\triangle$ AEC, since $\angle A = 45°$, being half a right angle, and $\angle E = 67\frac{1}{2}°$, by construction, the angle ACE must be the supplement of their sum, which is $67\frac{1}{2}°$. Hence $\angle ACE = \angle AEC$, and the side $AC = AE$. From each take AB, and we have $AC - AB = AE - AB = BE =$ the given quantity. Q. E. D.

Calculation.—In $\triangle$ EBC, Case 1, find $BC = AB$; then $AC = AE = AB + BE$.

Limits.—The difference BE may be any quantity whatever.

PROBLEM IV.—In a plane triangle are given one angle, the side opposite thereto, and the *difference* between the other two sides, to determine the triangle.

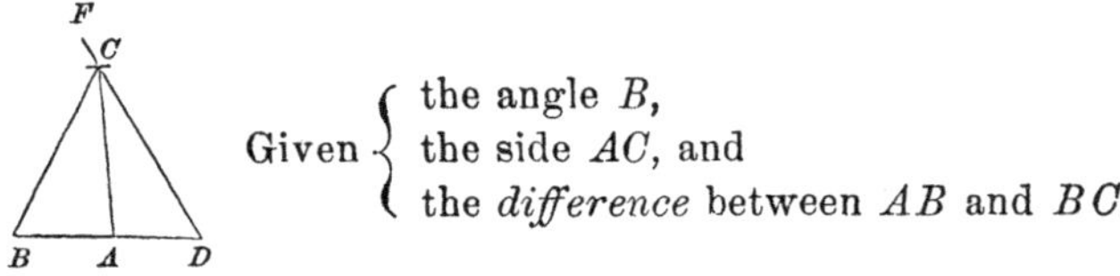

Analysis.—Let ABC represent the required triangle. Produce

BA, making $BD = BC$; then AD is the given difference. Join DC. Then, since $BD = BC$, we have the angles BDC and BCD equal, and each *half the supplement* of the given angle B. Hence the triangle DAC is determined, and we have this

Construction.—Draw an indefinite line, and on it lay $DA =$ the given difference between AB and BC. Make the angle ADF equal to *half the supplement* of the given angle B; and to DF apply $AC =$ the given side. Make the angle $DCB = \angle ADC$, and BAC will be the triangle required.

Demonstration.—We have to prove that the angle B is equal to the given angle, and that $AD = BC - BA$. Since the angles BCD and BDC are equal by construction, and each half the supplement of the given angle, the angle B must be equal to the given angle, and the side $BC = BD$. Take BA from each; then $BC - BA = BD - BA = AD$, the given difference. Q. E. D.

Calculation.—1. In $\triangle ADC$, Case 2, find $\angle ACD$; then $\angle ACB = \angle BCD - \angle ACD = \angle BDC - \angle ACD$.

2. In $\triangle ABC$, Case 1, find AB and BC.

Limits.—AC may be *any quantity greater than* AD.

Problem V.—In a plane triangle are given one angle, the side opposite thereto, and the *sum* of the other two sides, to determine the triangle.

Fig. 1.

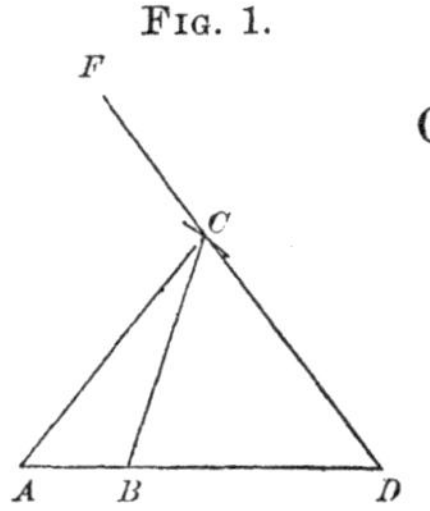

Given, the angle ABC, the side AC, and the *sum* of AB and BC.

Fig. 2.

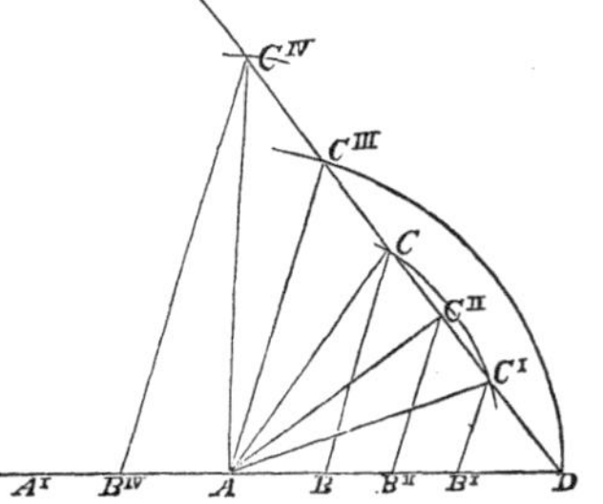

Analysis.—Let ABC (Fig. 1) represent the required triangle. The line AB being fixed in position, produce it until the produced part $BD = BC$; then AD will be equal to the given sum of AB and BC. Now, since $BD = BC$, the angles BCD and BDC are equal, and each half the given *exterior* angle ABC (25 I. cor. 6*). Hence, in the $\triangle ADC$ we know AC, AD, and the angle $D = \frac{1}{2}$ the given angle; whence this

* I. 32.

Construction.—Draw $AD =$ the given sum of the sides AB and BC. Make the angle $ADF =$ *half* the given angle, and to DF apply $AC =$ the given side. Then make the angle $DCB =$ the $< ADC$, and ABC will be the required triangle.

Demonstration.—We have to prove that the angle $ABC = 2\,ADF$, and that $AB + BC = AD$. By 25 I. cor. 6*, $< ABC =$ angles $BDC + BCD = 2\,BDC = 2\,ADF$.

Also (12 I.†) $BC = BD$. Add each to AB, and we have $AB + BC = AB + BD = AD$. Q. E. D.

Calculation.—In $\triangle\ ADC$, the first formed, find, Case 2, the angle ACD, whence we have angle CAD; then, in $\triangle\ ABC$, Case 1, find AB and BC.

Limits, and the *discussion* of the *general problem.*

1. In applying the given side AC to the line DF (see Fig. 2) in the above construction, if the arc were continued to the right, it would cut the line DF in another point C', and, by drawing $C'B'$ parallel to CB, or, which is the same thing, by making the angle $DC'B' = < ADC'$, we would have another triangle, $AB'C'$, fulfilling all the required conditions of the problem; for $AC' = AC$, $< AB'C' = < ABC$, and $AB' + B'C' = AB' + B'D = AD$. Hence the other parts in the triangles ABC and $AB'C'$ are equal,—that is, $AB = B'C'$, $BC = AB'$, angle $ACB = < C'AB'$, $< CAB = < AC'B'$; also $< AC'D = 180° - < AC'C = 180° - ACD$, and the angles $AC'D$ and ACD are supplements of each other.

In the preceding calculation, by Case 2, the angle *directly* obtained is the *acute* angle ACD, the *supplement of which* is the angle $AC'D$. When we use the angle ACD in the subsequent part of the calculation, we obtain the sides AB and BC, of which AB is *less* than BC. When we use its supplement $AC'D$, we obtain AB' and $B'C'$, of which AB' is *greater* than $B'C'$.

2. On DF, let fall the perpendicular AC'', and draw $C''B''$ parallel to BC; then AC'' is the *inferior limit* of the length of the given side; for, if the length of AC were given *less* than the perpendicular AC'', it could not be applied from A to the line DF, and the problem would be impossible.

When the given side is *just equal* to the perpendicular AC'', the $\triangle\ AB''C''$ becomes *isosceles*, having $AB'' = B''C''$, and each equal to half the given line AD. For the $< AC''B'' = 90° - B''C''D = 90° - ADC'' = DAC'' = B''AC''$. Hence $AB'' = B''C'' = B''D = \frac{1}{2}\,AD$.

* I. 32. † I. 6.

3. AD is the *superior limit* of the length of the given side. Apply to DF, $AC''' = AD$; then angle $AC'''D = \angle ADC = \angle BCD$. Hence AC''' is parallel to BC; but the side AB then vanishes, or becomes equal to 0, and the triangle closes in the line AC'''.

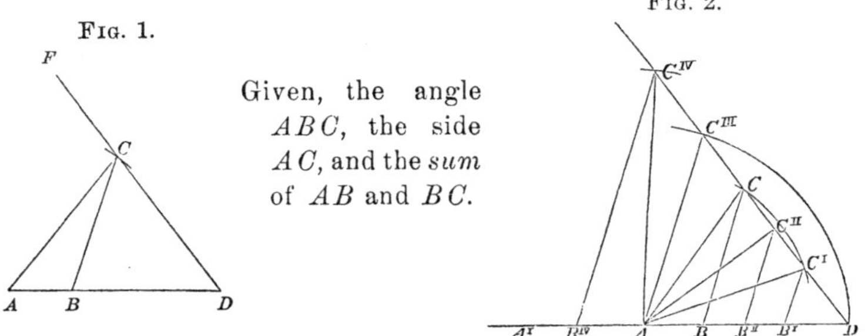

FIG. 1. FIG. 2.

Given, the angle ABC, the side AC, and the *sum* of AB and BC.

4. If to DF we apply a line AC^{iv}, *greater* than AD, and draw $C^{iv}B^{iv}$ parallel to CB, it will make the angle $A'B^{iv}C^{iv}$, with DA produced, equal the given angle ABC; then $B^{iv}D = B^{iv}C^{iv}$, and AD becomes the *difference* of the sides of the triangle $AB^{iv}C^{iv}$, in which an angle B^{iv}, equal to the *supplement* of ABC, is known, and the side AC^{iv}, which leads exactly to Problem IV., page 23, where the angle B^{iv}, the side AC^{iv}, and the *difference* between AB^{iv} and $B^{iv}C^{iv}$ are given.

PROBLEM VI.—In a parallelogram are given the angles, one diagonal, and the sum of all the sides, to determine the parallelogram.

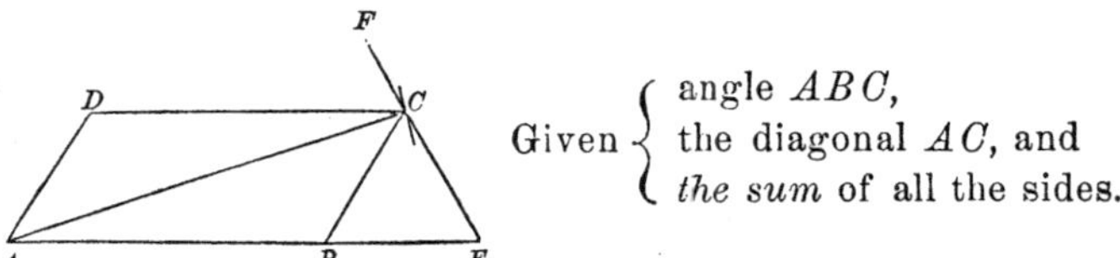

Given { angle ABC, the diagonal AC, and *the sum* of all the sides.

Analysis.—Let $ABCD$ represent the required parallelogram. Produce the side AB until the produced part $BE = BC$; then the side $AE = AB + BC =$ half the given perimeter, and is known. Hence, in the triangle ABC we have the angle B, the opposite side AC, and the *sum* of the sides AB and BC, to construct the triangle ABC just as in the last problem. Then complete the parallelogram $ABCD$, and it will be the one required.

The *construction, demonstration, calculation,* and *limits* are precisely as in the last problem.

NOTE.—When the angle B is a right angle, the parallelogram becomes a rectangle.

Problem VII.—In a plane triangle are given the base, the *difference* of the angles at the base, and the *sum* of the other two sides, to determine the triangle.

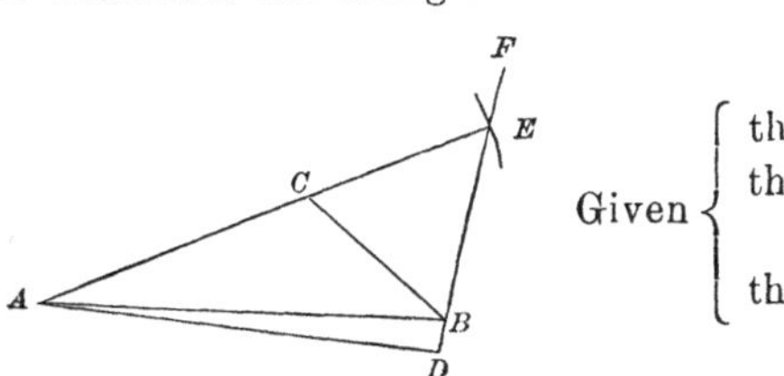

Given $\left\{\begin{array}{l}\text{the base } AB, \\ \text{the difference between angles} \\ \quad BAC \text{ and } ABC, \text{ and} \\ \text{the } sum \text{ of } AC \text{ and } CB.\end{array}\right.$

Analysis.—Suppose ABC to be the required triangle; and having neither side given in position, produce the *longer* of the unknown sides AC until the produced part $CE = CB$. Then $AE =$ the given sum of AC and CB. Join EB, and on it, produced, let fall the perpendicular AD. Then $< ABC - < CAB = (ABE - CBE) - CAB = ABE - AEB - EAB = ABE - (AEB + EAB) =$ (I. 25 cor. 6*) $ABE - ABD = 180° - ABD - ABD = 180° - 2\,ABD = 2\,(90° - ABD) = 2\,BAD$. Hence BAD is half the given difference of the angles. Whence this

Construction.—Draw $AB =$ the given base, and make the angle $BAD = \frac{1}{2}$ the given difference of the angles at the base. Through B draw a line DF perpendicular to AD, to which apply $AE =$ the given sum of the sides. Make the angle $EBC = < BEA$, and ABC will be the required triangle.

Demonstration.—Since the angles EBC and BEC are equal by construction, the sides CB and CE are equal. To each add AC; then $AC + CB = AC + CE = AE$, the given sum. Also, by the analysis $ABC - BAC = 2\,BAD =$ the given difference by construction. Q. E. D.

Calculation.—1. In triangle ABD, the first formed, Case 1, find $< ABD$; then $ABE = 180° - ABD$.

2. In triangle ABE, Case 2, find angles AEB and BAE; then $ACB = 2\,AEB$.

3. In triangle ACB, Case 1, find the sides AC and CB.

Limits.—The given difference may be taken any quantity less than a right angle; and the sum of the sides, any quantity greater than the base.

* I. 32.

PROBLEM VIII.—In a plane triangle, having the base and perpendicular height given, it is required to inscribe a square, of which one side shall be coincident with the base of the triangle.

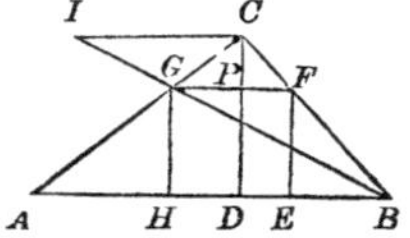

Given AB and CD.

Analysis.—Suppose the square $EFGH$ to be inscribed in the triangle ABC. Join BG, and produce it to meet CI, drawn parallel to AB, in I. Then, by similar triangles, $BF:BC::FE:CD$. Also, $BF:BC::FG:CI$. Hence, $FE:CD::FG:CI$. But $FE = FG$, being sides of the same square. Wherefore $CI = CD$; and we have this

Construction.—Draw $AB =$ the given base, and from *any* point in it erect the perpendicular $DC =$ the given perpendicular, and complete the triangle ABC. Through C, draw CI parallel to AB, and equal to CD. Join BI, cutting AC in G. Draw GF parallel to AB, and GH and FE each parallel to CD; then $EFGH$ will be a square inscribed in ABC.

Demonstration.—Since CD is perpendicular, and GF parallel, to AB, GH and FE are perpendicular to AB and GF, and hence $EFGH$ is rectangular. It now remains to show that the sides are equal. By similar $\triangle$s, $BC:CD::BF:FE$; and $BC:CI::BF:FG$. But the first three terms in these two proportions are equal. Hence the fourth terms are equal, and $FE = FG = GH = HE$. Q. E. D.

Calculation.—By similar $\triangle$s, $CD:AB::CP:GF$ or PD. By composition, $CD + AB:AB::CP + PD$ or $CD:GF = \dfrac{AB \times CD}{AB + CD}$ $=$ the side of the square required.

Limits.—1. With the same base AB, and the same perpendicular height CD, the point D may be taken *anywhere* in the line AB, and the side of the inscribed square will always be the same length $= \dfrac{AB \times CD}{AB + CD}$.

2. Taking the general problem of inscribing a square in a given triangle, neither of the angles adjacent to the base AB can be *greater* than a right angle.

3. In a scalene *acute*-angled triangle, or a *right*-angled triangle, any one of the three sides may be made the base; and, as the side

of the inscribed square is always equal to the *product* of the *base* and *perpendicular*, divided by *their sum*, it is evident that there may be three different squares inscribed, of different sizes, by making each side successively the *base*, and the perpendicular let fall on it from the opposite angle, the height.

4. When the triangle is *isosceles*, there can be but two squares of *different sizes*, those two squares being equal of which the equal sides are coincident with a side of the square.

In an *equilateral* triangle, either side may be made the base, and the squares will all be equal.

In an *obtuse-angled* triangle, the side opposite the obtuse angle *must* be the base, and there can be but one square inscribed.

5. A rectangle whose sides are in *any ratio* to each other may be inscribed in a given triangle in like manner with the square, by taking CI to CD in the ratio of the required length to the breadth, and proceeding in all respects as directed for the square where the ratio of the sides is a *ratio of equality*. If it is desired to have $GF = 2\ GH$, make $CI = 2\ CD$. If GF is to be a *half* or a *third* of GH, make CI equal to a *half* or a *third* of CD. And, generally, representing the ratio of the length of the rectangle to the breadth by m to n, take CI a fourth proportional to m, n, and CD, so that we will have $m : n :: CD : CI$, and draw BGI. Then will GH or $FE : GF :: CD : CI :: m : n$.

PROBLEM IX.—In a given triangle, to inscribe a rhombus which shall have given angles.

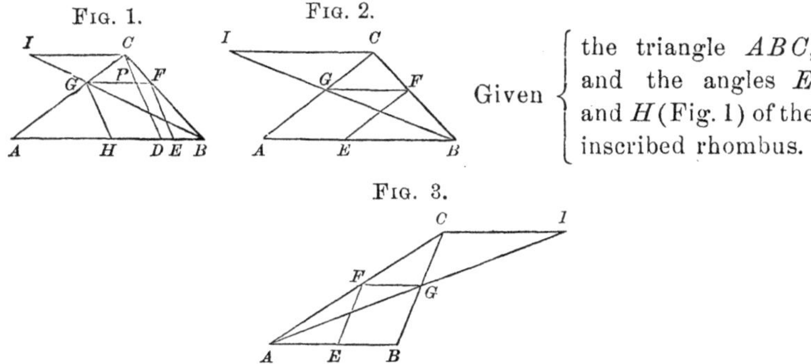

Given { the triangle ABC, and the angles E and H (Fig. 1) of the inscribed rhombus.

Analysis.—Suppose $EFGH$ (Fig. 1) to be a rhombus inscribed in the triangle ABC, having the angles E and H of the given magnitude. Draw CD parallel to FE or GH; then the angle $D =$ the given angle E. Hence CD is determined in the triangle ACD. Join BG, and produce it, to meet a line drawn through C, parallel

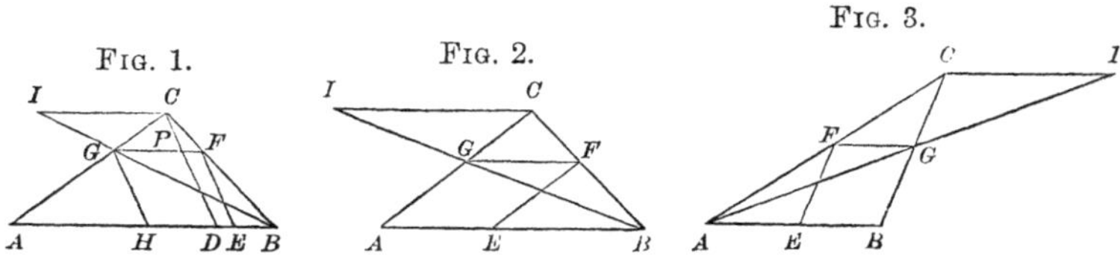

to AB, in I. Now, as in last problem, we have $BF : BC :: FE : CD$; and $BF : BC :: FG : CI$. Whence, since $FG = FE$, we have $CI = CD$, and the *construction*, *demonstration*, and *calculation* are precisely as in the last problem. GF or $FE = \frac{AB \times CD}{AB + CD}$.

Limits.—1. When one of the given angles of the rhombus (Fig. 2) is equal to the angle BAC of the given triangle, then CD will coincide with CA, and we have $BF : BC :: FE : CA$, and $BF : BC :: FG : CI$. Hence $CI = CA$, and CI must be made equal to CA, and CA must be used instead of CD in the demonstration and calculation. Then GF or $FE = \frac{AB \times CA}{AB + CA}$.

2. When one of the given angles of the rhombus (Fig. 3) is equal to the angle ABC of the triangle, then CD will coincide with CB, and CI must be drawn equal to CB, and CB must be used instead of CD in the demonstration and calculation. GF or $FE = \frac{AB \times CB}{AB + CB}$.

3. A parallelogram whose angles are given, and its sides in *any ratio* to each other, may be inscribed in a given triangle, in like manner, by taking CI in the same ratio to CD (Fig. 1), to CA (Fig. 2), or to CB (Fig. 3) that the sides are to have to each other. See *Limit* 5 to last problem.

PROBLEM X.—Having given the lengths of three perpendiculars drawn from a point within an equilateral triangle to its three sides, to determine the triangle.

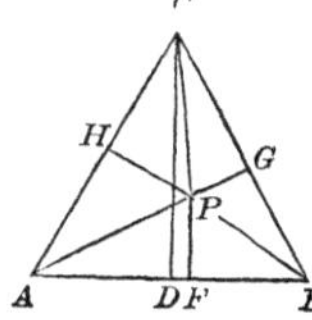

Given $\begin{cases} \text{the three perpendiculars } PF, PG, \text{ and } PH, \\ \text{and the triangle } ABC \\ \textit{to be equilateral.} \end{cases}$

Analysis.—In the quadrilateral $AFPH$, the angles at F and H being right angles, the angles at A and P *must*, together, be equal to two right angles, or 180°. But the angle A, being an angle of an equilateral triangle, is 60°. Hence the angle FPH is 120°. In like manner it may be shown that in the quadrilateral $BFPG$ the angle FPG is equal to 120°; and in the quadrilateral $CHPG$ the angle $GPH = 120°$. Whence this

Construction.—Draw *any line*, as AB, on which erect the perpendicular FP equal to one of the given perpendiculars. At P draw lines making with PF angles equal to 120°, and on them lay PG and PH respectively equal to the other two given perpendiculars. Through the points G and H draw lines perpendicular to PG and PH respectively, meeting each other in C, and meeting the line to which FP is perpendicular, in A and B; then will ABC be the required triangle.

Demonstration.—We have to prove only that the angles A, B, and C are equal. In the quadrilateral $AFPH$, the angles F and H are right angles by construction, and $< FPH = 120°$, hence angle $A = 60°$. In the same way, since $< FPG = 120°$, $< B = 60°$; and since $GPH = 120°$, angle $C = 60°$; and hence the angles A, B, and C are all equal. Q. E. D.

Calculation.—Join AP, BP, and CP, and on AB let fall the perpendicular CD. Now, the triangle ABC is made up of the three triangles APB, BPC, and APC, whose bases AB, BC, and AC are all equal. Hence $AB \times CD =$ (the double area of ABC) $= AB \times PF + AB \times PG + AB \times PH =$ (the double areas of the three constituent triangles APB, BPC, and APC). That is, $AB \times CD = AB \times (PF + PG + PH)$, or $CD = PF + PG + PH =$ the sum of the three given perpendiculars. Wherefore, in the right-angled triangle ACD, having CD, find, Case 1, $AC = AB = BC$, the side required. Then $\frac{1}{2} AB \times CD =$ area ABC.

PROBLEM XI.—Having given the lengths of the three lines, drawn from the three angles of a plane triangle to the middle of the opposite sides, to determine the triangle.

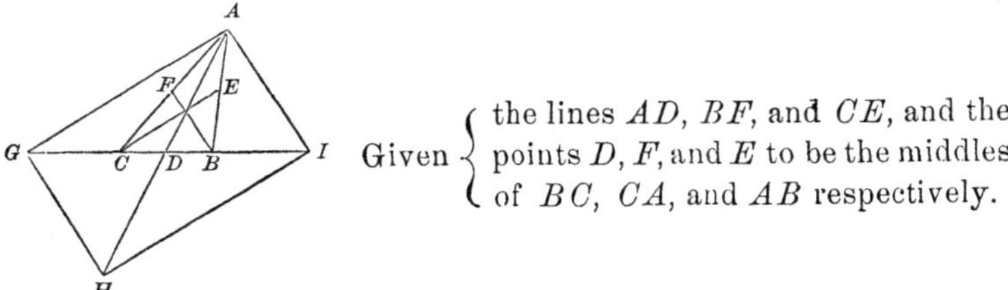

Given $\left\{\begin{array}{l}\text{the lines } AD,\ BF, \text{ and } CE, \text{ and the}\\ \text{points } D,\ F, \text{ and } E \text{ to be the middles}\\ \text{of } BC,\ CA, \text{ and } AB \text{ respectively.}\end{array}\right.$

Analysis.—Suppose ABC to be the triangle required. Produce the side BC both ways, meeting lines drawn parallel to BF and CE in I and G respectively. Then, in similar triangles CAI and CFB, since $CA = 2\,CF$, we have $AI = 2\,BF$, and $CI = 2\,BC$, whence $BI = BC$. Also in similar triangles BCE and BGA, since $BA = 2\,BE$, we have $AG = 2\,CE$, and $BG = 2\,BC$, whence $GC = BC$. Complete the parallelogram $AGHI$; then $GH = AI = 2\,BF$, and $HI = AG = 2\,CE$; whence this

Construction.—With the *doubles* of the three given distances describe the triangle AGH, and complete the parallelogram $AGHI$. Draw the diagonal GDI, and trisect it in the points B and C. Join CA and AB, bisect them in F and E, and join CE and BF; then will ABC be the required triangle.

Demonstration.—We have to prove only that $CE = \frac{1}{2}AG$, and that $BF = \frac{1}{2}GH = \frac{1}{2}AI$. Since, by construction, $BC = CG = BI$, and $BE = EA$, and $CF = FA$, we have, by parallel lines, $CE = \frac{1}{2}AG$, and $BF = \frac{1}{2}AI = \frac{1}{2}GH$. Q. E. D.

Calculation—By 14 IV. cor.,* $AH^2 + GI^2 = 2\,AG^2 + 2\,GH^2$. Hence $GI = \sqrt{2\,AG^2 + 2\,GH^2 - AH^2}$, and $BC = \frac{1}{3}GI$. Then—

Method 1.—In triangle AGI, Case 4, find the angles. Then $\angle BCE = BGA$, and $\angle CBF = CIA$. In triangle BCE, Case 3, find BE, then $AB = 2\,BE$; also in triangle CBF, Case 2, find CF, then $AC = 2\,CF$.

Method 2.—By 14 IV.,† $AC^2 + AB^2 = 2\,AD^2 + 2\,CD^2$; that is, $4\,AF^2 + 4\,BE^2 = 2\,AD^2 + 2\,CD^2$. (1.)

Also $AC^2 + CB^2 = 2\,CE^2 + 2\,BE^2$; that is, $4\,AF^2 + 4\,CD^2 = 2\,CE^2 + 2\,BE^2$. (2.) By subtracting Equation (2) from Equation (1) we have $4\,BE^2 - 4\,CD^2 = 2\,AD^2 + 2\,CD^2 - 2\,CE^2 - 2\,BE^2$. By

* B. II. † A. II.

transposition, $6\,BE^2 = 6\,CD^2 + 2\,AD^2 - 2\,CE^2$. Whence $BE = \sqrt{\frac{3\,CD^2 + AD^2 - CE^2}{3}}$. And $AB = 2\,BE$. Also, from Equation (2),

$$AF = \sqrt{\frac{2\,CE^2 + 2\,BE^2 - 4\,CD^2}{4}} = \tfrac{1}{2}\sqrt{(2\,CE^2 + 2\,BE^2 - 4\,CD^2)}.$$

Then $AC = 2\,AF$, and we have all the sides.

Limits.—Of the three given lines AD, BF, and CE, each must be less than the sum of the other two in order that the triangle AGH may be possible.

PROBLEM XII.—In a trapezoid, two of whose angles are right angles, and whose diagonals intersect each other at right angles, are given the base, and the diagonal drawn from the extremity of the base which is adjacent to a right angle, to determine the trapezoid.

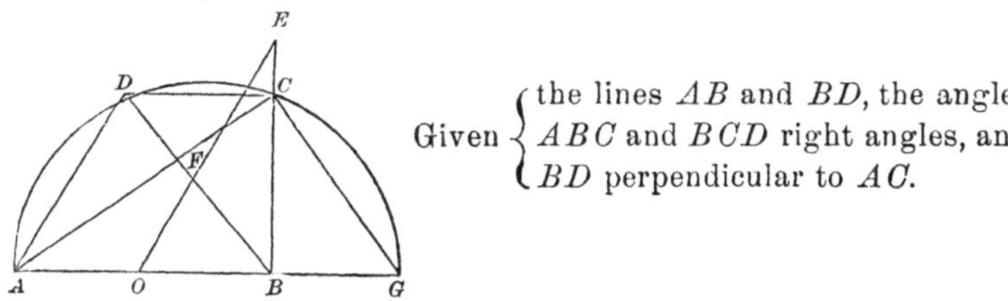

Given $\begin{cases} \text{the lines } AB \text{ and } BD, \text{ the angles} \\ ABC \text{ and } BCD \text{ right angles, and} \\ BD \text{ perpendicular to } AC. \end{cases}$

Analysis.—Let $ABCD$ represent the required trapezoid, having B and C right angles. Parallel to BD draw CG, meeting AB, produced, in G. Then ACG is a right angle, being $= AFB$, and $CG = BD$. Hence, by similar triangles GBC and GCA, $GB : GC :: GC : GA$. Wherefore $GC^2 = (BD^2) = GA \times GB$, and we must produce the given line AB, so that $GA \times GB = GC^2 = BD^2$.* Whence this

Construction.—Draw $AB =$ the given base, and erect the perpendicular $BE =$ the given diagonal BD. Bisect AB in O, join OE, and produce OB until $OG = OE$. On AG describe a semicircle, cutting BE in C. Join AC and CG, and parallel to CG and AB draw BD and CD, respectively, intersecting in D. Then $ABCD$ is the trapezoid required.

Demonstration.—The angle ABC being a right angle by construction, and $\angle AFD = \angle ACG =$ a right angle, being in a semicircle, BCD is a right angle, and the diagonals AC and BD intersect each other at right angles. It remains only to prove that BD or $CG = BE$. Now, $BE^2 = OE^2 - OB^2 = OG^2 - OB^2 = (OG + OB) \times$

* VI. 29.

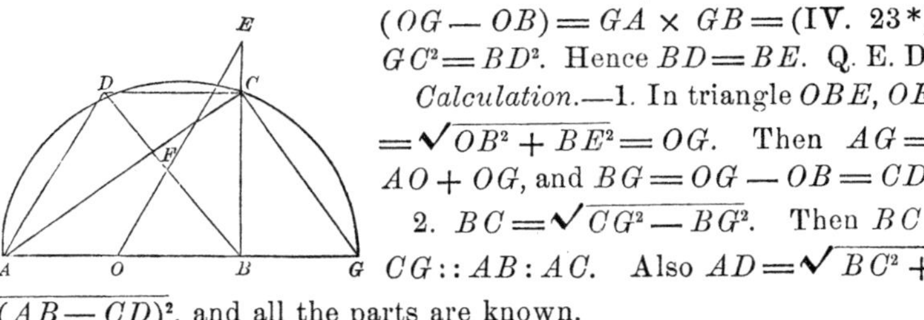

$(OG - OB) = GA \times GB =$ (IV. 23*) $GC^2 = BD^2$. Hence $BD = BE$. Q. E. D.

Calculation.—1. In triangle OBE, $OE = \sqrt{OB^2 + BE^2} = OG$. Then $AG = AO + OG$, and $BG = OG - OB = CD$.

2. $BC = \sqrt{CG^2 - BG^2}$. Then $BC : CG :: AB : AC$. Also $AD = \sqrt{BC^2 + (AB - CD)^2}$, and all the parts are known.

Problem XIII.—In a triangle are given one angle, and the lengths of the two lines drawn from the middle of each including side to the opposite angle, to determine the triangle.

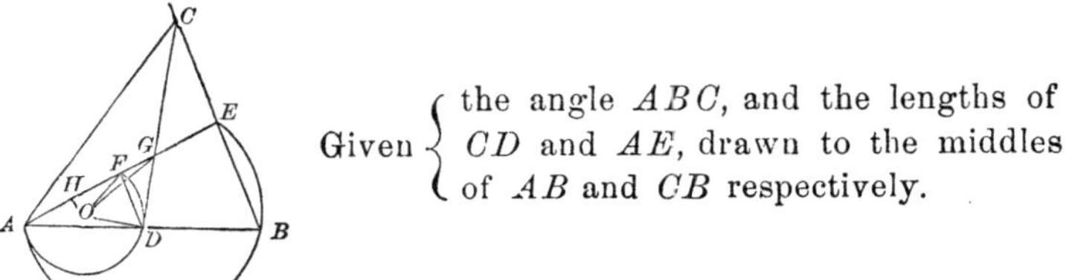

Given $\begin{cases} \text{the angle } ABC, \text{ and the lengths of} \\ CD \text{ and } AE, \text{ drawn to the middles} \\ \text{of } AB \text{ and } CB \text{ respectively.} \end{cases}$

Analysis.—Suppose ABC to be the required triangle, in which CD and AE are the given bisecting lines. Draw DF parallel to BC; then, since $AD = DB$, we have $AF = FE$. Now, in similar triangles ABE and ADF, since $AB = 2\,AD$, we have $BE = (EC) = 2\,DF$. Again, in similar triangles EGC and FGD, since $EC = 2\,DF$, we have $CG = 2\,GD$, and $EG = 2\,FG$. Hence $GD = \frac{1}{3}\,CD$, $FG = \frac{1}{3}\,FE$, and $EG = \frac{2}{3}\,EF = \frac{1}{3}\,AE$. Whence this

Construction.—Make $AE =$ *one* of the given lines, of which take $EG =$ one-third, and $EF =$ one-half, then $EG = 2\,FG$. On AE and AF describe arcs (III. Prob. 16†) to contain the given angle; and to the arc on AF apply $GD = \frac{1}{3}$ the *other* given line. Draw ADB, and join DG and BE, and produce them to meet in C. Then join AC, and ABC will be the triangle required.

Demonstration.—Join DF. Since the angles ADF and ABE are equal by construction, DF is parallel to BE; hence, as $AF = FE$, we have $AD = DB$, and $BE = 2\,DF$. It remains to prove that $BE = EC$, and $DC = 3\,GD$.

In similar triangles EGC and FGD, since $GE = 2\,FG$ by con-

* VI. 8, Cor. † III. 33.

truction, we have $EC = 2\,DF$, and $CG = 2\,GD$. Hence $EC =$ BE, each being equal to $2\,DF$, and $CD = CG + GD = 3\,GD$. Q. E. D.

Calculation.—By Book III. Prob. 16,* O being the centre of the arc ADF, the angle AFO is equal to the difference between the given angle and a right angle. Join OG, and let fall the perpendicular OH; then $FH = HA = \frac{1}{2}\,AF = \frac{1}{4}\,AE$.

1. In triangle OFH, Case 1, find $OF = OD$. Angle $OFG = 180° - < OFH$.

2. In triangle OFG, Case 3, find angle FOG and the side OG.

3. In triangle ODG, Case 4, find the angle DOG. Then, when the given angle is an *obtuse* angle, the angle $DOF = < DOG - < FOG$. But when the given angle is *acute*, the angle $DOF = < DOG + < FOG$. And angle DAF (or BAE) $= \frac{1}{2}\,DOF$ (III. 18†).

4. In triangle ABE, Case 1, find AB and BE. Then $BC =$ $2\,BE$.

5. In triangle ABC, Case 3, find AC. Then all the parts are determined.

Limits.—1. GD must be greater than GF, in order to apply it to the arc ADF. Hence a *third* of one of the given lines must be *greater* than *one-sixth* of the other, or, the *shorter* line must be *greater* than *half the longer one.*

2. When the given angle is a right angle, the arcs ABE and ADF become semicircles, the centre O comes to H, OF and OD each $= \frac{1}{2}\,AF$, and $OG = \frac{1}{2}\,AF + FG$. Hence, in triangle DOG, and angle DOG, half of which is BAE, then proceed to find AB, BE, and BC, as above; when $AC = \sqrt{AB^2 + BC^2}$.

Problem XIV.—Through a given point to draw a line parallel to a given line.

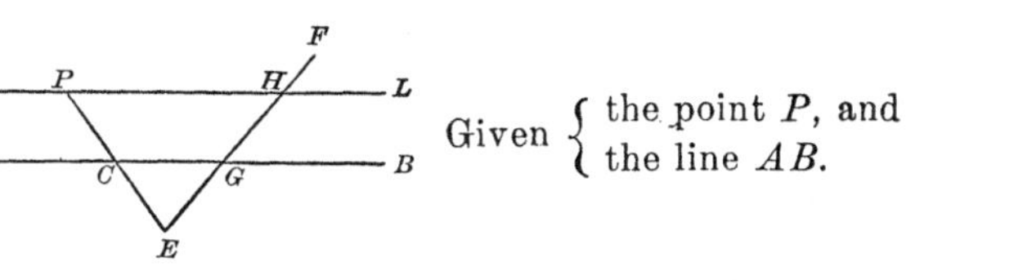

Given { the point P, and the line AB.

Analysis.—Suppose IPL to be drawn parallel to AB. From *any* point, as E, below AB, draw two lines, as ECP and EGH; then (IV. 16‡) $EC : CP :: EG : GH$. Whence this

* III. 33. † III. 20. ‡ VI. 2.

Construction.—Draw, from the given point to the given line AB, *any* line PC, and produce it till $CE = PC$. Then from E draw *any* line EGF, and lay $GH = GE$. Through P and H draw the line IPL, which will be parallel to AB.

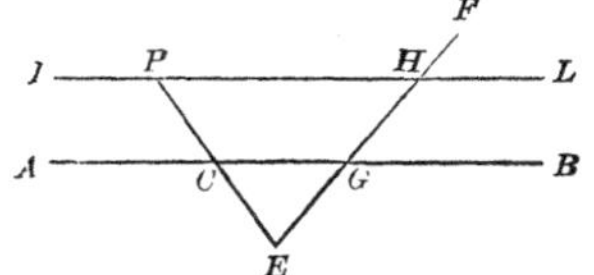

Demonstration.—Since, by construction, $EC = CP$, and $EG = GH$, $IPHL$ is parallel to AB (IV. 16*).

Problem XV.—In a plane triangle are given one angle, the side opposite thereto, and the ratio of the other two sides, to determine the triangle.

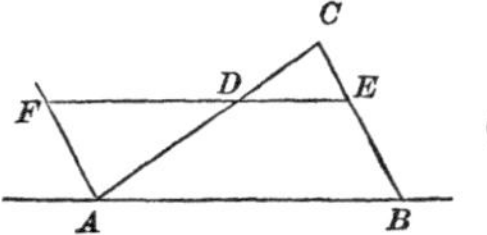

Given { the angle C,
the side AB, and
the ratio of AC to CB as m to n.

Analysis.—Suppose the triangle ABC constructed as required. Parallel to AB draw *any* line EDF; then (IV. 15†) $CD : CE :: AC : CB :: m : n$ in the given ratio. Draw AF parallel to BC, then $EF = AB$. Whence this

Construction.—Draw two lines CA and CB, making at their intersection C, the given angle. Lay $CD = m$, and $CE = n$. Join ED, and produce it, making EDF equal the given side. Draw FA and AB parallel to BC and EF, respectively; then ABC will be the triangle required.

Demonstration.—By parallel lines, $AB = EF =$ the given side by construction; also, $AC : CB :: CD : CE :: m : n$, by construction, in the given ratio. Q. E. D.

Calculation.—1. In the triangle CDE, Case 3, find the angles CDE and CED; then $CAB = CDE$, and $CBA = CED$.

2. In the triangle ABC, Case 1, find the sides AC and BC.

Example.—Take angle $C = 90°$, $AB = 65$, and $CA : CB$ as 4 to 3; then AC will be 52, and BC 39.

* VI. 2. † VI. 2.

PROBLEM XVI.—In a plane triangle, having given all the angles, and the perimeter, to determine the triangle.

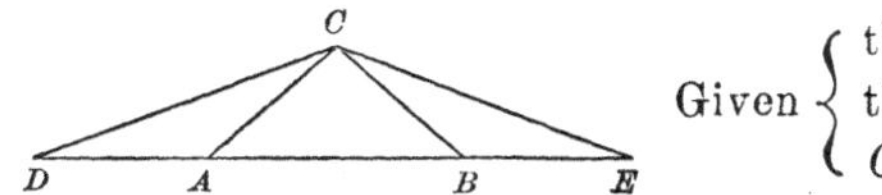

Given $\begin{cases} \text{the angles } A, B, \text{ and } C, \text{ and} \\ \text{the perimeter } AB + BC + \\ CA \text{ in } one\ sum. \end{cases}$

Analysis.—Let ABC represent the required triangle. First obtain the perimeter in one line by producing AB both ways, and making $AD = AC$, and $BE = BC$; then DE is equal to the given perimeter. Join DC and EC; then (I. 11 and 25, cor. 6*) the angle $D = \frac{1}{2}$ the given angle A, and the angle $E = \frac{1}{2}$ the given angle B. Whence this

Construction.—Draw the line $DE =$ the given perimeter, make the angle $EDC =$ half the given angle A, and the angle $DEC =$ half the given angle B. Then make the angle $DCA =$ the angle D, and the angle $ECB =$ the angle E, and ABC will be the required triangle.

Demonstration.—By 12 I.,† $AC = AD$, and $BC = BE$; hence $AB + BC + CA = BE + AB + AD = DE =$ the given perimeter by construction.

Also (I. 25, cor. 6‡) the angle $BAC =$ twice the angle D, and the angle $ABC =$ twice the angle E; hence A and B are equal to the given angles. Q. E. D.

Calculation.—1. In $\triangle DEC$, Case 1, find DC.

2. In $\triangle DAC$, Case 1, find $AC = AD$.

3. In $\triangle ABC$, Case 1, find AB, and $BC = BE$.

PROBLEM XVII.—Upon a given base to construct a right-angled triangle, having given the length of the perpendicular let fall from the right angle on the hypothenuse.

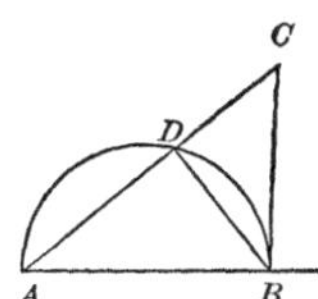

Given $\begin{cases} \text{the base } AB, \\ \text{the distance } BD, \text{ and} \\ \text{the angle } ADB \text{ a right angle.} \end{cases}$

Analysis.—Let ABC represent the required triangle. Since ADB is a right angle, the point D is in a semicircle described on AB. Whence this

* I. 5 and 32. † I. 6. ‡ I. 32.

Construction.—On the given base AB describe a semicircle, and erect a perpendicular BC indefinitely. In the semicircle apply the chord BD of the given length of the perpendicular, and draw ADC. Then ABC is evidently the triangle required.

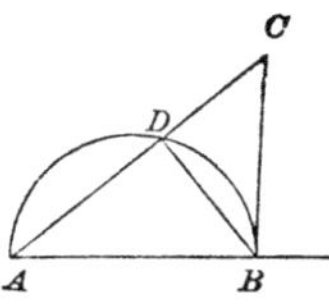

Calculation. — $AD = \sqrt{AB^2 - BD^2}$. Then $AD : AB :: AB : AC$; and $AD : DB :: AB : BC$.

Example.—Take $AB = 15$, and $BD = 12$; then $AD = 9$, $BC = 20$, and $AC = 25$.

Problem XVIII.—Through a given point to draw a straight line making a given angle with a given straight line.

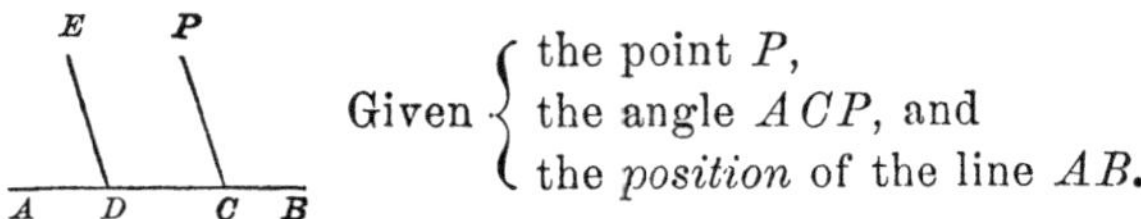

Given $\begin{cases} \text{the point } P, \\ \text{the angle } ACP, \text{ and} \\ \text{the } \textit{position} \text{ of the line } AB. \end{cases}$

Analysis.—Suppose CP drawn to make with AB the angle ACP equal the given angle. From *any* point, as D, in AB, draw DE parallel to CP, then (I. 20, cor. 3*) the angle $ADE = ACP$.

Construction.—From *any* point, as D, in AB, draw DE, making the angle $ADE =$ the given angle ACP; then through the given point P draw PC parallel to DE, and it will be the line required, for (I. 20, cor. 3†) the angle $ACP =$ the angle ADE.

Calculation.—Since the point P and the line AB are given in *position*, the length of the perpendicular let fall from P on AB is known. Designating the foot of that perpendicular by F, in the triangle PFC, Case 1, find PC and CF.

* I. 29. † I. 29.

PROBLEM XIX.—Given, the base of a triangle, an angle at the base, and a point in the base through which the diameter of the circumscribing circle, drawn from the vertex, passes, to determine the triangle.

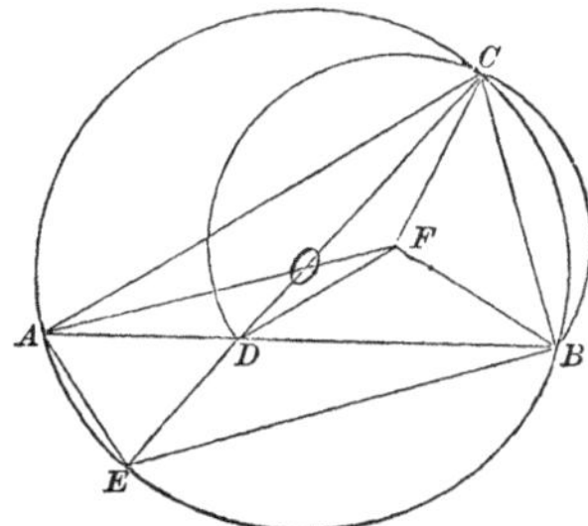

Given $\begin{cases}\text{the base } AB\text{, the angle } BAC\text{,} \\ \text{and the point } D \text{ through which} \\ \text{the diameter } COE \text{ passes.}\end{cases}$

Analysis.—Suppose ABC to be the required triangle, circumscribed by a circle whose diameter is CDE. Since the point D is given, the product of AD and BD is known. Join AE; then the angle $BCE = \angle BAE$ (III. 18, cor. 1*) = the *complement* of the given angle BAC. Whence this

Construction.—Make $AB =$ the given base, $BD =$ the given distance of the point D from B, and the angle $BAC =$ the given angle; then (III. Prob. 16†) on BD describe a segment BCD to contain an angle equal to the *complement* of the given angle BAC, cutting AC in C. Join BC and CD. Draw AE perpendicular to AC, meeting CD, produced, in E. On CE, as a diameter, describe a circle whose centre is O; then ABC will be the required triangle.

Demonstration.—We have to prove only that the circle described on CE as a diameter passes through the point B. By III. 18, cor. 1,‡ the angle $CBA = \angle CEA$, and angle $ABE = \angle ACE$; hence $CBE = CBA + ABE = CEA + ACE = 90° =$ an angle in a semicircle, and therefore the circle described on CE passes through the point B. Q. E. D.

Calculation.—Draw the radii FB, FC, and FD, and join AF.

1. In triangle BFD (III. 18§) the angle $DFB = 2\,DCB$; find, Case 1, $DF = CF$. Now, angle $BDF = \frac{1}{2}(180° - BFD)$, and angle $ADF = 180° - BDF$.

2. In triangle ADF, Case 3, find angle DAF, and AF. Then angle $FAC = \angle BAC - \angle DAF$.

3. In triangle FAC, Case 2, find AC; in triangle BAC, Case 3, find BC; and in triangle ECB, Case 1, find CE; when all the parts are known.

* III. 21. † III. 33. ‡ III. 21. § III. 20.

Limits.—When the given angle BAC is a right angle or an obtuse angle, the problem fails

PROBLEM XX.—In a plane triangle are given the vertical angle, and the radii of the *inscribed* and *escribed* circles, the latter touching the base and the sides produced, to determine the triangle.

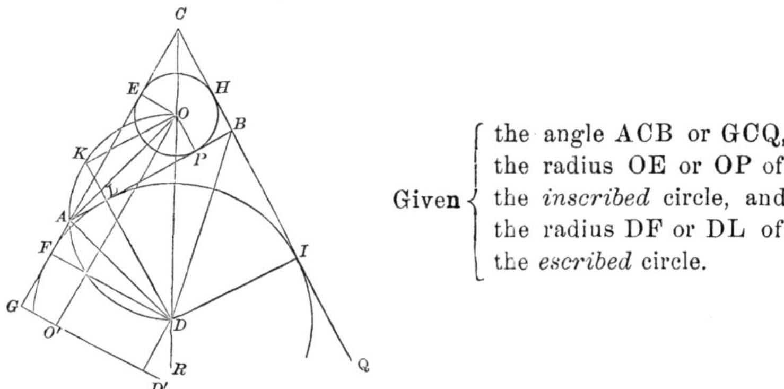

Given { the angle ACB or GCQ, the radius OE or OP of the *inscribed* circle, and the radius DF or DL of the *escribed* circle.

Analysis.—Suppose ACB to represent the required triangle, of which the sides CA and CB are produced indefinitely to G and Q; O the centre of the *inscribed* circle, and D the centre of the *escribed* circle. Draw COD, and it will bisect the angle GCQ (III. Prob. 15*); also AO bisects the angle BAC, and DA and DB bisect the angles GAB and ABG respectively. Hence the angle OAD is a right angle. Draw the radii OE and DF perpendicular to CG, and OP and DL perpendicular to AB. Then the circles $O - E,P$; $D - F,L$, are evidently given in *magnitude* and *position*, and the base $ALPB$ is a common tangent to them at P and L. Whence this

Construction.—Draw CR, bisecting the given angle GCQ. At *any* point, as G, in CG, erect the perpendicular $GD' =$ the given radius DF, and on it lay $GO' =$ the other given radius OE. Through D' and O parallel to CG, draw lines intersecting CR in the points D and O, which will be the centres of the given circles. Let fall the perpendiculars DF and OE; describe the circles $D - F,L,I$, and $O - P,E,H$. On OD describe a semicircle, in which apply DLK the sum of the given radii OP and DL; join OK, and draw $ALPB$ parallel thereto; then ABC will be the required triangle.

* IV. 4.

Demonstration.—When we prove that $ALPB$ is a common tangent to the two circles, it is all evident by the analysis. (See Prob. XIII. of "The Circle.") Since $ALPB$ is parallel to OK by construction, and OP parallel to DK, because both are perpendicular to OK or AB, we have $DL + OP = DL + LK = DK =$ the sum of the radii. Hence DL and OP are respectively the given radii, and, being perpendicular to AB, $ALPB$ is the common tangent. (III. 9.*)

Calculation.—1. In the triangles CFD and CEO, Case 1, find CD and $CF = CI$; also CO. Then $DO = CD - CO$, and the angle $FDC = 90° - FCD$.

2. In triangle KDO, Case 2, find angle KDO; then angle $FDA = \frac{1}{2} FDL = \frac{1}{2}(FDC - KDO)$.

3. In triangle FDA, Case 1, find $AF = AL$. Then $AC = CF - AF$, and angle $BAC = 180° - 2\,FAD$.

4. In triangle ABC, Case 1, find BC and AB.

Scholium.—The angles OAB and DAB being respectively halves of the angles CAB and BAG, the angle DAO is a right angle, and the point A is in a semicircle on DO, the distance between the centres of the given circles.

PROBLEM XXI.—Through a given point in the perpendicular let fall from the right angle of a given right-angled triangle upon the hypothenuse, to draw a line, cutting one side, and the other side produced, so that said line shall be bisected by the given hypothenuse.

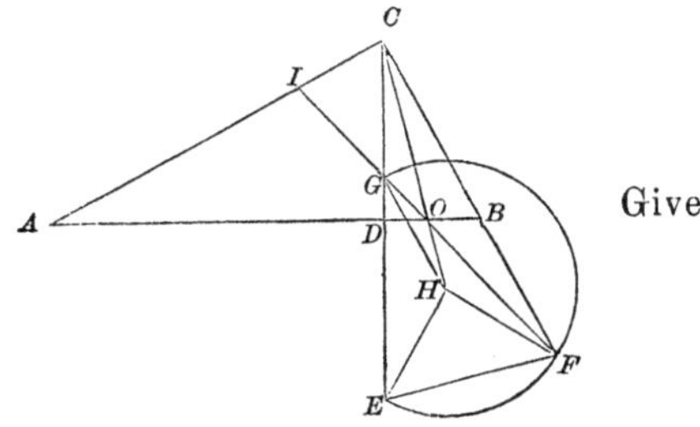

Given $\left\{\right.$ the right-angled triangle ABC, the point G in the perpendicular CD, and the line OGI to be equal to OF, F being in CB produced.

Analysis.—Suppose $IGOF$ drawn so that $OI = OF$. Produce CD until $DE = CD$, and join EF. Because the angle ICF is a right angle, and $CD = DE$, a circle described with O as a centre and radius OI or OF will pass through C and E. Therefore the

* III. 16.

angle IFE or GFE = the angle ICE or ICD. (III. 18, cor. 1.*) Whence this

Construction.—On the line GE (III. Prob. 16†) describe a segment of a circle to contain an angle equal to ACD, cutting CB, produced, in F. Through F and G draw the line $FOGI$, and it will be the line required.

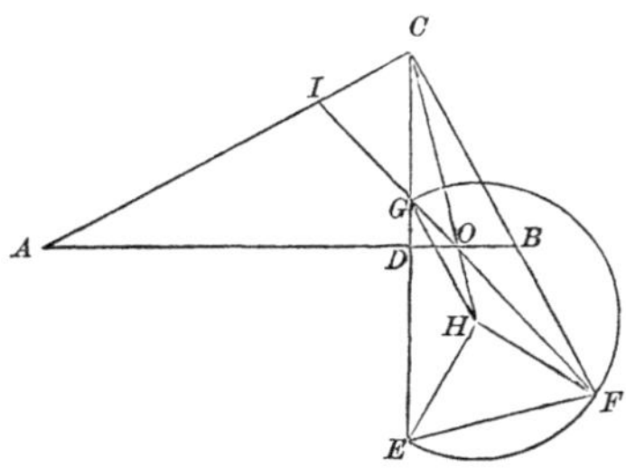

Demonstration.—Since the angle GFE or IFE = the angle ACD or ICD, and DO is perpendicular from the middle of the chord CE, O is the centre and IOF the diameter of a circle passing through the points I, C, F, and E. Whence $OI = OF$. Q. E. D.

Calculation.—Join the centre H with the points E, F, G, and C; then the angle $GHE = 2\,GFE = 2\,ACD = 2\,ABC$. 1. In $\triangle\,GHE$, having $GE = GD + CD$, Case 1, find $HG = HE = HF$. Angle $CGH = 180° - HGE$.

2. In $\triangle\,CGH$, Case 3, find CH and angle GCH; then angle $HCF = DCB - GCH$.

3. In $\triangle\,HCF$, Case 2, find CF; then $BF = CF - CB$.

4. In $\triangle\,CFE$, Case 3, find $< CFE$; then angle CFI = angle $CFE - GFE$.

5. In $\triangle\,ICF$, Case 1, find IF and IC; then $IO = \frac{1}{2}\,IF$.

Limits.—The distance of the given point G from C may be *any quantity not greater than* CD. When $CG = CD$, the required line will pass through D, and be equal to CE.

PROBLEM XXII.—In a plane triangle are given the two sides, and the length of the line bisecting the included angle, to determine the triangle.

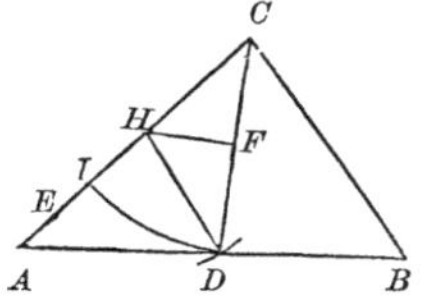

Given { the sides AC and CB, and the length of the line CD *bisecting* the angle CAB. }

Analysis.—Let ACB represent the required triangle, in which the line CD bisects the angle ACB. Draw DH parallel to BC; then

* III. 21. † III. 33.

(I. 20, cor. 2*) the angle $HDC = BCD = ACD$ by the conditions of the problem. Hence $HC = HD$. Also, IV. 17,† and by parallel lines, we have $AC : CB :: AD : DB :: AH : HC$. Whence this

Construction.—Draw the line $CA =$ the longest given side, on which lay $CE =$ the shortest given side, and $CI =$ the given line bisecting the included angle. Divide AC in H (IV. Prob. 1‡) so that AH may be to HC as AC is to CE. With the centre C and radius CI describe an arc, to which apply $HD = HC$. Join HD, CD, and AD, and produce AD to meet a line drawn through C, parallel to HD, in B. Then ACB is the required triangle.

Demonstration.—We have (I. 11 and 20, cor. 2§) the angle $HCD = HDC = BCD$. Hence CD bisects the angle ACB. Then $AC : CB :: AD : DB :: AH : HC ::$ (by construction) $AC : CE$. That is, $AC : CB :: AC : CE$; wherefore $CB = CE$. Q. E. D.

Calculation.—On CD let fall the perpendicular HF; then $CF = FD = \frac{1}{2} CD$. By construction, $AC : CB :: AH : HC$. By composition, $AC + CB : CB :: AH + HC \, (AC) : HC = \dfrac{AC \times CB}{AC + CB}$. Now,

1. In triangle HCF, Case 2, find angle $HCF = ACD$; then angle $ACB = 2\,ACD$.

2. In triangle ACB, Case 3, find AB; then $AC : CB :: AD : DB$. By composition, $AC + CB : CB :: (AD + DB) \, AB : DB = \dfrac{AB \times CB}{AC + CB}$. Then $AD = AB - BD$. Or we may obtain AD from the proportion just used,—$AC + CB : AC :: AD + DB \, (AB) : AD = \dfrac{AB \times AC}{AC + CB}$.

Limits.—The bisecting line CD may be *any* quantity *less* than the *longer given side.*

* I. 29. † VI. 3. ‡ VI. 10. § I. 5 and 29.

Problem XXIII.—In a plane triangle are given the ratio of the sides, and the segments of the base made by a perpendicular let fall thereon from the opposite angle. (See Prob. II., "Analysis by Algebra.")

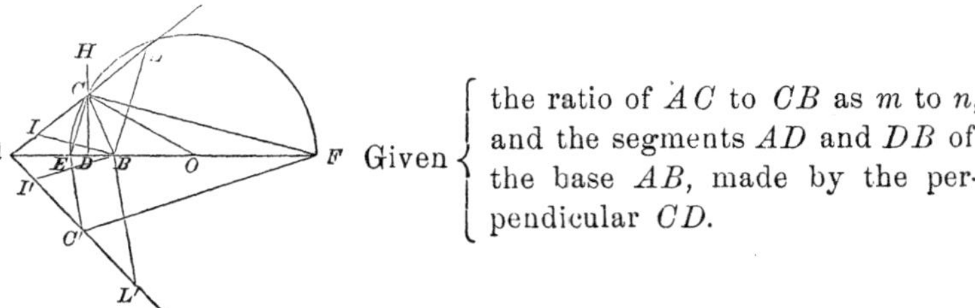

Given $\left\{ \text{the ratio of } AC \text{ to } CB \text{ as } m \text{ to } n, \text{ and the segments } AD \text{ and } DB \text{ of the base } AB, \text{ made by the perpendicular } CD. \right.$

Analysis.—Let ABC represent the required triangle, of which the segments AD and DB of the base AB are given, and AC to CB as m to n, of which m is the greater. Produce AC indefinitely to G, and bisect the angles ACB and BCG by the lines CE and CF, respectively meeting AB, and AB produced, in E and F. Draw BI parallel to CF, and BL parallel to CE, meeting AG in I and L. Then (I. 20, cors.*) angle $CLB = ACE = ECB = CBL$; hence $CL = CB$, and (IV. 15†) $AC : CL\ (CB) :: AE : EB :: m : n$. Also (I. 20, cors.*) angle $CIB = GCF = BCF = CBI$; hence $CI = CB$, and (IV. 15†) $AF : FB :: AC : CI$ or $CB :: m : n$. Wherefore $AC : CB :: AE : EB :: AF : FB :: m : n$.

Also angle $ECF = ECB + BCF = \frac{1}{2}(ACB + BCG) =$ a right angle, and the point C is in the semicircle described on EF. Whence this

Construction.—Draw an indefinite line, and on it lay $AD =$ the greater given segment, and $DB =$ the less. At D erect the perpendicular DH indefinitely. To divide the line AB, and AB produced, as required, in E and F,‡ draw a line AM, making *any* angle with AB, and on AM lay $AC' = m$, the *greater* term of the given ratio, and $C'I'$, $C'L'$, each equal to n, the less. Join BL', BI', and parallel to them, respectively, draw $C'E$, $C'F$, the latter meeting AB produced in F. On EF describe a semicircle whose centre is O, cutting the perpendicular DH in C. Join CA, CE, CB, CO, and CF, and ACB will be the required triangle.

Demonstration.—By IV. 15,§ $AE : EB :: AC' : C'L' :: m : n$; and $AF : FB :: AC' : C'I' :: m : n$. Hence, by the analysis, it is evident that CE bisects the angle ACB, CF bisects the exterior angle

* VI. 2. † I. 29. ‡ See Chauvenet's Geometry, III. 26 and 27. § VI. 2.

BCG; $AC:CB::AE:EB::m:n$; and the angle ECF is a right angle, and therefore the point C is in a semicircle described on EF. Q. E. D.

Calculation.—By construction, $AL'\ (m+n):AC'\ (m)::AB:$ $AE=\frac{m}{m+n}\times AB$; then $EB=AB-AE$. Also $AI'\ (m-n)$: $I'C'\ (n)::AB:BF=\frac{n}{m-n}\times AB$; and $AF=AB+BF$. Also $EF=EB+BF$, and $OC=OE=\frac{1}{2}EF$. Now, $ED=EB-BD$, and $OD=OE-ED$; then $DC^2=OC^2-OD^2$, and $AC=\sqrt{AD^2+DC^2}$, also $BC=\sqrt{BD^2+DC^2}$.

Scholium 1.—If, instead of having the *segments* of the base given, the *whole base* and the *perpendicular height* or *area* of the triangle are known, then from *any* point in the line of the base erect a perpendicular, as DC, of its given length, and through C draw a line parallel to the base; then either of the points in which such parallel line cuts the semicircle will be the vertex of a triangle fulfilling the conditions of the problem. The calculation will be evident from the construction and preceding calculation.

Scholium 2.—The vertex of the triangle may be in a *given line of any kind* which intersects the semicircle ECF. (See Prob. XXX. of "The Circle," Case 2.)

Scholium 3.—Since, by the analysis, we have CI, CB, and CL all equal, C is the centre of a semicircle passing through the points I, B, and L, and IBL is a right angle. Hence the angle $ACE=ALB=\frac{1}{2}(CLB+CBL)=$ (I. 25, cor. 6*) $\frac{1}{2}ACB=ECB$, and CE bisects the angle ACB.

Scholium 4.—Since, by the analysis, $AF:FB::AE:EB$, we have $AF:AE::FB:EB$. By division, $AF-AE\ (2\,OE):AE::FB-EB\ (2\,OB):EB$; that is, $AE:EB::OE:OB::m:n$. Again, by division, $AE-EB:AE::OE-OB\ (EB):OE$. Whence the radius EO *is a fourth proportional* to $AE-EB$, EB, and AE.

Scholium 5.—Since $AE-EB:EB::AE:EO$, we have, by composition, $AE:EB::AO:EO$. Alternately, $AE:AO::EB:EO$; by division, $EO:AO::BO:EO$; inversely, and putting OC for EO, we have $AO:OC::OC:BO$. Whence (IV. 20†) the triangles OAC and OCB are similar, and the angle $OCB=OAC$. Now (I. 11 and 25, cor. 6‡), the angle $ACE=OEC-OAC=OCE-OCB=BCE$. Hence CE bisects the angle ACB, and we have (IV. 17§) $AC:CB::AE:EB::m:n$, which is another method of demonstration.

* I. 32. † VI. 6. ‡ I. 5 and 32. § VI. 3.

NOTE.—The geometrical property involved in this problem is of great practical utility in the construction of various problems, and the student should endeavor to be familiar with it; wherefore the following *general* problem is annexed.

PROBLEM XXIV.—Having two points given, it is required to find a circle, such that if lines be drawn from the two given points to *any* point in the circumference of this circle, the lines so drawn shall always have to each other a given ratio as m to n.

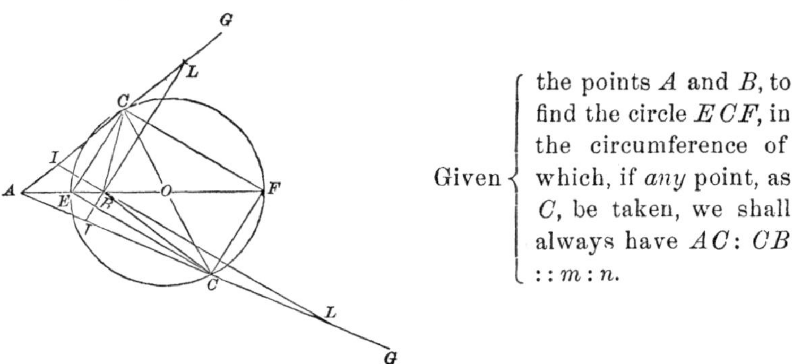

Given $\begin{cases}\text{the points } A \text{ and } B, \text{ to find the circle } ECF, \text{ in the circumference of which, if } any \text{ point, as } C, \text{ be taken, we shall always have } AC : CB :: m : n.\end{cases}$

Analysis.—Let A and B represent the given points, and C, above AF, *any point such* that $AC : CB :: m : n$ in the given ratio. Produce AC indefinitely to G; bisect the angles ACB and BCG by the lines CE and CF respectively, meeting AB, and AB produced, in E and F. Draw BI parallel to CF, and BL parallel to CE, meeting AG in I and L. Then (I. 20, cors.*) angle $CLB = ACE = ECB = CBL$; hence $CL = CB$, and (IV. 15†) $AC : CL\ (CB) :: AE : EB :: m : n$.

Also (I. 20, cors.*) angle $CIB = GCF = BCF = CBI$; hence $CI = CB$, and (IV. 15†) $AF : FB :: AC : CI$ or $CB :: m : n$. Wherefore $AE : EB :: AC : CB :: AF : FB :: m : n$.

Also angle $ECF = ECB + BCF = \frac{1}{2} ACB + \frac{1}{2} BCG =$ a right angle, and the point C is in a circle described on EF as a diameter.

Construction.—As in last problem, divide AB so that $AE : EB :: m : n$, and take AF so that $m : n :: AF : FB$, that is, by division, $m - n : n :: AF - FB\ (AB) : FB = \dfrac{n}{m-n} AB$. On EF describe a circle whose centre is O, and it will be the circle required.

Demonstration.—Take any point C *above* the diameter. Draw CA, CE, CB, CF, produce AC to G, draw BI parallel to CF, and

* I. 29. † VI. 2.

BL parallel to CE, meeting AG in I and L respectively. By construction, $AE : EB :: m : n$, and $AF : FB :: m : n$; hence, by the analysis, it is evident that CE bisects ACB, CF bisects the exterior angle BCG, $AC : CB :: m : n$, and the angle ECF is a right angle, and the point C is in the circumference of a circle on EF. If the point C be taken at *any point* in the circumference *below* the diameter, the same reasoning applies.

NOTE.—See Scholiums 3, 4, and 5 to last problem.

Calculation.—As in Problem XXIII., find AE and AF; then the radius $OE = \frac{1}{2}(AF - AE)$.

Corollary.—The lines AE and EB will subtend the same visual angle from every point in the circumference $ECFC$.

Scholium.—If the given ratio of AC to CB is a *ratio* of *equality*, that is, if $m = n$, then $AE = BE$, and the value of $FB = \frac{n}{m - n} \times AB$ becomes $\frac{n}{0} AB$, which, as the divisor is zero, is *infinite*; hence the radius EO is *infinite*. But a circle described through E with an *infinite radius* is a straight line perpendicular to AB, extending above and below. Hence a perpendicular through the middle of the base, extending both ways, will be the line required.

PROBLEM XXV.—In a right-angled triangle are given the perimeter and radius of the inscribed circle, to determine the triangle.

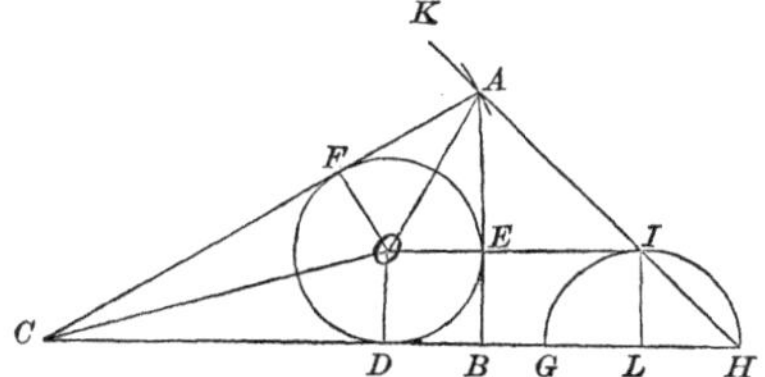

Given $\begin{cases} \text{the perimeter of } ABC \\ \text{and radius } OD. \end{cases}$

Analysis.—Let ABC represent the required triangle, and DEF its inscribed circle, of which O is the centre. Join OA, OC, and let fall on the sides the three perpendiculars OD, OE, and OF. Then (III. Prob. 15*) $CF = CD$, and $AF = AE$. Also, $BDOE$ is evidently a square, whose side is the given radius. Wherefore we have $CA = CF + AF = CD + AE$. To each of these equals add the equals DB and BE respectively, and we have $CA + DB = CD + AE + BE$. But the sum of these equals makes up the whole

* IV. 4.

given perimeter of the triangle. Therefore $CA + DB = \frac{1}{2}$ the given perimeter; that is, the hypothenuse $CA =$ half the *given perimeter minus the given radius*, and the sum of the two sides CB and $BA =$ *half the given perimeter plus the radius*. Whence this

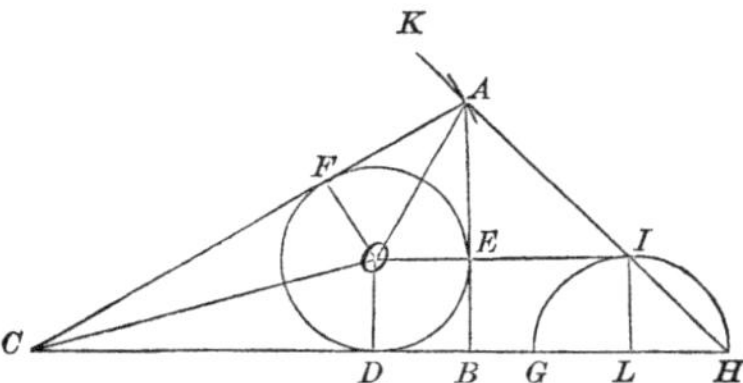

Construction.—Make $CL = \frac{1}{2}$ the given perimeter. With centre L and given radius, describe the semicircle GIH; then $CG =$ the length of the required hypothenuse CA, and $AH =$ the sum of the sides AB and BC. (See Prob. V., this section.) Erect the perpendicular LI, draw HIK, and to it apply $CA = CG$, let fall the perpendicular AB, which will be parallel to IL, and hence, since $IL = LH$, we have $AB = BH$, and consequently $CB + AB = CB + BH = CH$; whence the truth of the construction is manifest. In the triangle ABC (III. Prob. 15*) inscribe the circle DEF.

Calculation.—In triangle CAH, Case 2, find angle $ACH = ACB$.

In triangle ACB, Case 1, find the sides CB and BA.

PROBLEM XXVI.—In a plane triangle are given one angle, and the lengths of the three perpendiculars let fall from the angular points upon the opposite sides, to determine the triangle.

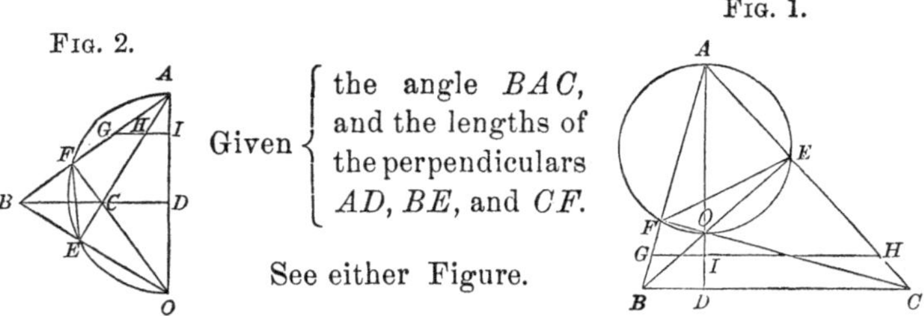

Analysis.—Let ABC, in either figure, represent the required triangle; then, since the double area of a triangle is equal to the *base* multiplied by the *perpendicular height*, we have $BC \times AD = AB \times CF = AC \times BE$. Putting each of these equations into a proportion, we have $AB : BC :: AD : CF$; $AC : BC :: AD : BE$; and $AB : AC :: BE : CF$, which proportions are immediately deduci-

* IV. 4.

ble also from the similar right-angled triangles BAD, BCF; ACD, BCE; and ABE, ACF. Hence the sides are *inversely* proportional to the given perpendiculars falling upon them.

In AB and AC take $AG = BE$, and $AH = CF$, and draw GH. Then, by the last of the above proportions, $AB : AC :: BE\ (AG) : CF\ (AH)$. Hence (IV. 16*) GH is parallel to BC, and the triangles ABC and AGH are equiangular and similar. But the triangle AGH is known in all its parts; therefore, since AD is known, the triangle ABC is known in all its parts. Whence this

Construction.—At A make an angle equal to the given angle, and on its including sides AB and AC, lay $AG =$ the perpendicular to fall on AC, and $AH =$ the perpendicular to fall on BC. Join GH, and on GH produced, if necessary, let fall the perpendicular AI, and produce it, making $AD =$ the perpendicular drawn from the point of the given angle to the opposite side BC. Through D, parallel to GH, draw BC, and on AB and AC, respectively, let fall the perpendiculars CF and BE; then ABC will be the required triangle.

Demonstration.—The triangle ABC, being similar to AGH, and having $AD =$ the given perpendicular from the point A, has, from the analysis, the proper angles and sides, and it remains only to show that BE and CF are of the given lengths.

In the similar triangles ABD and BCF, $AB : BC :: AD : CF$; but AD is made of the given length; hence, since the perpendiculars are *inversely* as the sides upon which they fall, CF is of the given length. Also, in similar triangles, ACD and BCE, $AC : BC :: AD : BE$; therefore, since AD is the given length, BE is the given length also. Q. E. D.

Calculation.—In the triangles AFC and AEB, Case 1, find AC and AB; then, in similar triangles BCF and ABD, we have $AD : CF :: AB : BC$.

NOTE.—See Theorem XV.

* VI. 2.

PROBLEM XXVII.—In a plane triangle are given the base, the line that bisects the vertical angle, and the diameter of the circumscribing circle, to determine the triangle.

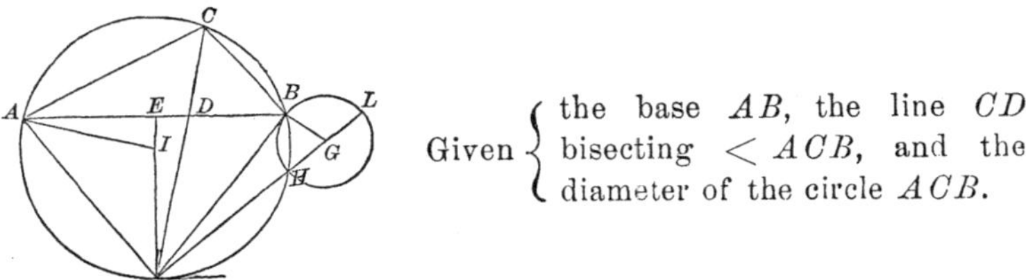

Given $\begin{cases} \text{the base } AB, \text{ the line } CD \\ \text{bisecting } < ACB, \text{ and the} \\ \text{diameter of the circle } ACB. \end{cases}$

Analysis.—Let ACB represent the required triangle, and $ACBF$ the circumscribing circle, of which the centre is I. Produce CD to F, and join FB. Then (III. 18*), since angle $ACF = BCF$, the arc $AF = BF$, and angle $ABF = BCF$. Hence (I. 25, cor. 2†) the triangles FBD and FCB are similar, and we have $FD : FB :: FB : FC$, in which FB and CD, part of the line CF, are known. Whence this

Construction.—About *any* point I as a centre, with the given radius, describe a circle, in which apply $AB =$ the given base, and draw EIF perpendicular to AB; then (III. 6‡) $AE = EB$, and arc $AF = FB$. Join FB, and draw BG perpendicular to FB and equal to half the given bisecting line CD. With centre G and radius GB, describe the circle BHL, and draw $FHGL$; then $HL = 2\,BG =$ the bisecting line. Apply $FDC = FL$, join AC and CB, and ABC will be the required triangle.

Demonstration.—We have to prove only that $CD = HL$. Now, the triangles FBD and FCB are similar by the analysis; hence $FC : FB :: FB : FD$. But (IV. 30§) $FL\,(FC) : FB :: FB : FH$; hence $FD = FH$, and $CD = FC - FD = FL - FH = HL = 2\,BG =$ the given length of the bisecting line. Q. E. D.

Calculation.—Join AI and AF. We have $EI = \sqrt{AI^2 - AE^2}$, $EF = EI + IF$, $FB^2 = FE^2 + EB^2$, $FG = \sqrt{FB^2 + BG^2}$; then $FL = FG + GB = FC$, and $FH = FG - GB = FD$. Also $ED = \sqrt{FD^2 - FE^2}$, $BD = BE - ED$, and $AD = AE + ED$. Now, in the similar triangles FBD and FCB, we have $FB : BD :: FC : BC$. Also, in similar triangles FAD and FCA, we have $FA : AD :: FC : AC$, and all the lines are known.

NOTE.—If the vertical angle were given instead of the diameter of the

* III. 21. † I. 26. ‡ III. 3. § III. 6.

circumscribing circle, we would (III. Prob. 16*) on AB describe a circle whose segment ACB shall contain the given angle, and then proceed precisely as in the above problem.

PROBLEM XXVIII.—In a plane triangle are given the perpendicular height, the line bisecting the base, and the line bisecting the vertical angle, to determine the triangle.

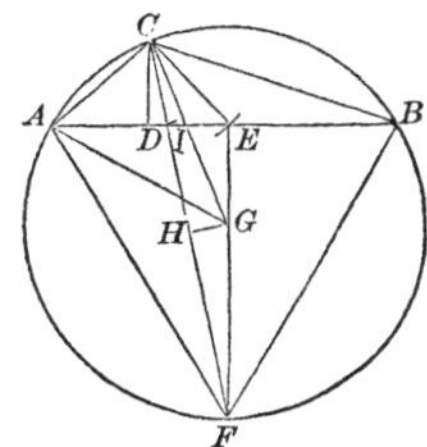

Given { the perpendicular CD, the line CE bisecting AB, and the line CI bisecting $< ACB$.

Analysis.—Suppose ACB to be the required triangle. Because the angle $ACI = BCI$, the line CI produced will bisect the arc AFB of a circle circumscribing the triangle ACB. At E erect a perpendicular to AB, and it will pass through the centre of the circle and the middle of the arc AFB. Hence CI produced, and the perpendicular EF, will meet at F. Bisect the chord CF by the perpendicular HG, and G will be the centre of the circumscribing circle. Whence this

Construction.—Draw any line, as AB, indefinitely, in it take any point D, and erect the perpendicular $DC =$ the given perpendicular. From C apply $CE =$ the given line bisecting the base, and $CI =$ the given line bisecting the vertical angle. At E erect a perpendicular to meet CI, produced, in F, bisect CF by the perpendicular HG, join CG, and with G as centre, and radius GC or GF, describe the circle $ACBF$. Join CA and CB, also AG, AF, and BF; then ABC will be the required triangle.

Demonstration.—Since arc $AF = BF$, the angle $ACI = BCI$, and CI bisects the vertical angle. Q. E. D.

Calculation. — We have $DE = \sqrt{CE^2 - CD^2}$, and $DI = \sqrt{CI^2 - CD^2}$; then $EI = DE - DI$.

1. In similar triangles CDI and FEI, we have $DI : CD :: IE : EF$, and $DI : CI :: EI : FI$; then $FC = FI + IC$.

2. In similar triangles FCA and FAI, we have $FC : FA :: FA :$

* III. 33.

FI; hence $FA=\sqrt{FC \times FI}$, and $EA=\sqrt{FA^2-EF^2}$; then $AB=2\,EA$. Also $AI=AE-EI$, and $BI=AE+EI$.

3. Again, in similar triangles FCA and FAI we have $FA:AI::FC:CA$; and in similar triangles FBI and FCB we have $FB:BI::FC:CB$. Lastly, $FE:FI::FH:FG$ or $GA=$ the radius of the circle; when all the lines are known.

Problem XXIX.—In a plane triangle are given the vertical angle, the line bisecting it, and the difference between the including sides, to determine the triangle.

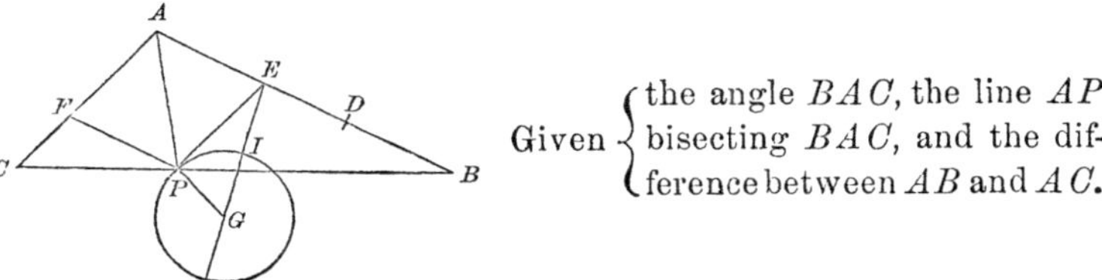

Given $\begin{cases}\text{the angle } BAC, \text{ the line } AP \\ \text{bisecting } BAC, \text{ and the dif-} \\ \text{ference between } AB \text{ and } AC.\end{cases}$

Analysis.—Let BAC represent the required triangle. Take $AD=AC$; then $BD=$ the given difference. Draw PF parallel to AB, and PE parallel to AC; then, since the angle $PAF=PAE$, $AEPF$ is a rhombus, having all its sides equal and determined. Hence $ED=AD-AE=AC-AF=CF$. Now, in similar triangles CFP and PEB, we have $CF\,(ED):FP\,(PE)::PE:EB=(ED+DB)$. Hence $ED \times (ED+DB)=PE^2$. Whence this

Construction.—Construct the rhombus $AEPF$, all of whose parts are known, and on PE erect the perpendicular $PG=\frac{1}{2}$ the given difference between AB and AC. With the centre G and radius GP, describe a circle, meeting EG, joined and produced, in I and H. Make $EB=EH$, and draw BPC; then will $AB-AC=2\,PG=$ the given difference, and BAC will be the triangle required.

Demonstration.—By IV. 30*, we have $EH\,(EB):EP::EP:EI$. Also, by similar triangles PEB and CFP, we have $EB:EP::PF\,(EP):CF$ or ED. Hence $ED=EI$, and $AB-AC=EB-CF=EB-ED=EH-EI=IH=2\,PG=$ the given difference.

Calculation.—In triangle AFP, Case 1, find $AF=AE=EP$;

* III. 36.

then $EG = \sqrt{EP^2 + PG^2}$, and $EH = EG + GP = EB$. Also, $EI = EG - GP = ED = CF$. Hence we have $AB = AE + EB$, and $AC = AF + CF$, and all the parts are known.

Remarks.—1. When the angle BAC is a right angle, $AEPF$ will be a square.

2. This problem is the same as having a point P, equidistant from the lines AB and AC, given in position, to draw through P a line so that AB and AC shall have a given difference. See Problem LXIII. of this division.

PROBLEM XXX.—To draw through a given point, between two lines forming any given angle with each other, a line, such that the parts of it intercepted between the point and the given lines shall have a given ratio to each other.

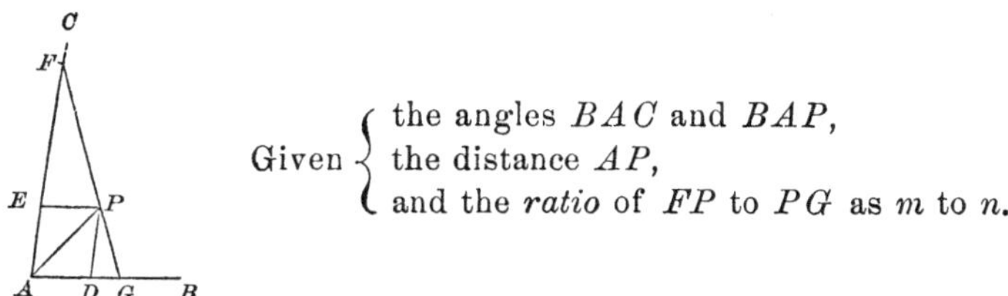

Given $\begin{cases} \text{the angles } BAC \text{ and } BAP, \\ \text{the distance } AP, \\ \text{and the } \textit{ratio} \text{ of } FP \text{ to } PG \text{ as } m \text{ to } n. \end{cases}$

Analysis.—Let AB and AC be the two lines given in position, P the given point between them, and FPG the required line, having $FP : PG :: m : n$.

Draw PE parallel to AB, and PD parallel to AC. Then, in similar triangles FEP and PDG, we have $FE : PD\ (EA) :: FP : PG :: m : n$. Whence this

Construction.—Having drawn PE and PD parallel to AB and AC respectively, take EF a fourth proportional (IV. Prob. 2*) to n, m, and AE, so that we shall have $n : m :: AE : EF$. Through F and P draw FPG, and it will be the line required.

Demonstration.—$FP : PG :: FE : EA$ or $PD :: m : n$ in the given ratio. Q. E. D.

Calculation.—1. In the triangle APD, find, by Case 1, $PD = AE$, and $AD = EP$; then $n : m :: AE : EF = \frac{m \cdot AE}{n}$. Also $m : n :: AD : DG = \frac{n \cdot AD}{m}$. Whence we have $AF = AE + EF$, and $AG = AD + DG$.

* VI. 12.

2. In triangle GAF, Case 3, find FG; then $FA : FG :: FE : FP$, and $FA : FG :: EA : PG$.

Scholium.—If FP is to be *equal* to PG, make $EF = EA$; if FP is to be 3 PG, make $EF = 3\,EA$, etc.

PROBLEM XXXI.—In a right-angled triangle are given the perimeter, and the perpendicular let fall from the right angle on the hypothenuse, to determine the triangle.

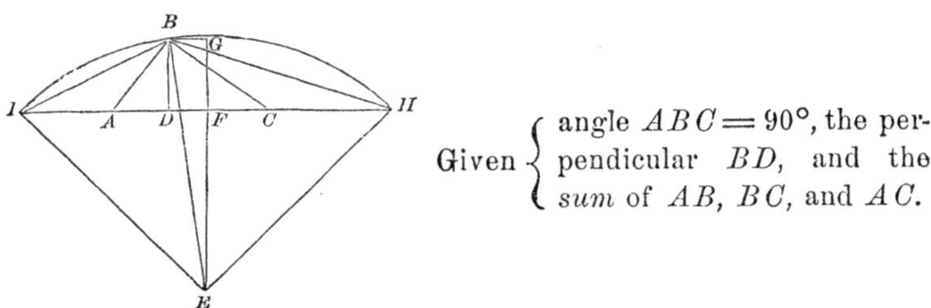

Given { angle $ABC = 90°$, the perpendicular BD, and the *sum* of AB, BC, and AC.

Analysis.—Let ABC represent the required triangle. Produce AC both ways, making $AI = AB$, and $CH = CB$; then IH is known, being equal to the given perimeter. Join IB and BH; then (I. 25, cor. 6*) the angle $AIB = ABI = \frac{1}{2} BAC$, and $\angle CHB = CBH = \frac{1}{2} ACB$. Hence $HIB + IHB = \frac{1}{2}(BAC + ACB) = \frac{1}{2}$ of $90° = 45°$. Therefore angle $IBH = 180° - 45° = 135°$, and the angle at the centre of a circle passing through I, B, and H must = twice $(180° - \angle IBH)$ = twice $(180° - 135°) = 90°$. Whence this

Construction.—Bisect the given perimeter IH in F; erect the perpendicular $FE = FH$ or FI. Join EH and EI, and, with centre E and radius EI or EH, describe the arc IBH. Produce EF, making FG = the given perpendicular; through G draw GB parallel to IH, intersecting the arc at B. Join IB and BH; make the angle $IBA = HIB$, and the angle $HBC = IHB$; then ABC will be the required triangle.

Demonstration.—It is evident (I. 12†) that $AB + BC + AC = AI + CH + AC = IH$, and that $BD = FG$ = the given perpendicular. It remains to prove that the angle ABC is a right angle. Since, by construction, $FE = FI = FH$, *each of the acute angles* in the triangles EFI and EFH is $45°$; hence the angle $IEH = 90°$. The angle $EIB = EBI$ (I. 11‡), and angle $AIB = ABI$ by construction; hence angle $EBA = EIF = 45°$. Also angle $EBH =$

* I. 32. † I. 6. ‡ I. 5.

EHB, and angle $CBH = CHB$; hence angle $EBC = EHF = 45°$. Wherefore angle $ABC = EBA + EBC = 90°$. Q. E. D.

Calculation.—$EI^2 = EF^2 + FI^2 = 2\,FI^2$, and $EI = FI\sqrt{2} = EB$; $EG = EF + FG$; $BG = \sqrt{EB^2 - EG^2} = DF$; then $ID = IF - DF$, and $HD = IF + DF$. Now, in triangle DIB, Case 3, find the angle DIB; then angle $DAB = 2\,DIB$, and angle $ACB = 90° - DAB$. In triangle ADB, Case 1, find AB and AD; and in BDC, Case 1, find BC and CD; then $AC = AD + DC$.

Limit.—The perpendicular may be any quantity not greater than the *difference* between EB and EF, or EI and EF; that is, not greater than the difference between half the perimeter multiplied by the square root of 2, and half the perimeter. If $a =$ perpendicular height and $p =$ semi-perimeter, a must be less than $p\,(\sqrt{2} - 1)$.

PROBLEM XXXII.—In a plane triangle are given the two sides, and the length of the line drawn from the included angle to the centre of the inscribed circle, to determine the triangle.

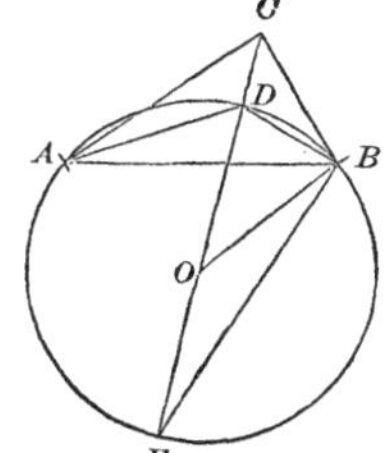

Given { the sides AC and CB, and the distance CD to the centre of the inscribed circle.

Analysis.—Let ABC represent the required triangle, of which D is the centre of the inscribed circle (the circle not drawn in the figure); then (III. Prob. 15*) the line AD bisects the angle A, BD the angle B, and CD the angle C. Let a circle be described about the triangle ADB, meeting CD, produced, in E, and join EB. Then (III. 18 and I. 25, cor. 6†) the angles $BED + BDE = BAD + (BCD + DBC) =$ the sum of half the angles A, C, and B respectively, which is equal to $90°$. Hence the angle DBE is a right angle. Also, since $\angle CAD = DAB = DEB$, and $\angle ACD = BCE$, the triangles ACD and BCE are similar, and we have $CD : CA :: CB : CE$; whence this

Construction.—Take $CD =$ the given distance, and on *it*, produced, take CE a fourth proportional to CD and the two given sides CA and CB. On DE describe a circle of which the centre is

* IV. 4. † III. 20 and I. 32.

O, and to the circle apply $CA =$ *one* of the given sides, and $CB =$ the *other*. Draw AD, DB, EB, and OB, and ABC will be the required triangle, as is evident from the analysis.

Calculation.—By construction, $CD : CA :: CB : CE = \frac{CA \times CB}{CD}$. Then $DE = CE - CD$, $OD = \frac{1}{2} ED = OB$, and $OC = OD + CD$.

Now, in triangle OBC, Case 4, find the angle OCB; then angle $ACB = 2\ OCB$.

In triangle ACB, Case 3, find angles A and B and the side AB.

In triangle ADB, Case 1, find AD and DB, then all the parts are known.

Corollary.—If D be the centre of a circle inscribed in a triangle, CD, produced, will pass through the centre of the circle circumscribing the triangle ADB.

PROBLEM XXXIII. (By a civil engineer).—Having a plane triangle given, it is required to draw a line cutting two of the sides, so that a point in that line shall be a given distance from one of the angles at the base, and the part of the line intercepted between such point and each side shall be equal to the lower segment cut from that side.

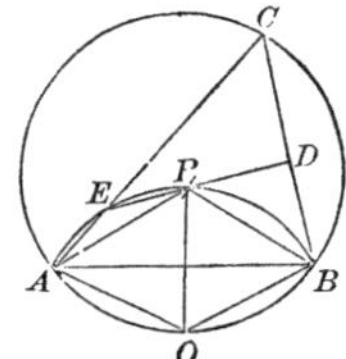

Given $\begin{cases} \text{all the angles, the side } AB\text{, and the} \\ \text{distance } BP\text{, to draw } EPD\text{, so that} \\ PD = DB\text{, and } PE = EA. \end{cases}$

Analysis.—Suppose the problem constructed, as in the figure. About the triangle ACB describe a circle, and let O be the middle of the arc AB. Join OA, OB, and OP. Now, if $OP = OB$, the angle OPD will evidently be equal to OBD; also, since $OB = OA$, the angle OPE will be equal to OAE. Whence this

Construction.—Having described a circle about the given triangle ABC, and taken O, the middle of the arc AB, with the centre O and radius equal to OA or OB, describe an arc, in which apply BP $=$ the given distance of the point from B. Then draw the line

DPE, making the angle $BPD =$ the angle DBP; then will $PD = DB$, and $PE = EA$.

Demonstration.—Since angle $BPD = DBP$ by construction, we have $PD = DB$. Hence it only remains to prove that $PE = EA$.

Now, the angles OBP and OPB being equal, also BPD and DBP, we have angle $OBD = OPD$; also the angles OBC and OAC, or OBD and OAE, together (III 18, cor. 4*) are equal to 180°. Hence $OBD + OAE = OPD + OPE$ (the sum of each pair of angles being equal to 180°) $= OBD + OPE$. Therefore $OAE = OPE$. Taking the equal angles OAP and OPA from these, we have the angle $EAP = EPA$, and therefore $PE = EA$. Q. E. D.

Calculation.—The angle $AOB = 180° - C$; hence we have each of the angles OAB and $OBA = \frac{1}{2}$ the given angle C. In triangle ABO, Case 1, find $OB = OP = OA$. In triangle OBP, find angle OBP. Now, the angle $OBC = OBA + ABC$, and the angle $PBD = OBC - OBP = BPD$. In triangle BPD, Case 1, find BD and PD. Again, in triangle ABP, Case 3, find AP, and angle BAP; then angle $EAP = BAC - BAP$. Lastly, in triangle EAP, Case 1, find $PE = EA$; then $EC = AC - EA$, and $CD = CB - BD$.

PROBLEM XXXIV.—In a rhombus are given the perimeter, and the sum of the diagonals, to find the diagonals.

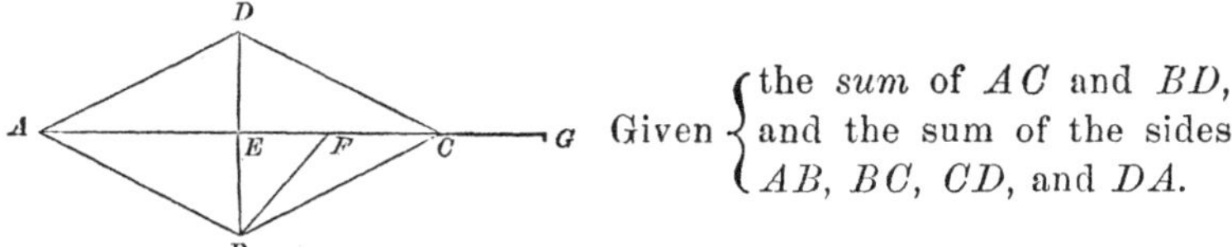

Given { the *sum* of AC and BD, and the sum of the sides AB, BC, CD, and DA.

Analysis.—Let $ABCD$ be the required rhombus, whose diagonals intersect each other at E. Since in a rhombus the sides are all equal, each side is one-fourth of the given perimeter. Also the diagonals cross each other at right angles.

Now, take $EF = EB$, then $AF = \frac{1}{2}$ the sum of the diagonals, and the angle $EFB = 45°$. Whence this

Construction.—In the indefinite line AG, take $AF = \frac{1}{2}$ the given sum of the diagonals. At F, draw a line making with FA an angle $= 45°$, to which line apply $AB = \frac{1}{4}$ the given perimeter. To AG apply $BC = BA$, draw CD parallel to AB, and AD parallel to BC. Draw the other diagonal BD; then $ABCD$ will be the required rhombus.

* III. 22.

Demonstration.—The sides are evidently all equal, and their sum equal to the given perimeter. Now, in the triangle BEF, since the angle E is $90°$, and $EFB = 45°$ by construction, angle $EBF = 45°$, and we have $EF = EB$. Hence $AC + BD = 2\ AE + 2\ BE = 2\ AE + 2\ EF = 2\ AF =$ the given sum of the diagonals. Q. E. D.

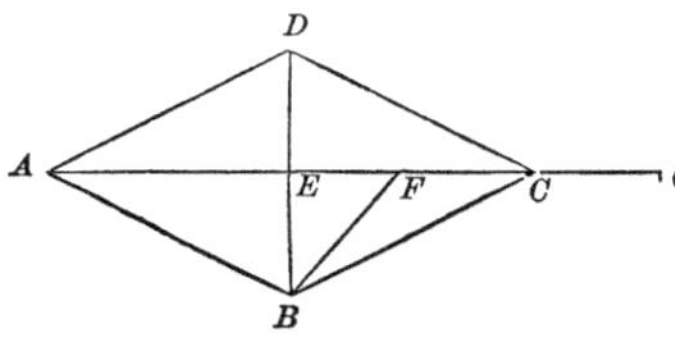

Calculation.—1. In the triangle AFB, Case 2, find angle $BAF = BAE$.

2. In the triangle BAE, Case 1, find AE and EB; then $AC = 2\ AE$, and $BD = 2\ EB$.

Limit.—Twice the square of one-fourth of the perimeter must not be *greater* than the square of half the sum of the diagonals. When twice the square of one-fourth the given perimeter is just *equal* to the square of half the sum of the diagonals, the rhombus becomes a square.

PROBLEM XXXV.—In a right-angled triangle are given the hypothenuse, and the difference between the two lines drawn from its extremities to the centre of the inscribed circle, to determine the triangle.

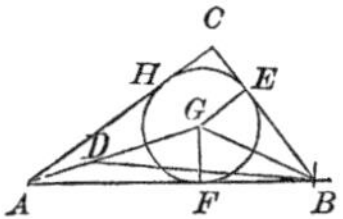

Given $\begin{cases} AB, \text{ the angle } C \text{ a right angle, and the} \\ \textit{difference} \text{ between } AG \text{ and } GB, G \text{ being} \\ \text{the centre of the inscribed circle.} \end{cases}$

Analysis.—Let ABC be the right-angled triangle required. In AG take $GD = GB$, and join BD; then $AD =$ the given difference between AG and GB. Now, since AG and BG bisect the angles BAC and ABC respectively (III. Prob. 15*), the angles $GAB + GBA = \frac{1}{2}(BAC + ABC) = 45°$. Hence $AGB = 135°$, and GDB and GBD each equal half the supplement of $135° = \frac{1}{2}\ 45° = 22\frac{1}{2}°$. Whence this

Construction.—Draw an indefinite line AG, in which take $AD =$ the given difference. Make the angle $GDB = 22\frac{1}{2}°$, and apply $AB =$ the given hypothenuse. Make the angle $DBG = 22\frac{1}{2}°$. Also, make the angle $GAC = GAB$, and the angle $GBC = GBA$; then ABC is the triangle required.

Demonstration.—Since $GD = GB$ (I. 12†), we have $AG - GB$

* IV. 4. † I. 6.

$= AG - GD = AD =$ the given difference. Also, since, by construction, the angles BAC and ABC are bisected by the lines AG and BG respectively, G is the centre of the inscribed circle.

Again, since the angles GDB and GBD, by construction, each $=$ $22\frac{1}{2}°$, the angle $AGB = 135°$; therefore the angles GAB and GBA together $= 45°$; and we have $CAB + CBA = 2\,(GAB + GBA) =$ $90°$. Whence ACB is a right angle. Q. E. D.

Calculation.—1. In triangle ADB, Case 2, find angle $DAB =$ GAB. Then angle $BAC = 2\,GAB$.

2. In triangle ABC, Case 1, find AC and BC. In triangle AGB, Case 1, find AG, GB, and GF.

Limits.—The given difference may be any quantity *less* than the hypothenuse.

When the difference is nothing, the triangle is isosceles.

PROBLEM XXXVI.—In a plane triangle are given the base, an angle at the base, and the *sum of the squares* of the other two sides, to determine the triangle.

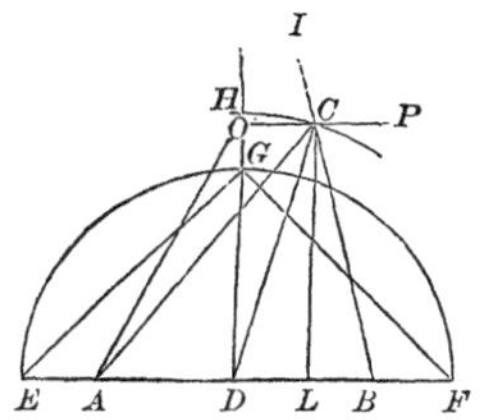

Given $\begin{cases} \text{the base } AB, \\ \text{the angle } ABC, \text{ and} \\ \text{the } sum \text{ of } AC^2 + BC^2 = m^2. \end{cases}$

Analysis.—Let ABC represent the required triangle. Bisect the given base AB in D, join CD, erect the perpendicular $DH = DC$, and join AH. Now (IV. 14*), we have $AC^2 + BC^2 = 2\,AD^2 +$ $2\,DC^2 = 2\,AD^2 + 2\,DH^2 = 2\,AH^2$. Hence the given line m, whose square is equal to the sum of the squares of AC and BC, must be the hypothenuse of an isosceles right-angled triangle, one of the equal sides of which is equal to AH. Whence this

Construction.—Draw $AB =$ the given base, bisect it in D, and make the angle $ABI =$ the given angle. With the centre D, and radius equal to half the given line m, describe a semicircle, meeting AB, produced both ways, in E and F. Then $EF = m$. At D erect the perpendicular DG, join GE and GF, and to DG, produced, apply $AH = EG$. Also, to BI apply $DC = DH$. Join AC, BC,

* II. A.

and CD, and let fall the perpendicular CL; then will ABC be the required triangle.

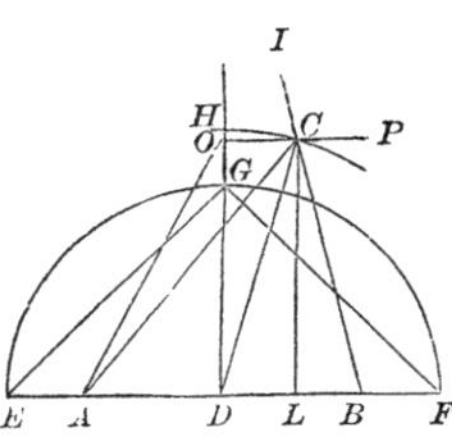

Demonstration.—We have to prove only that $AC^2 + BC^2 = EF^2 = m^2$. By IV. 14,* we have $AC^2 + BC^2 = 2\,AD^2 + 2\,DC^2 = 2\,AD^2 + 2\,DH^2 = 2\,AH^2 = 2\,EG^2$, by construction, $= EG^2 + GF^2 = EF^2 = m^2$. Q. E. D.

Calculation.—$DC^2 = DH^2 = AH^2 - AD^2 = EG^2 - AD^2 = \frac{1}{2}\,EF^2 - AD^2$. Hence DC is known. In triangle DBC, Case 2, find angle BDC and side BC; then $AC = \sqrt{EF^2 - BC^2}$. Or, in $\triangle\ ADC, \angle ADC = 180° - BDC$; by Case 3, find AC.

NOTE.—If the perpendicular height CL is given, instead of the angle ABC, lay $DO =$ the given perpendicular, draw OP parallel to AB, to which apply $DC = DH$, and join AC and CB. The *demonstration* is the same as before. In the *calculation*, find DC^2 as before, then $DL = \sqrt{DC^2 - CL^2}$, $AL = AD + DL$, and $BL = AD - DL$. Then $AC = \sqrt{AL^2 + LC^2}$, and $BC = \sqrt{BL^2 + LC^2}$.

Limits in the last case.—The square of the given perpendicular must not be greater than half the square of the given line, diminished by the square of half the base. That is, DO must not be greater than DH. If DO is just equal to DH, the triangle ABC will be *isosceles*.

PROBLEM XXXVII.—In a plane triangle are given the base, an angle at the base, and the *difference* between the squares of the other two sides, to determine the triangle.

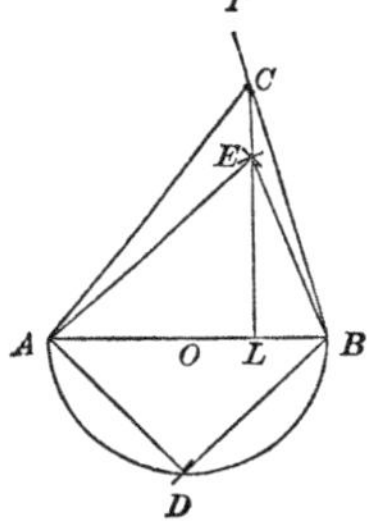

Given $\begin{cases} \text{the base } AB, \\ \text{the angle } ABC, \text{ and} \\ \text{the } \textit{difference} \text{ of } AC^2 \text{ and } BC^2 \text{ equal } m^2. \end{cases}$

Analysis.—Let ABC represent the required triangle, and let fall the perpendicular CL on the base. Then $AC^2 - AL^2 = CL^2 =$

* II. A.

$BC^2 - BL^2$. Whence $AC^2 - BC^2 = AL^2 - BL^2 = m^2$, and we have this

Construction.—On the given base AB describe a semicircle, in which apply $AD = m$, and join BD. With A and B as centres, and radii equal to AB and BD respectively, describe arcs intersecting in E. Make the angle $ABI =$ the given angle, and through E draw LEC perpendicular to AB, meeting BI in C. Join AC and BC, and ABC will be the triangle required.

Demonstration.—We have to prove only that $AC^2 - BC^2 = AD^2 = m^2$. We have $AC^2 - AL^2 = CL^2 = BC^2 - BL^2$. Hence $AC^2 - BC^2 = AL^2 - BL^2$. Also, $AE^2 - AL^2 = EL^2 = BE^2 - BL^2$. Hence $AE^2 - BE^2 = AL^2 - BL^2$. Wherefore we have $AC^2 - BC^2 = AE^2 - BE^2 =$ (by construction) $AB^2 - BD^2 = AD^2 = m^2$. Q. E. D.

Corollary.—The difference of the squares of two lines drawn from *any point* in the perpendicular LC, or LC produced either way, to the points A and B, will always be equal to AD^2, or $AC^2 - BC^2 = AL^2 - BL^2$.

Calculation.—$BE = BD = \sqrt{AB^2 - AD^2}$. Then, in triangle AEB, Case 4, $AB : AE + BE :: AE - BE : 2\,OL$, O being the centre of the semicircle. Hence we have $AL = AO + OL$, and $BL = AO - OL$.

Now, in triangle BLC, Case 1, find CL; then $AC = \sqrt{AL^2 + CL^2}$, and $BC = \sqrt{BL^2 + CL^2}$.

NOTE.—If the perpendicular LC were given instead of angle ABC, make the perpendicular $LEC =$ the given perpendicular, and join CA and CB. The *demonstration* and *calculation* will remain the same.

Limit.—The length of m must always be less than the base; that is, AD must be less than AB.

PROBLEM XXXVIII.—"In a city are three steeples visible from a station, at which the angle between the first and second steeples was taken, also the angle between the second and third, the second steeple being nearest the station. The three respective distances between the three steeples are known. It is required to find their respective distances from the station or place of observation."—From Gummere's Surveying.

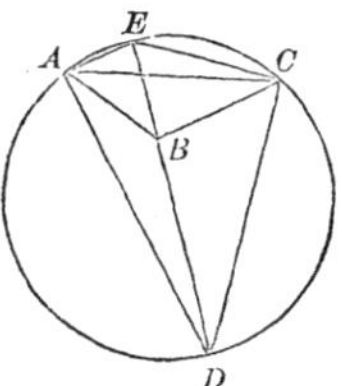

Given $\begin{cases} \text{the distances } AB,\ BC, \text{ and } AC \\ \text{the angles } ADB \text{ and } BDC, \text{ and} \\ \text{consequently } ADC. \end{cases}$

Analysis.—Let A, B, and C represent the positions of the first, second, and third steeples respectively, and D the station. Through the three points A, D, and C pass the circumference of a circle. Join DB, and produce it to meet the circumference in E, and join AE and CE. Then (III. 18, cor. 1*) the angle $ACE =$ the given angle ADB between the first and second steeples, and the angle $CAE =$ the given angle BDC between the second and third steeples. Whence this

Construction.—With the three given distances form the triangle ABC, having B nearest to D, make the angle $ACE =$ the given angle between the first and second steeples, and $CAE =$ that between the second and third steeples. About the triangle AEC (III. Prob. 13†) describe a circle; join EB, and produce it to meet the circumference at D, the station required. Join DA and DC.

Demonstration.—By III. 18, cor. 1,‡ the angle $ADB = ACE$, and $BDC = CAE$, which are the given angles by construction. Q. E. D.

Calculation.—1. In triangle BAC, Case 4, find angle BAC; then angle $BAE = BAC + CAE$. In triangle EAC, Case 1, find AE.

2. In triangle BAE, Case 3, find angles ABE and $AEB = AED = ACD$; then angle $ABD = 180° - ABE$.

3. In triangle ABD, Case 1, find DA and DB, and in triangle ADC, Case 1, find DC.

NOTE.—If B is *farther* from D than A and C are, construct the triangle ABC with B *above* the line AC. Join BE, and produce it to D, and proceed as before, only, in this case, the angle BAE will equal $BAC - CAE$.

* III. 21. † IV. 5. ‡ III. 21.

PROBLEM XXXIX.—"There are given the heights of two columns, the distance of the foot of the lower column from the middle of the base of a statue in a straight line between them, and the distances from the head of the statue to the top of each column, to determine the distance between the tops of the columns; also the height of the statue."—Gummere's Surveying.

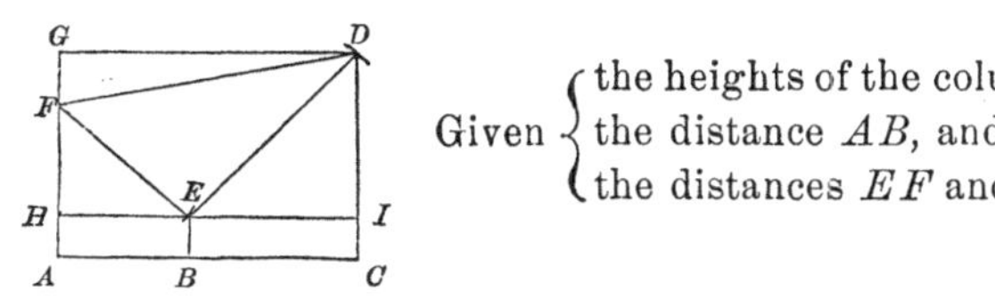

Given $\begin{cases} \text{the heights of the columns } AF \text{ and } CD, \\ \text{the distance } AB, \text{ and} \\ \text{the distances } EF \text{ and } ED. \end{cases}$

Analysis.—Suppose AF and CD, in the annexed figure, to represent the columns, and BE the statue. Through D, the top of the higher column, draw DG parallel to AC, to meet AF, produced, in G. Then $AG = CD$, and we have the line GD given in position, and AB, AF, FE, and ED given in length; whence this

Construction.—On an indefinite horizontal line lay AB equal the distance of the foot of the lower column from the middle of the base of the statue. At A and B erect perpendiculars, and lay $AF =$ the height of the lower column, and $AG =$ that of the higher. Now, the head of the statue is in the perpendicular erected at B. To this perpendicular apply FE equal the given distance from the top of the lower column to the head of the statue, and BE will be the statue's height. Through G draw a line parallel to AB, and the top of the higher column must be in this line. To this line apply ED equal to the given distance from the head of the statue to the top of the higher column; let fall the perpendicular DC on the horizontal line, and join DF, which is evidently the distance between the tops of the columns required.

Calculation.—Through E draw HEI parallel to ABC. Then $HE = AB$, and $FH = \sqrt{FE^2 - AB^2}$, $AH = AF - FH = BE =$ the height of the statue $= IC$. And $DI = DC - IC$. Also $EI = \sqrt{ED^2 - DI^2}$. Now, $HI = HE + EI = AB + EI = GD$. And $GF = AG - AF$. Hence we have $DF = \sqrt{GD^2 + GF^2} =$ the required distance between the tops of the columns.

PROBLEM XL.—In a plane triangle are given the base and perpendicular height, to determine it, such that *one* of the angles at the base shall be double the *other*.

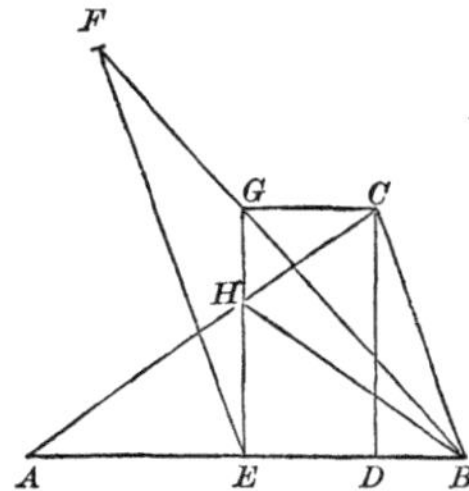

Given $\begin{cases} \text{the base } AB \text{ of the triangle } ABC, \\ \text{the height } CD, \text{ and} \\ \text{angle } ABC \text{ to be equal to } 2\,BAC. \end{cases}$

Analysis.—Draw BH bisecting the angle ABC; then each of the angles ABH and HBC will be equal to the angle BAC. Through H draw EHG parallel and equal to the perpendicular CD, and join CG, which will be parallel to AB. Then (I. 11 and 28*) we have $AH = BH$, $AE = BE$, and $GC = ED$. Also, since BH bisects the angle ABC (IV. 17†), we have $AB : BC :: AH : HC ::$ $AE : ED$ or CG. That is, $AB : BC :: AE : CG$. But $AB = 2\,AE$, therefore $BC = 2\,CG$. Draw EF parallel to BC, meeting BG, produced, in F. Then the triangles BEF and GCB will be similar, and, since the side $BC = 2\,CG$, we have $FE = 2\,BE = AB$. Whence this

Construction.—Draw $AB =$ the given base, bisect it in E, erect the perpendicular $EG =$ the given height, join BG, and to it, produced, apply $EF = AB$. Draw BC parallel to EF, GC parallel to AB, join AC and BH, and let fall the perpendicular CD; then ABC will be the required triangle.

Demonstration.—In similar triangles FEB and BCG, we have $FE\ (AB) : BC :: EB : CG :: AE : ED :: AH : HC$; that is, $AB :$ $BC :: AH : HC$. Hence (IV. 17‡) BH bisects the angle ABC, and we have angle $CBH = ABH =$ (I. 11§) BAH, and $ABC = 2\,BAC$. Q. E. D.

Calculation.—1. In triangle BEG, Case 3, find angle $EBG =$ EBF. 2. In triangle EBF, Case 2, find angle $EFB = GBC$; then angle $ABC = EBG + GBC$, and angle $BAC = \frac{1}{2}\,ABC$.

3. In triangle ABC, Case 1, find AC and BC; then $ED = CG$ $= \frac{1}{2}\,BC$, and $AD = AE + ED$, and $BD = AE - ED$.

Scholium.—As $EF = AB$ is greater than EB, the arc described with EF can cut BG, or BG produced, in but one point, and the

* I. 5 and 34. † VI. 3. ‡ VI. 3. § I. 5.

angle EFB can have but one value. Also, since $AB = 2\,ED + 2\,DB = 2\,CG + 2\,BD = BC + 2\,BD$, we have constructed the right-angled triangle CDB, having given the base CD, and the *sum of the hypothenuse* BC, and twice the perpendicular BD (equal to AB).

PROBLEM XLI.—In an isosceles triangle are given the angle between the equal sides, and the three distances from a point, either *within* or *without* the given triangle, to the three angles, to determine the triangle.

FIG. 1.

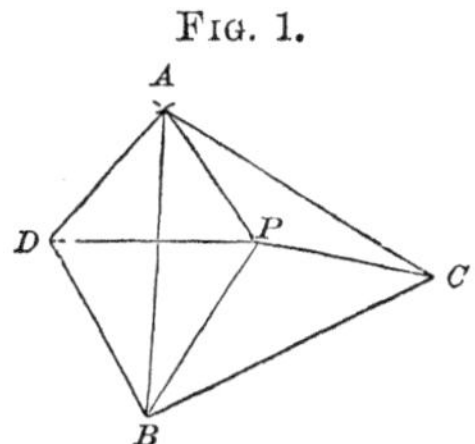

Given, the angle ABC, the sides BA and BC to be equal, and the distances PA, PB, and PC, in either Figure.

FIG. 2.

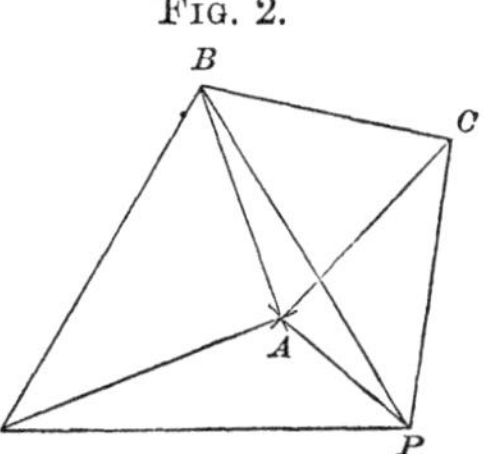

Analysis.—Suppose ABC (referring to *either* figure) to represent the required triangle, with ABC the given angle, and $BA = BC$. At B make the angle $PBD =$ the given angle ABC, lay $BD = BP$, and join AD. Then, comparing the triangles ABD and CBP, the angle $ABD = PBD - PBA = ABC - PBA = CBP$; also the including sides AB and BD, in one triangle, are equal to CB and BP in the other. Hence (I. 6*) $AD = PC$. Whence this

Construction.—Draw $BP =$ the given distance from the point to the angle B, make the angle $PBD =$ the given angle, lay $BD = BP$, and join PD. With P and D as centres, and radii equal to the other two given distances respectively, describe arcs, intersecting in a point A on the *opposite* side of DP from B, when the point P is to be *within* the triangle (Fig. 1), but on the *same* side of DP that B is when the point P is to be *without* the triangle (Fig. 2). Join AD, AP, and AB. Make the angle $ABC = DBP$, lay $BC = AB$, and join AC and PC; then ABC will be the required triangle, PC being equal to AD, as is shown in the analysis.

Calculation.—1. In triangle DBP, Case 1, find DP. 2. In triangle APD, Case 4, find angle APD; then (in Fig. 1) angle $APB = APD + DPB = APD + \frac{1}{2}$ the supplement of the given angle ABC or PBD. And (in Fig. 2) angle $APB = DPB - APD = \frac{1}{2}$ the supplement of the given angle ABC or PBD,— APD.

* I. 4.

3. In triangle APB, Case 3, find $AB = BC$. 4. In triangle ABC, Case 1, find AC.

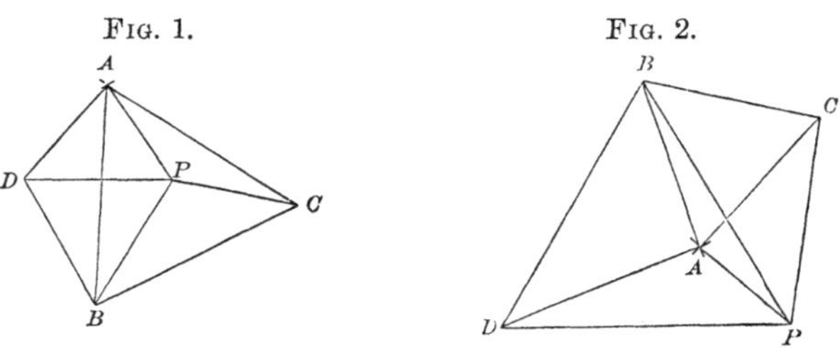

Limits.—1. $PA + PC$ must be greater than PB, the longest given distance.

2. The angle ABC may be made on *either* side of AB; but if made on the other side, the letters P and D must be transposed.

Scholium.—When the triangle ABC is to be *equilateral*, the *analysis*, *construction*, and *calculation* are precisely the same as in the above problem, using the same letters; only in this case it is simpler to describe *equilateral triangles* on PB and BA, as $PD = PB$ and $AC = AB$.

Problem XLII.—Given the three distances from a point, either *within* or *without* a square, to the three *nearest* corners, to determine the square.

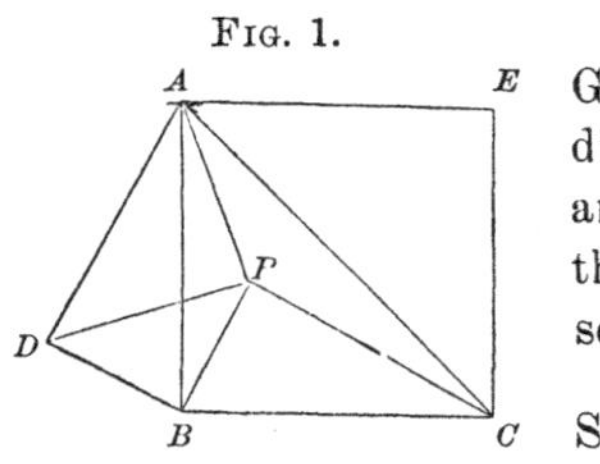

Given the three distances PA, PB, and PC, drawn to the corners of the square.

See either Figure.

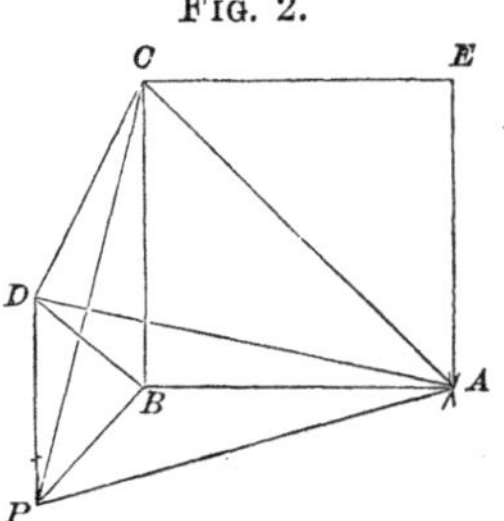

Analysis.—Join AC. Then in the isosceles triangle ABC (either figure), the angle ABC, being a right angle, is given, and the three distances PA, PB, and PC; hence we have to construct the triangle ABC *precisely* as in the last problem, using the same letters, and then complete the square $ABCE$, which will be the one required.

Property in relation to a square.—If from *any* point lines be drawn to the four corners of a square, the *sum of the squares* of the lines drawn to the extremities of *one* diagonal, will be equal to the *sum of the squares* of the lines drawn to the extremities of the *other* diagonal.

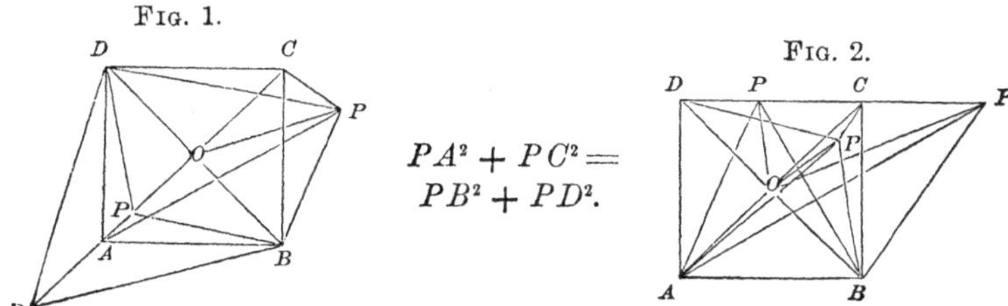

Demonstration.—In Figure 1, give attention to the point P *without the square*, or *in the diagonal*, or *in the diagonal produced, separately*, and the lines drawn *from the point under consideration* to the corners A, B, C, and D of the square; then we have (IV. 14*) $PA^2 + PC^2 = 2\,PO^2 + 2\,AO^2 = 2\,PO^2 + 2\,DO^2 = PB^2 + PD^2$. In like manner, in Figure 2, giving attention to the point P *within the square*, or *in the side*, or *in the side produced, separately*, we have $PA^2 + PC^2 = 2\,PO^2 + 2\,AO^2 = 2\,PO^2 + 2\,DO^2 = PB^2 + PD^2$. When the point P is in the diagonal, or the diagonal produced, it will be observed that $PA =$ the difference between PO and AO, and $PC =$ their sum. Hence $PA^2 + PC^2 = (PO \backsim AO)^2 + (PO + AO)^2 = 2\,PO^2 + 2\,AP^2$. Q. E. D.

PROBLEM XLIII.—From a given point in the base of a given right-angled triangle to draw a line to a point in the hypothenuse, such, that if a line be drawn from this last point, parallel to the base, to meet the perpendicular, these two last lines may be equal.

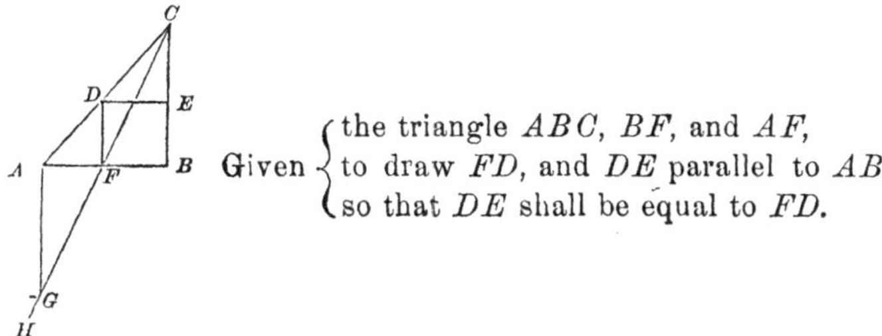

Analysis.—Suppose FD drawn so that DE parallel to AB shall be equal to FD. Draw CFH, meeting a line through A, parallel to

* II. A.

FD, in G. Then, by similar triangles CDE and CAB, we have $CD : DE :: CA : AB$; also, in similar triangles CDF and CAG, we have $CD : DF :: CA : AG$; whence $DE : DF :: AB : AG$; and in order to have $DF = DE$ we must have $AG = AB$; wherefore this

Construction.—To CF, produced, apply $AG = AB$, draw FD parallel to AG, and DE parallel to AB, and they will be the equal lines required.

Demonstration.—By similar triangles, $CA : AB :: CD : DE$; also, $CA : AG\ (AB) :: CD : DF$. Hence $DE = DF$. Q. E. D.

Calculation.—1. In triangle BFC, Case 3, find angle BCF and CF; then angle $ACG = BCA - BCF$.

2. In triangle ACG, Case 2, find CG; then $CG : AG\ (AB) :: CF : FD = DE = \dfrac{AB \times CF}{CG}$.

Scholium 1.—D is the centre of a circle that would pass through F, and be tangent to the perpendicular BC, at E.

Scholium 2.—If it is desired that $DF = 2\,DE$, apply $AG = 2\,AB$; and so of any other ratio; take AG to AB in the same ratio that it is desired to have DF to DE, and then proceed as in the problem.

NOTE.—The same *analysis*, *construction*, *demonstration*, and *calculation*, also *scholium* 2, apply when the angle B is *acute* or *obtuse*.

PROBLEM XLIV.—It is required to find a point in the perpendicular of a given right-angled triangle, such that two lines drawn from said point, one to a *given point* in the base, the other parallel to the base and meeting the hypothenuse, shall be equal.

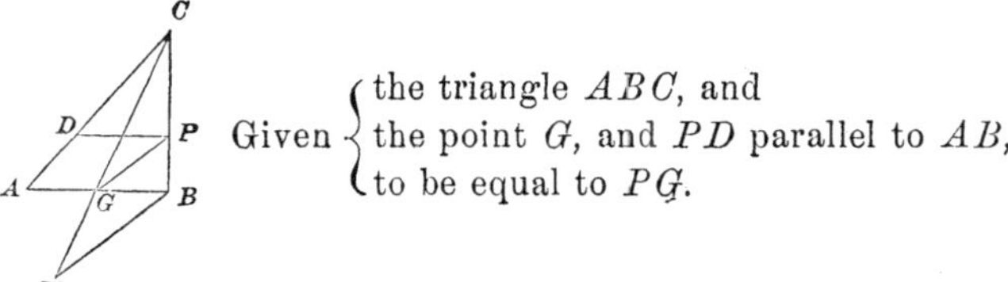

Given { the triangle ABC, and the point G, and PD parallel to AB, to be equal to PG.

Analysis.—Suppose PD drawn parallel to AB and equal to PG. Join CG, and produce it to meet BH, drawn parallel to PG, in H. Then, by similar triangles CPD and CBA, we have $CP : CB ::$

$PD:BA$. Also, by similar triangles CPG and CBH, we have $CP:CB::PG:BH$. Hence $PD:PG::AB:BH$, and, in order that PD shall equal PG, we must take $BH=AB$. Whence this

Construction.—Join CG, and to it, produced, apply $BH=BA$. Draw GP parallel to BH, and PD parallel to AB; then $PG=PD$.

Demonstration.—By similar triangles, $CB:CP::BA:PD$; also, $CB:CP::BH(BA):PG$. Hence $PG=PD$. Q. E. D.

Calculation.—1. In triangle BCG, Case 3, find angle BCG and CG.

2. In triangle BCH, Case 2, find CH. Then, by similar triangles BCH and PCG, $CH:CG::BH(AB):PG=PD$.

Scholium.—If it is desired that PG shall be *equal to* $2\,PD$, apply $BH=2\,AB$; and so of any other ratio, take BH to AB in the same ratio that it is desired to have PG to PD, and proceed as in the problem.

NOTE.—The same *analysis*, *construction*, *demonstration*, and *calculation*, as also *scholium*, apply when the angle B is *acute* or *obtuse*.

PROBLEM XLV.—In a plane triangle are given the base, an angle at the base, and the *sum* of the side opposite this angle, and the perpendicular height of the triangle, to determine the triangle.

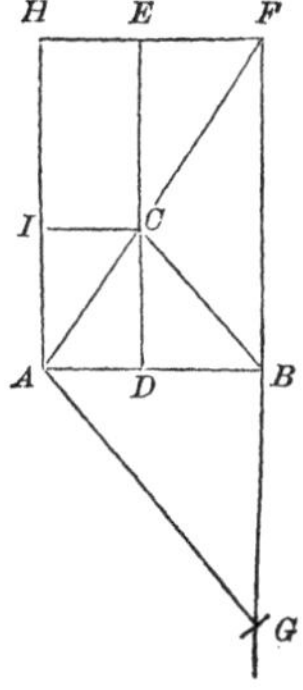

Given { in the triangle BAC the base AB, the angle BAC, and the *sum* of BC and CD.

Analysis.—Suppose ABC to represent the required triangle. Produce the perpendicular DC, making $CE=CB$; then DE is known, being equal to the given sum of BC and CD. Through E, parallel to AB, draw a line, meeting AC, produced, in F; and through A, draw a line parallel to BC to meet FB, joined and produced, in G. Now, by similar triangles FCE and ACD, $FC:CA::EC:CD$. By composition, $FC:FA::EC:ED$. Also, by similar triangles FCB and FAG, $FC : FA :: CB$ or $EC : AG$. Therefore $AG = ED$; and we have this

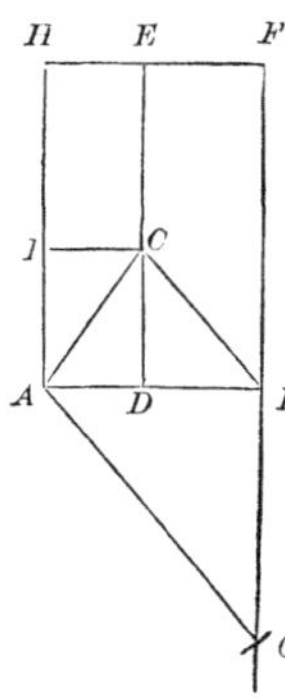

Construction.—Draw $AB =$ the given base, and make the angle BAF equal to the given angle. Erect the perpendicular $AH =$ the given sum of the side and perpendicular, draw HF parallel to AB, join FB, and to it, produced, apply $AG = AH$. Draw BC parallel to AG, then ABC will be the triangle required, having $BC + CD = AH$.

Demonstration.—Through C, parallel to AH, draw DCE, and draw CI parallel to AB; then we must prove that $BC = CE$. By parallel lines we have $AF : FC :: AH : HI = CE$. Also, $AF : FC :: AG$ or $AH : BC$; whence $BC = CE$. And $BC + CD = EC + CD = ED = AH$. Q. E. D.

Calculation.—1. In triangle AHF, Case 1, find AF.

2. In triangle AFB, Case 3, find angle AFB and FB.

3. In triangle AFG, Case 2, find FG; then $FG : FB :: AG : BC$. Also $DC = DE - BC$, and $DE : DC :: AF : AC$.

PROBLEM XLVI.—In a plane triangle are given one angle, an adjacent side, and the angle made with the side opposite the given angle by a line drawn from the middle point of said side to the given angle, to determine the triangle.

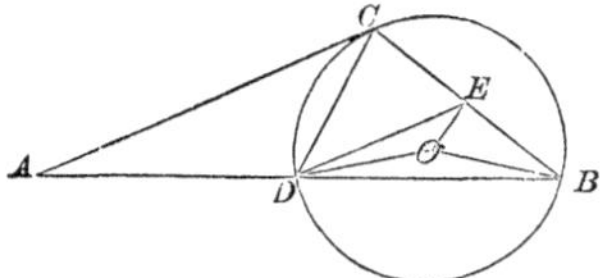

Given $\begin{cases} \text{the angle } ACB, \text{ the side } BC, \\ \text{and the angle } BDC, D \text{ being} \\ \text{the middle of } AB \end{cases}$

Analysis.—Suppose the figure constructed. Draw DE parallel to AC. Then, since $BD = DA$, we have $BE = EC$. Also, since the side BC and the angle BDC are known, the segment BDC is known (III. Prob. 16*). Whence this

Construction.—Make $BC =$ the given side, and $BCA =$ the given angle of the triangle. On BC (III. Prob. 16†) describe a circular segment to contain the angle to be made with the side opposite the given angle. Bisect BC in E, and through E, parallel to CA, draw a line to meet the segment in D. Join BD and CD, and produce BD, to complete the required triangle ABC.

* III. 33. † III. 33.

Demonstration.—Since BC, and the angles ACB and BDC, are the given quantities by construction, we have to prove only that D is the middle point of AB.

Now, since ED is parallel to AC, and $BE = EC$ by construction, we have (IV. 15*) $BD = DA$. Q. E. D.

Calculation.—From O, the centre of the circular segment, draw OB, OD, and OE; then, 1. In right-angled triangle BOE, angle $BOE = BDC$; hence, Case 1, find OE, and $OB = OD$; then the angle OED is equal to the difference between the given angle BCA or BED, and $90°$.

2. In triangle OED, Case 2, find ED; then $CA = 2\,ED$.

3. In triangle ACB, Case 3, find AB; then $BD = \frac{1}{2} AB = AD$.

Problem XLVII.—In a right-angled triangle are given the *difference* between the base and perpendicular, and also the *difference* between the base and hypothenuse, to determine the triangle.

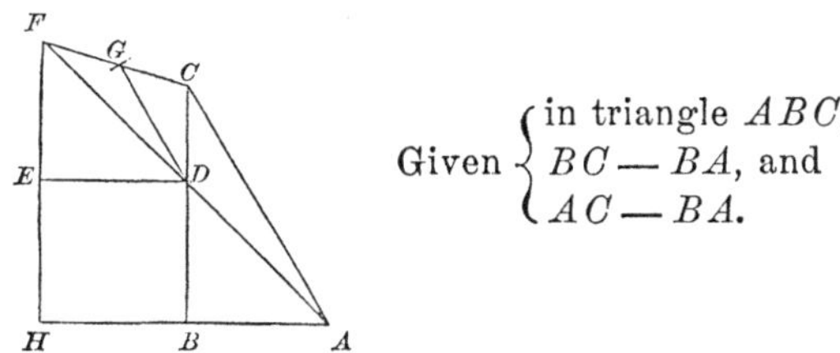

Given $\begin{cases} \text{in triangle } ABC, \\ BC - BA, \text{ and} \\ AC - BA. \end{cases}$

Analysis.—Let ABC represent the required triangle. Take $BD = BA$, then DC is known, and the angle $BAD = \frac{1}{2}$ a right angle. Also take $AH = AC$, then BH is known. Draw HF parallel to BC, and DE parallel to AB; then, since $BD = BA$, we have $HF = AH = AC$, and $EF = ED = HB$. Draw DG parallel to AC, meeting FC in G; then we have, by similar triangles, $FA : FD :: AH : DE$, also $FA : FD :: AC$ or $AH : DG$; wherefore $DG = DE$, and we have this

Construction.—Draw DE and EF at right angles, and each equal to the given difference between the hypothenuse and base, and DC parallel to EF, and equal to the given difference between the base and perpendicular. Join FD and FC, and to FC apply $DG = DE$. Through C, parallel to DG, draw a line meeting FD, produced, in A; and through A, parallel to DE, draw a line meeting FE and CD, produced, in H and B respectively. Then ABC will be the triangle required, having $AC - BA = DE$, and $BC - BA = DC$.

* VI. 2.

Demonstration.—Since, by construction, $EF = ED$, we have $BD = BA$, and $HF = AH$. Hence $BC - BA = BC - BD = DC$, which is equal to the given difference by construction. Also, by similar triangles, $FD : FA :: DE : AH$; $FD : FA :: DG$ or $DE : AC$; whence $AC = AH$. Hence $AC - BA = AH - BA = BH = DE =$ the given difference by construction. Q. E. D.

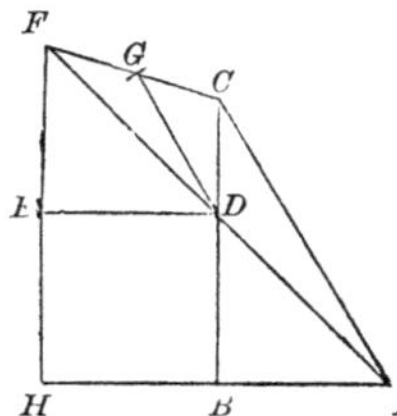

Calculation.—$FD = \sqrt{2\,ED^2}$. 1. In triangle FDC, Case 3, find angles, and FC. 2. In triangle FGD, Case 2, find FG; then, by similar triangles, $FG : FC :: DG$ or $DE : AC = AH$.

Now, $AB = AH - BH = AC - DE$, and $BC = BD + DC = AB + DC$.

PROBLEM XLVIII.—In a right-angled triangle are given the *sum* of the base and perpendicular, and also the *sum* of the base and hypothenuse, to determine the triangle.

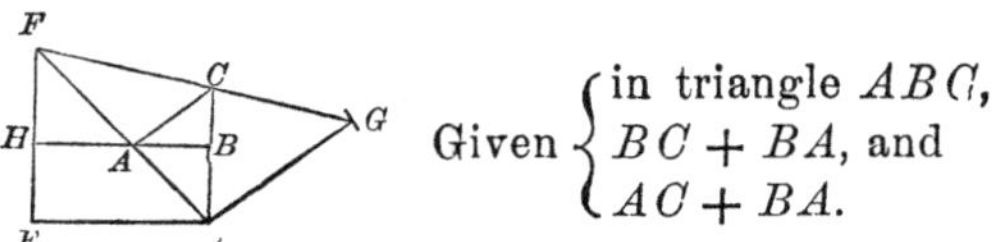

Given $\begin{cases} \text{in triangle } ABC, \\ BC + BA, \text{ and} \\ AC + BA. \end{cases}$

Using the term *sum* instead of *difference*, and to FC, *produced*, applying $DG = DE$, as in the above figure, the *analysis*, *construction*, *demonstration*, and *calculation* will be in the precise terms given for the preceding problem, making proper modification in regard to produced lines.

PROBLEM XLIX.—In a plane triangle are given the perpendicular height, and the radii of its inscribed and circumscribed circles, to determine the triangle.

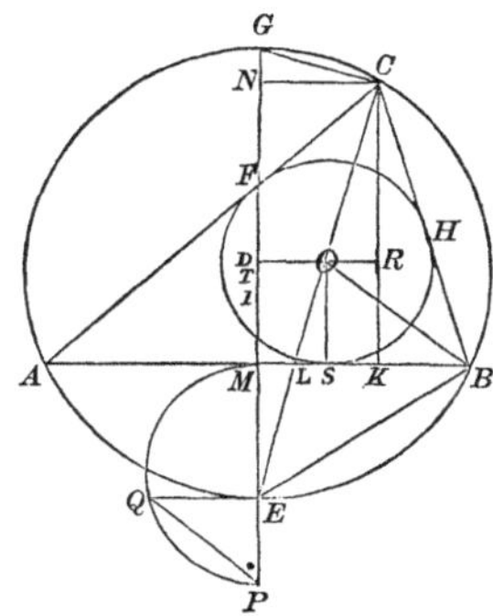

Given $\left\{ \text{in triangle } ABC, \text{ the perpendicular } CK, \text{ and the radii } OS \text{ and } ET, \text{ or } TG, \text{ of its inscribed and circumscribed circles.} \right.$

Analysis.—Let ABC represent the required triangle, CK the given perpendicular, O the centre of the inscribed circle, and T the centre of the circumscribed. Join BO and CO, producing the latter to meet the circumscribed circle in E, and draw the diameter $EMTG$. Also, draw ROD and CN parallel to AB, and join EB. Now (I. 25, cor. 6*), the angle $EOB = OCB + OBC = ACO + OBA$ $=$ (III. 18, cor. 1†) $ABE + OBA = OBE$. Hence the triangle EOB is isosceles, and we have $EO = EB$. Again, by similar triangles EBL and ECB, $EL : EB$ or $EO :: EB$ or $EO : EC$. By division, $EO - EL : EO :: EC - EO : EC$; that is, $OL : EO ::$ $CO : EC$, or $EC : EO :: CO : OL :: CR : OS$. Hence $EC \times OS =$ $EO \times CR$, and $EC = \frac{EO \times CR}{OS}$.

In similar triangles EDO and ENC, $ED : EO :: EN : EC$; that is, $EM + OS : EO :: EM + CK : \frac{EO \times CR}{OS}$. Or, by dividing the second and fourth terms by $\frac{EO}{OS}$, we have $EM + OS : OS :: EM +$ $CK : CR$. From which $EM = \frac{OS \times (CK - CR)}{CR - KR} = \frac{OS^2}{CK - 2\,OS}$ That is, EM is a third proportional to the *excess* of the *perpendicular height* of the triangle *above* the *diameter* of the *inscribed circle*, and the *radius* of the same circle. Whence this

Construction.—With the given radius ET describe the circle $AEBC$, and draw the diameter ETG. Take $EI =$ the diameter of

* I. 32. † III. 21.

the inscribed circle, draw EQ perpendicular to EI and equal the radius of the inscribed circle, and $IP =$ the given height of the required triangle. Through P and Q describe the semicircle PQM, and draw AMB parallel to EQ. Take $MN = IP$, the given height, draw NC parallel to AB, and join AC and BC; then will ABC be the required triangle. Join EC and EB, and take $EO = EB$; then will O be the centre of the inscribed circle. Let fall the perpendiculars OS and CK, join CG, and draw ROD parallel to AB.

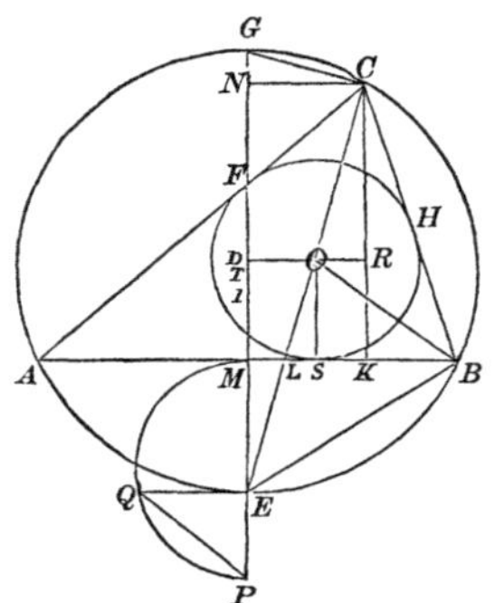

Demonstration.—$PE = IP - IE = CK - 2\,OS$. Also, $EQ = OS$. Hence $EM = \frac{EQ^2}{EP} = \frac{OS^2}{CK - 2\,OS}$. Whence the construction becomes evident from the analysis.

Calculation.—$EM = \frac{OS^2}{CK - 2\,OS}$; EB (IV. 23*) $= \sqrt{EG \times EM} = EO$; $CR = CK - OS$; EC, by the analysis, $= \frac{EO \times CR}{OS}$; $EN = EM + CK$; $EN : EC :: EM : EL = \frac{EC \times EM}{EN}$; $ML = \sqrt{EL^2 - EM^2}$; $AL = AM + ML$; $BL = BM - ML$; $AB = AL + BL$. Lastly, $EL : LB :: EB : BC = \frac{LB \times EB}{EL}$, and $EL : LA :: EA$ or $EB : AC = \frac{LA \times EB}{EL}$, when all the parts are known.

* VI. 8, cor.

PROBLEM L.—From the vertical angle of a given triangle to draw a line to a point in the base, such that the square of this line shall be equal to the rectangle of the segments of the base made by that point.

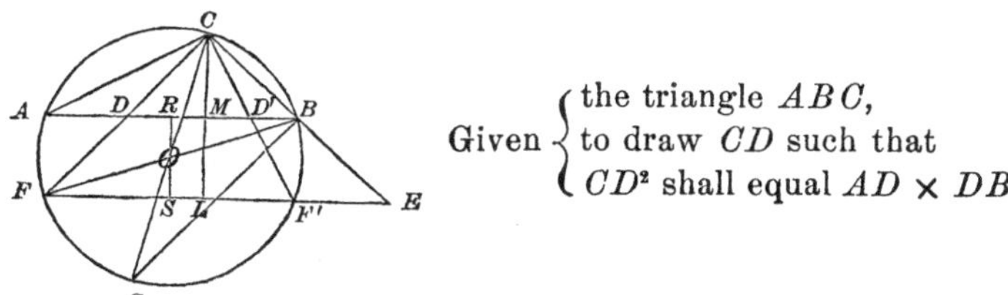

Given $\begin{cases} \text{the triangle } ABC, \\ \text{to draw } CD \text{ such that} \\ CD^2 \text{ shall equal } AD \times DB. \end{cases}$

Analysis.—About the given triangle ACB describe a circle whose centre is O, and, supposing CD to be the required line, produce it to meet the circumference at F. Then (IV. 28*) we have $AD \times DB = CD \times DF$. And, by the conditions of the problem, $AD \times DB = CD^2$; whence $CD \times DF = CD^2$, or $DF = CD$. Whence this

Construction.—Produce CB, making $BE = CB$. Through E draw a line parallel to AB, meeting a circle through ABC in F and F', and draw CDF and CDF'; then CD or CD' will be the line required.

Demonstration.—Since $BE = CB$, we have, by parallel lines, $DF = CD$, and $D'F' = CD'$. Hence $CD^2 = CD \times DF = AD \times DB$, and $CD'^2 = CD' \times D'F' = AD' \times D'B$. Q. E. D.

Calculation.—Draw the diameter COG, join GB, OB, and OF, and draw CML and ROS perpendicular to AB; then $CM = ML = ROS$. 1. In right-angled triangle CGB, the angle $G =$ the given angle CAB; find GC, Case 1. Then, in similar triangles GCB, ACM, $GC : CB :: AC : CM = ML = RS$. And $BM = \sqrt{BC^2 - CM^2}$, $EL = 2\,BM$, and $RM = BR - BM = \frac{1}{2}AB - BM = SL$.

2. $OR = \sqrt{OB^2 - BR^2}$, $OS = RS - OR = CM - OR$, and $FS = F'S = \sqrt{OF^2 - OS^2}$.

3. $ES = EL + SL$, $FE = ES + FS$, $F'E = ES - FS$, $BD = \frac{1}{2}FE$, $BD' = \frac{1}{2}EF'$, $AD = AB - BD$, $AD' = AB \sim BD'$, $CD = \sqrt{AD \times DB}$, and $CD' = \sqrt{AD' \times D'B}$.

Scholium 1.—When EF does not meet the circle, the problem is impossible.

Scholium 2.—If we want $AD \times DB : CD^2 :: m : n$, make $BE : BC :: m : n$, draw EF parallel to AB, and draw CDF or $CD'F'$; then CD or

* III. 35.

CD' will be the line required. For $n:m::BC:BE::CD:DF::CD':D'F'$. Hence $DF=\frac{m}{n}\cdot CD$, and $D'F'=\frac{m}{n}\cdot CD'$. Also $AD\times DB=CD\times DF=\frac{m}{n}\cdot CD^2$, and $AD'\times D'B=CD'\times D'F'=\frac{m}{n}\cdot CD'^2$. Whence $AD\times DB:CD^2::m:n$, and $AD'\times D'B:CD'^2::m:n$.

PROBLEM LI.—Given, a point *between* two parallel lines, and a point *in* one of these parallels, to find another point in this parallel, from which, if a line be drawn through the given point *between* the parallels to meet the other parallel, such line shall be equal to the distance from the point thus found to the given point *in* that parallel.

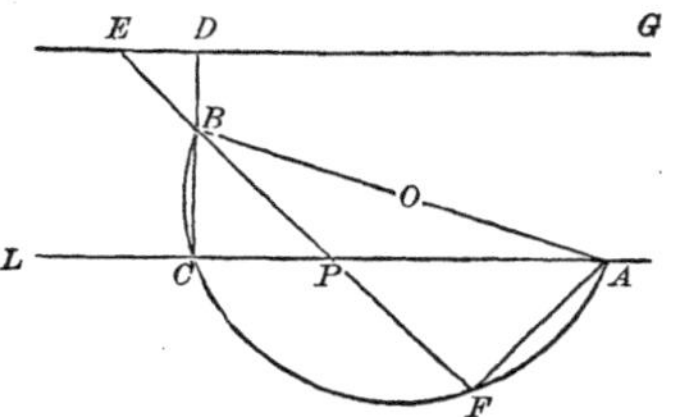

Given { the parallels EG and LA, and the points B and A, to find P such that PBE shall be equal to PA.

Analysis.—Suppose P to be the required point from which, if PBE be drawn, it will equal PA. Join AB, and on it describe the semicircle ACB, of which the centre is O. Draw CBD, which will be perpendicular to each of the two parallels, and hence, as the point B is given, BC and BD are known. Produce EP to F, and join AF. Then, by parallel lines and similar triangles, we have $CD:PE::CB:BP::AF:AP$; that is, $CD:PE::AF:AP$; hence, since $PE=AP$, $AF=CD$. Whence this

Construction.—On AB, the line joining the given points, describe a semicircle ACB. Through B draw CBD, apply $AF=CBD$, and draw $FPBE$, and P will be the required point from which PBE is drawn equal to AP, as is evident from the analysis.

Calculation.—1. Having the parallels and the points B and A given, and $AF=CBD$, which is given, the two triangles ACB and AFB, Case 2, are known in all their parts. Subtract the angle BAC from BAF, and we have the angle $CAF=$ (III. 18, cor. 1*) CBP.

2. In triangle CBP, Case 1, find CP and PB; then $PA=CA-CP=PBE$, and $BE=PE-PB$.

* III. 21.

Scholium 1.—If it is desired to have $PE : PA :: m : n$, take $m : n :: CD : AF$, and proceed as above.

Scholium 2.—If we take $m : n :: BD : AF$, and draw $FPBE$, we have $EB : PA :: m : n$, as is evident by comparing the similar triangles BED and APF.

PROBLEM LII.—Given, a right-angled triangle,—it is required to draw a line from the acute angle at the base to meet the perpendicular let fall from the right angle on the hypothenuse, in a point such that, if a line be drawn from this point, parallel to the base, to meet the perpendicular of the given triangle, these two lines thus drawn shall have to each other a given ratio.

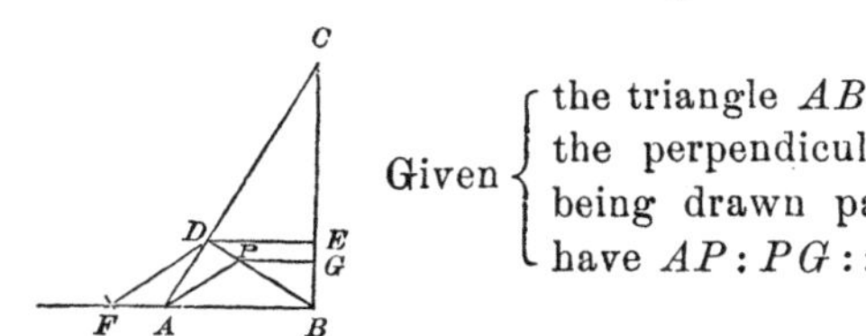

Given { the triangle ABC, to find the point P in the perpendicular BD, such that, PG being drawn parallel to AB, we shall have $AP : PG :: m : n$.

Analysis.—In the given right-angled triangle ABC, suppose AP and PG to be the required lines, such that $AP : PG :: m : n$. Draw DE parallel to PG or AB, and DF parallel to AP. Then, by parallel lines, $DF : DE :: AP : PG :: m : n$. Whence this

Construction.—Draw DE parallel to AB, and take $n : m :: DE :$ (a fourth term) DF, and apply this distance to meet BA, produced, in F. Draw AP parallel to DF, and PG parallel to AB; then P is the point required.

Demonstration.—By parallel lines, $AP : PG :: DF : DE :: m : n$ by construction. Q. E. D.

Calculation.—1. By similar triangles CBA and CDB, we have $CA : CB :: CB : CD = \frac{CB^2}{CA}$. Also, $CA : CD :: AB : DE$; then $n : m :: DE : DF$ by construction.

2. In triangle FDB, Case 2, find angle $BFD = BAP$.

3. In triangle BAP, Case 1, find AP and BP; then $m : n :: AP : PG$, and $PD = BD - BP$.

Scholium.—When AP is to be equal to PG, apply $DF = DE$.

PROBLEM LIII.—Given, the perimeter and perpendicular height of a triangle, to construct it, such that *one* of the angles at the base shall be *double* the *other.*

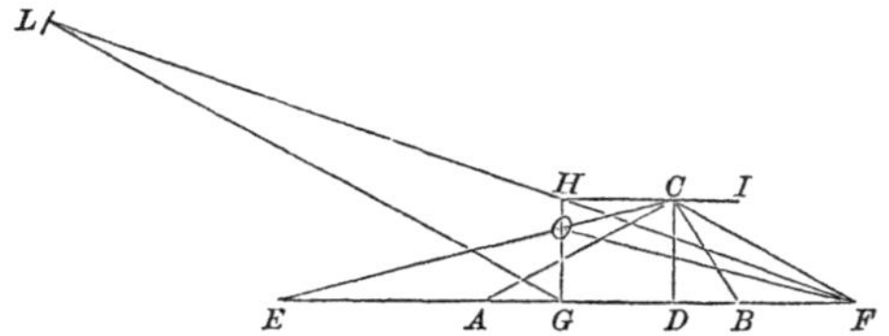

Given, the perimeter of ABC, the perpendicular height CD, and angle ABC to be equal to $2\,BAC$.

Analysis.—Let ABC represent the required triangle. Produce AB both ways, making $AE = AC$ and $BF = BC$, and join EC and FC. Then EF is known, being equal to the given perimeter. Draw ICH parallel to AB, meeting a perpendicular from G, the middle point of EF, in H. Then $GOH = CD$. Join FO. Then (I. 11, and cor. 6, 25*) angle $AFC = \frac{1}{2} ABC = BAC = 2\,FEC = 2\,EFO$. Hence FO bisects the angle EFC, and we have (IV. 17†, and parallel lines) $EF : FC :: EO : OC :: EG : GD$ or CH; that is, $EF : EG$ or $GF :: FC : CH$; whence, since $EF = 2\,GF$, we have $FC = 2\,CH$; and thence this

Construction.—Make $EF =$ the given perimeter; bisect it in G; erect $GH =$ the given perpendicular; draw HCI parallel to EF; join FH, and to it, produced, apply $GL = EF$; draw FC parallel to GL; join EC; make the angle $FCB = CFB$, and the angle $ECA = CEA$; let fall the perpendicular CD; then ABC will be the required triangle.

Demonstration.—It is evident that the perimeter and perpendicular are equal to the given quantities, and it remains to prove that the angle $ABC = 2\,BAC$. By similar triangles LGF and FCH, $LG\,(EF) : GF\,(EG) :: FC : CH\,(GD)$; that is, $EF : FC :: EG : GD ::$ (by parallel lines) $EO : OC$. Wherefore (IV. 17‡) FO bisects the angle EFC, and angle $EFC = 2\,EFO = 2\,FEO = BAC$. Also, $ABC = 2\,EFC = 2\,BAC$. Q. E. D.

Calculation.—1. In triangle FGH, Case 2, find angles, and FH.

2. In triangle FGL, Case 2, find angles; then angle $HFC = FLG$, and angle $DBC = 2\,BFC = 2\,(DFH + HFC)$, and $DAC = \frac{1}{2} DBC$ by the question.

3. In triangles ADC and BDC, Case 1, find AC, AD, BC, and BD; then $AB = AD + BD$.

Corollary.—Since angle $CFA = \frac{1}{2}\,CBA = CAB$, we have $CF =$

* I. 5 and 32. † VI. 3. ‡ VI. 3.

$CA = AE$. Hence, if we have the line EF given, and a line HI parallel thereto, we can find, as above, the points C and A such that FC, CA, and AE shall all be equal.

Scholium.—In the above solution we construct the triangle EFC, having the base EF and perpendicular CD given, so that one of the angles at the base is double the other; that is, angle $EFC = 2\ FEC$.

PROBLEM LIV.—Given, the vertical angle of a triangle, the angle made with the base by a line drawn from the vertical angle to the base, and the radii of the two circles inscribed in the two triangles into which the whole triangle is divided by this line, to determine the triangle.

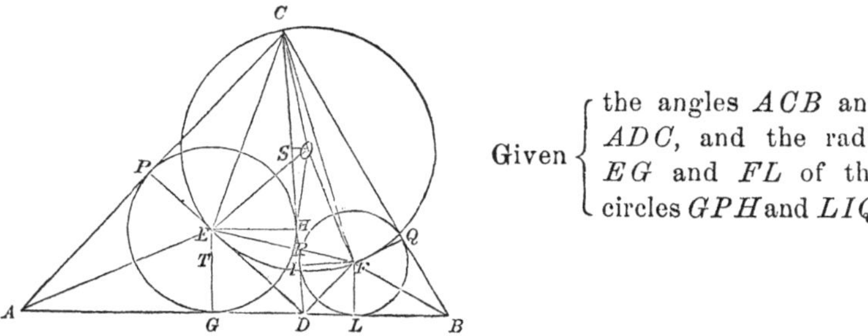

Given $\left\{\begin{array}{l}\text{the angles } ACB \text{ and} \\ ADC\text{, and the radii} \\ EG \text{ and } FL \text{ of the} \\ \text{circles } GPH \text{ and } LIQ.\end{array}\right.$

Analysis.—Suppose ABC to be the required triangle, having the angles ACB and ADC given. Join C with the centres E and F; then (III. Prob. 15*) angle $ECF = \frac{1}{2}$ the given angle ACB.

Construction.—Draw any line AB, and at any point D, in AB, make the angle $ADC =$ the given angle; and, touching the lines AB and DC with the given radii (see scholium next page), describe the circles E, GH and F, IL, E and F being the respective centres. Join EF, cutting the line CD in R, and on EF (III. Prob. 16†) describe a segment of a circle, of which O is the centre, to contain an angle equal to *half* the given vertical angle, cutting DC in C. Join CE and CF, also OC, OE, OR, and OF, and (III. Prob. 14‡) draw the lines CPA and CQB, touching the circles GH and IL in P and Q respectively. Then ABC will be the required triangle.

Demonstration.—By III. Prob. 15,§ angle $ACB = 2\ ECF$, and all the rest is manifest from the analysis and construction.

Calculation.—On CRD let fall the perpendicular OS.

1. In the triangles DGE and DLF, Case 1, find DE and $DG =$

* IV. 4. † III. 33. ‡ III. 17. § IV. 4.

DH, and FD and $LD = DI$; then $EF = \sqrt{DE^2 + FD^2}$, EDF being evidently a right angle, $IH = DH - DI$.

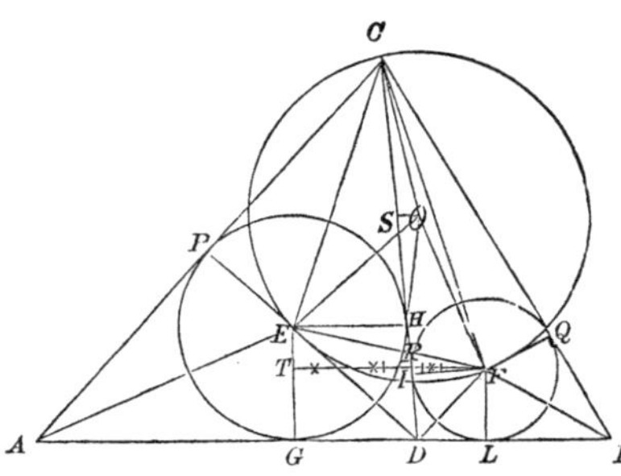

2. By similar triangles EHR and FIR, we have $EH : FI :: ER : FR$, and $EH : FI :: HR : IR$. By composition, $EH + FI : EH :: (ER + FR)$ or $EF : ER$, and $EH + FI : EH :: (HR + IR)$ or $IH : HR$; then $RF = EF - ER$, $IR = IH - HR$, and $DR = DH - HR$. In triangle EHR, Case 2, find angle $ERH = ERS = ERC$; hence the angles REC, $REO = REC - OCE$, and $EOF = 2\,ECF$, are known.

3. In triangle EOF, Case 1, find $OE = OC$. In triangle OER, Case 3, find angle ERO, and RO; then angle $SRO = ERO - ERS$.

4. In triangle SRO, Case 1, find RS and SO; then $CS = \sqrt{OC^2 - SO^2}$, $RC = RS + CS$, and $DC = DR + RC$.

5. In triangle ERC, Case 3, find EC, and angle $ECR = ECA$. In triangle FRC, Case 3, find FC, and the angle $FCR = FCB$; then angle $ACD = 2\,ECA$, and angle $DCB = 2\,FCB$.

6. In triangles ACD and DCB, Case 1, find AC and AD, and BC and BD; then $AB = AD + BD$.

Scholium.—To describe a circle, as required in the preceding problem, which shall have a given radius, and touch two lines given in position, as ADB and DC, it is only necessary to bisect the angles ADC and BDC by the straight lines DE and DF respectively; erect the perpendicular DI equal to the given radius, and parallel to AB, through I, draw a line to meet the bisecting lines in E and F, either of which will be the centre of a circle that will touch both lines.

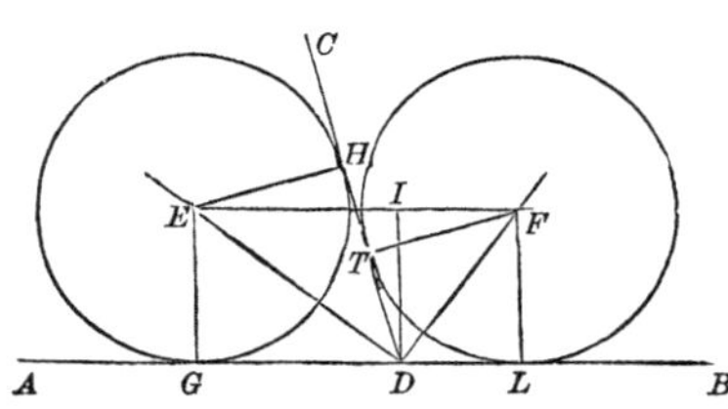

Demonstration.—Let fall the perpendiculars EG, EH, and FL, FT; then (I. 17*) the triangles DGE and DHE are equal in all

* I. 26.

their parts, as also the triangles DLF and DTF. Hence $EG = EH = DI$, and $FL = FT = DI$. The radii EG and FL may be of different lengths, and the construction modified accordingly.

Corollary.—The angle $EDF = EDC + CDF = \frac{1}{2}(ADC + CDB) =$ half of $180° = 90°$. Hence angle EDF is a right angle.

PROBLEM LV.—In a plane triangle are given the vertical angle, an adjacent side, and the length of a line drawn from the vertical angle to the base, dividing the base in a given ratio, to determine the triangle.

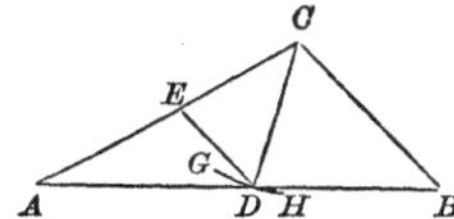

Given $\begin{cases} \text{the angle } ACB, \text{ the side } AC, \text{ and} \\ \text{the line } CD \text{ dividing } AB \text{ so that} \\ AD : DB :: m : n. \end{cases}$

Analysis.—Suppose ACB to be the required triangle. Draw DE parallel to CB; then (IV. 15*) $AE : EC :: AD : DB :: m : n$. Whence this

Construction.—Make ACB equal to the given angle, and AC equal to the given side. Divide AC (IV. Prob. 1†) in E so that $AE : EC :: m : n$. Draw ED parallel to CB, and to it apply $CD =$ the given line. Join CD and AD, and produce the latter to meet CB in B; then ACB is the triangle required.

Demonstration.—By IV. 15,‡ $AD : DB :: AE : EC :: m : n$; whence all is manifest. Q. E. D.

Calculation.—By construction, $m : n :: AE : EC$. By composition, $m + n : m :: AE + EC$ or $AC : AE = \frac{m \cdot AC}{m + n}$. Also, $m + n : n : AE + EC$ or $AC : EC = \frac{n \cdot AC}{m + n}$.

1. In triangle CED, the angle $CED = 180° - AED = 180° - ACB$ (I. 25, cor. 6§); find ED, Case 2; then $AE : ED :: AC : CB$.

2. In triangle ACB, Case 3, find AB. Then, since $m : n :: AD : DB$, we have, by composition, $m + n : m :: AB : AD = \frac{m \cdot AB}{m + n}$. Also, $m + n : n :: AB : DB = \frac{n \cdot AB}{m + n}$.

* VI. 2. † VI. 10. ‡ VI. 2. § I. 32.

PROBLEM LVI.—In a plane triangle are given the vertical angle, the perpendicular let fall on the base, and the *difference of the squares* of the perpendiculars let fall from the foot of this perpendicular upon the two sides of the triangle, to determine the triangle.

FIG. 1.

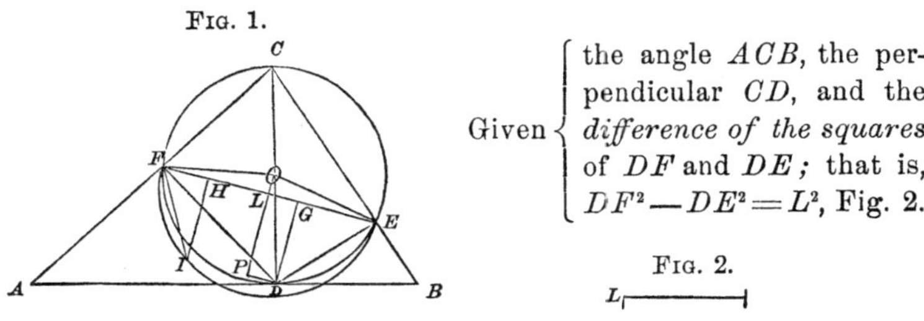

Given $\begin{cases}\text{the angle } ACB, \text{ the perpendicular } CD, \text{ and the} \\ \textit{difference of the squares} \\ \text{of } DF \text{ and } DE; \text{ that is,} \\ DF^2 - DE^2 = L^2, \text{ Fig. 2.}\end{cases}$

FIG. 2.

Analysis.—Suppose the triangle ACB constructed, as in the above figure. Bisect the given perpendicular CD in O, and about the centre O describe the circle $CFDE$, which, as CFD and CED are right angles, must pass through the points F and E. Join OF, OE, and FE, and on FE let fall the perpendicular DG; then the angle $FOE = 2\,ACB$; also (cor. to Prob. 36), $FG^2 - GE^2 = FD^2 - DE^2 = L^2$. Whence this

Construction.—With O as a centre, and radius equal to half the given perpendicular, describe a circle. At O make the angle $FOE =$ twice the given vertical angle, and join FE. On FE describe a semicircle, in which apply $FI = L$, and let fall the perpendicular IH. Bisect HE in G, draw GD parallel to HI, and draw the diameter DOC, which will be equal to the given perpendicular. Draw ADB perpendicular to DC, meeting CF and CE, joined and produced, in A and B; then ABC will be the required triangle.

Demonstration.—Join DF and DE, which will be (III. 18, cor. 2*) at right angles to AC and CB. The angle $ACB = \frac{1}{2}\,FOE =$ the given vertical angle. Also, $FD^2 - DE^2 =$ (cor. to Prob. 36) $FG^2 - GE^2 = (FG + GE) \times (FG - GE) = FE \times FH =$ (IV. 18, cor.†) $FI^2 = L^2$. Q. E. D.

Calculation.—1. In triangle FOE, Case 1, find FE; then $FH = \frac{FI^2}{FE} = \frac{L^2}{FE}$, and $EG = \frac{1}{2}\,EH = \frac{1}{2}\,(EF - FH)$.

2. On EF let fall the perpendicular OL, and produce it to meet DP, drawn parallel to FE, in P; then $DP = LG = EL - EG$. Also $OL = \sqrt{OE^2 - EL^2}$, $OP = \sqrt{OD^2 - DP^2}$, $DG = PL =$

* III. 31. † VI. 8, cor.

$PO - OL$, $DE = \sqrt{DG^2 + EG^2}$, $FD = \sqrt{DG^2 + GF^2}$, $CF = \sqrt{CD^2 - FD^2}$, and $CE = \sqrt{CD^2 - ED^2}$.

3. By similar triangles CFD, CDA, we have $CF : FD :: CD : DA$, and $CF : CD :: CD : CA$.

4. By similar triangles CED, CDB, we have $CE : ED :: CD : DB$, and $CE : CD :: CD : CB$, and $AB = DA + DB$.

Limit.—L must be less than FE.

PROBLEM LVII.—In a plane triangle are given the vertical angle, the perpendicular let fall on the base, and the *sum* of the perpendiculars let fall from the foot of this perpendicular upon the two sides of the triangle, to determine the triangle.

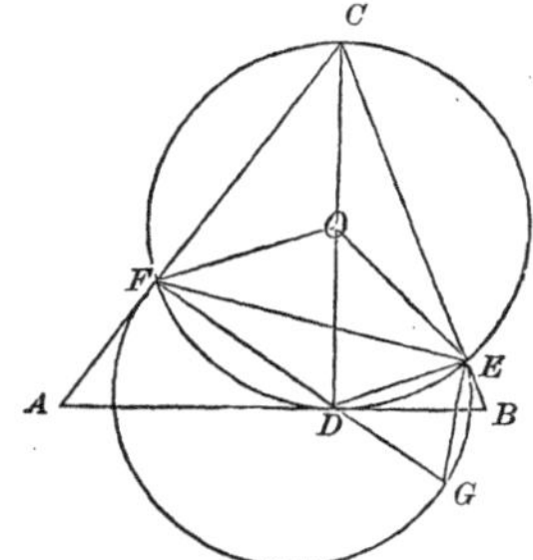

Given { the angle ACB, the perpendicular CD, and the *sum* of the perpendiculars FD and DE.

Analysis.—Suppose ABC to be the triangle required. On CD describe the circle $CFDE$, of which O is the centre, and join OF, OE, and FE; then the angle FOE = twice the given angle ACB, and hence the triangle FOE is known in all its parts. Produce FD, making $DG = DE$, and join EG; then $FG = FD + DE$ is known, and the angle $FGE = \frac{1}{2} FDE = \frac{1}{2}$ the supplement of the given angle ACB is known. Whence this

Construction.—With a radius equal to half the given perpendicular, describe a circle of which the centre is O. At O make an angle FOE = twice the given vertical angle, and join FE. On FE (III. Prob. 16*) describe a segment FGE to contain an angle equal to *half the supplement* of the given vertical angle, in which apply the line FDG = the given sum of the perpendiculars, D being the intersection of FG with the circumference of the circle first described. Draw DOC, GE, and DE. Also, through D, perpendicular to CD, draw AB, meeting CF and CE, joined and produced, in A and B respectively; then ABC will be the required triangle.

* III. 33.

Demonstration.—By III. 18, cor. 2,* DF and DE are perpendicular to AC and BC respectively. The angle $ACB = \frac{1}{2} FOE =$ the given vertical angle, angle FGE, by construction, equal to $\frac{1}{2} FDE$. Hence $DEG = \frac{1}{2} FDE = FGE$, and $DE = DG$; and we have $FD + DE = FG =$ the given sum. Q. E. D.

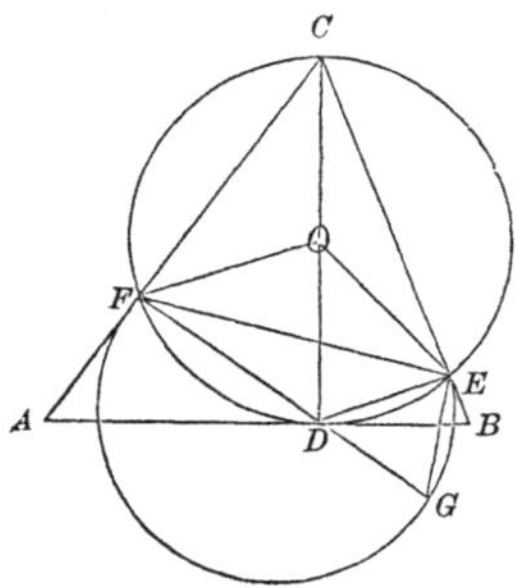

Calculation.—1. In triangle FOE, Case 1, find FE. 2. In triangle FGE, Case 2, find angles, and EG. In triangle EDG, Case 1, find ED; then $FD = FG - ED$. 3. In triangle FDE, all the angles being known, we have (III. 18, cor. 1†) angle $ACD = FCD = FED$, and angle $BCD = ECD = EFD$. In triangles ACD and BCD, Case 1, find CA and AD, CB and DB; then $AB = AD + DB$.

PROBLEM LVIII.—In a plane triangle are given the base, one angle at the base, and the segments into which the base is divided by a line bisecting the vertical angle, to determine the triangle.

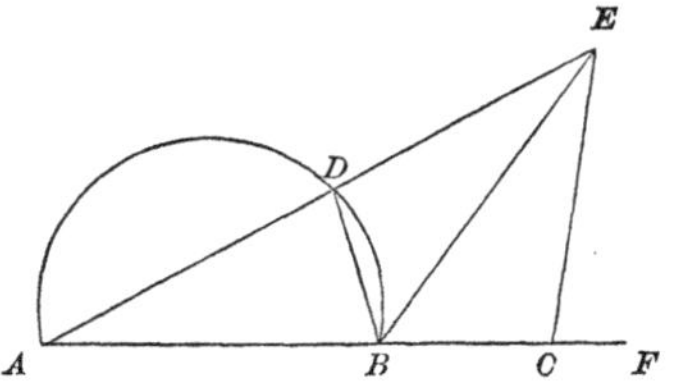

Given { in the triangle AEC, the base AC, the angle ACE, and the segments AB and BC, made by EB, bisecting the angle AEC.

Analysis.—Let AEC represent the required triangle. Make the angle $EBD = EBC$; then $BD = BC$, and angle $BDE = BCE =$ the given angle. Produce AC to F; then angle $ADB = FCE$, they being supplements of the equal angles BDE and BCE, the given angle. Whence this

Construction.—Draw an indefinite line AF, in which take AB and $BC =$ the given segments of the base, and make the angle $ACE =$ the given angle. On AB (III. Prob. 16‡) describe a segment to contain an angle equal to FCE, in which apply $BD = BC$. Join AD, and produce it to meet CE in E, and join EB; then ACE will be the required triangle.

Demonstration.—It remains to prove only that EB bisects the angle AEC. Since, by construction, $BD = BC$, and angle $ADB =$

* III. 31. † III. 21. ‡ III. 33.

FCE, we have angle $EDB = ECB$; wherefore $ED = EC$, and angle DEB or $AEB = CEB$, and EB bisects the angle AEC. Q. E. D.

Calculation.—1. In triangle ADB, Case 2, find angles, and AD.

2. In triangle ACE, Case 1, find AE and $EC = ED$; then $AD = AE - ED$.

3. In triangle ABE, Case 1, find BE, if desired.

PROBLEM LIX.—In a right-angled triangle are given the base, and the rectangle of the hypothenuse and perpendicular, to determine the triangle.

FIG. 1.

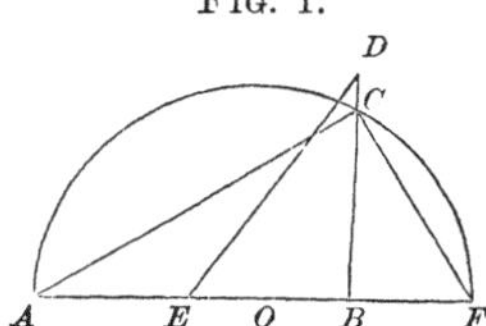

Given $\begin{cases}\text{in the triangle } ABC, \text{ the base } AB, \\ \text{and the rectangle } AC \times CB = \\ L^2, \text{ Fig. 2.}\end{cases}$

FIG. 2.

L ——————

Analysis.—Let ABC represent the required triangle. Produce BC, and take BD a third proportional to AB and L (IV. Prob. 2, cor.*); then $AB : L :: L : BD$, and $AB \times BD = L^2 = AC \times CB$ by the question. Draw CF at right angles to AC, meeting AB, produced, in F. Then, by similar triangles ABC and CBF, we have $AB : AC :: CB : CF$. Hence $AB \times CF = AC \times CB = AB \times BD$; from which $CF = BD$. Whence this

Construction.—Bisect the given base AB in E, and, having found BD as in the analysis, take $EF = ED$. On AF describe the semicircle ACF, and join AC and CF; then ABC will be the triangle required.

Demonstration.—By IV. 23,† we have $CF^2 = AF \times FB = (EF + EA) \times (EF - EA) = (ED + EB) \times (ED - EB) = ED^2 - EB^2 = BD^2$; hence $CF = BD$. By similar triangles ACB, BCF, we have $AB : AC :: CB : CF$ or BD; hence $AC \times CB = AB \times BD = L^2$. Q. E. D.

Calculation.—By construction, $EF = ED = \sqrt{EB^2 + BD^2}$; then $AF = EF + AE$, and $FB = EF - AE = EF - EB$, and (IV. 23‡) $FC = \sqrt{AF \times FB}$, $AC = \sqrt{AF \times AB}$, and $BC = \sqrt{AB \times FB}$.

* VI. 11. † VI. 8, cor. ‡ VI. 8, cor.

PROBLEM LX.—Having given the lengths of two lines at right angles to each other at a common extremity, to draw a line from the other extremity of *one* to meet the *other line produced* such that the parallel to this line so drawn intercepted between the two given lines, and drawn through the extremity of the line that was produced, shall be equal to the sum of this line and the part produced.

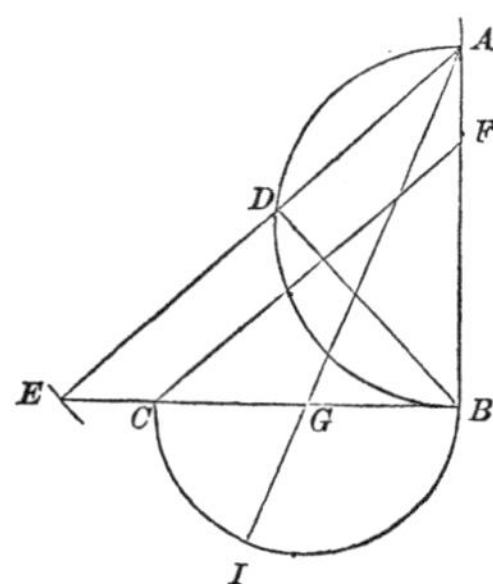

Given $\begin{cases}\text{the lines } AB \text{ and } BC \text{ at right angles} \\ \text{at } B, \text{ to draw } AE \text{ such that the} \\ \text{parallel } CF \text{ shall be equal to } BE.\end{cases}$

Analysis.—Suppose the problem constructed as in the figure. On AB describe the semicircle ADB. Then (IV. 30*) $AE:EB::EB:ED$. But, by similar triangles AEB and FCB, we have $AE:EB::CF$ or $EB:CB$; whence $ED=CB$; and we have this

Construction.—On the given lines AB and BC describe semicircles, G being the centre of the one on BC. Draw AGI; and to BC, produced, apply $AE=AGI$, and draw CF parallel to AE; then will $CF=BE$, and AE will be the line required.

Demonstration.—$AB^2=AG^2-GB^2=(AG+GB)\times(AG-GB)=(AG+GI)\times(AG-GB)=AI\times(AI-BC)=AE\times(AE-BC)$; that is, $AE:AB::AB:AE-BC$. But, by similar triangles ABE and ABD, we have $AE:AB::AB:AD$; hence $AE-BC=AD=AE-ED$. Whence $ED=BC$. Now (IV. 30†) $AE:EB::EB:ED$ or CB. Also, by similar triangles AEB and FCB, we have $AE:EB::CF:CB$; hence $CF=BE$. Q. E. D.

Calculation.—$AG=\sqrt{AB^2+BG^2}$, $AE=AI=AG+GI$, $AD=AG-BG$, $ED=BC$, $BE=\sqrt{AE\times ED}$; then $BE:EA::BC:CF$, and $BE:BA::BC:BF$.

* III. 36. † III. 36.

PROBLEM LXI.—From two given points to draw two right lines to meet in a line of *any kind* given in position, so that the *difference* of the *squares* of these lines shall be equal to a given quantity.

FIG. 1.

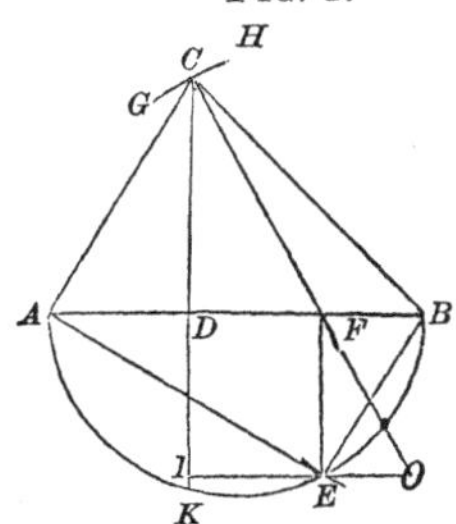

Given $\left\{\begin{array}{l}\text{the points } A \text{ and } B\text{, and the line } GH \\ \text{of any kind, to draw } BC \text{ and } AC \text{ to} \\ \text{meet in } GH \text{ such that } BC^2 - AC^2 = \\ L^2\text{, Fig. 2.}\end{array}\right.$

FIG. 2.

L ————

Analysis.—Suppose AC and BC to be the lines required. Join AB, and on it let fall the perpendicular CD. Take $DF = DA$; then $L^2 = BC^2 - AC^2 =$ (cor. to Prob. 36) $BD^2 - AD^2 = (BD + AD) \times (BD - AD) = AB \times BF$. Whence this

Construction.—On the given distance AB describe a semicircle, in which apply $BE = L$. Let fall the perpendicular EF on AB; bisect AF in D; erect a perpendicular at D to meet the given line GH in C, and join AC and BC, which will be the lines required.

Demonstration.—$BC^2 - AC^2 = BD^2 - AD^2 =$ (by analysis) $AB \times BF =$ (IV. 23*) $BE^2 = L^2$. Q. E. D.

Calculation.—$BF = \frac{BE^2}{AB}$, $AF = AB - BF$, $AD = \frac{1}{2} AF$, $DB = DF + BF$. Now, combining AD and DB with the given position and equation of the line GCH, AC and BC are determined.

Scholium.—If the given line GH is the arc of a circle whose centre is O and radius OC, produce CD to meet a line drawn through O parallel to AB, in I. Then, because the *position* of GH is given, the centre O is given, and DI and OI are known, as also the radius OC. Then $CI = \sqrt{OC^2 - OI^2}$, $CD = CI - DI$; also, $AC = \sqrt{AD^2 + CD^2}$, and $BC = \sqrt{DB^2 + CD^2}$.

* VI. 8, cor.

PROBLEM LXII.—"Having, at an unknown distance from the foot of a steeple, taken its elevation, I advanced sixty yards towards it on level ground, and then observed the angle of elevation, which was just the *complement* of the former. Advancing twenty yards still nearer, the angle of elevation was now *just double* of the first. Required, the altitude of the steeple, and the distances of the stations from its foot."

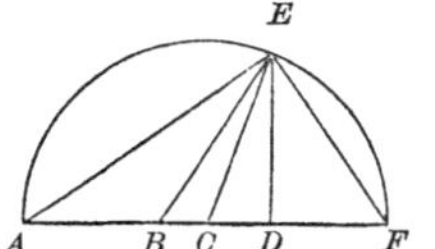

Given $\begin{cases} \text{the distances } AB \text{ and } BC, \text{ the angle} \\ EBD \text{ to be the } \textit{complement} \text{ of } EAD, \\ \text{and } ECD \text{ to be the } \textit{double} \text{ of } EAD, \\ \text{to find } ED. \end{cases}$

Analysis.—Let A, B, C represent the three stations at which the angles of elevation were taken, and DE the steeple. Since the angle at C is, by the problem, double that at A, it is evident that C is the centre and CA the radius of a semicircle AEF passing through the first station A and the top of the steeple E. Describe the semicircle, and join FE; then EBD, by the question, being the complement of EAD, is equal to AFE; and hence $BE = EF$, and $BD = DF$. Whence this

Construction.—On an indefinite right line lay $AB = 60$, and $BC = 20$. With centre C and radius CA describe the semicircle AEF. Bisect BF in D; erect the perpendicular DE, and join E with the points A, B, C, and F; then DE will represent the steeple.

Demonstration.—The angle ECD (III. 18*) $= 2\,EAD$; also (I. 5†), the triangles BDE and FDE are equal. Hence angle $EBD = EFD = EFA =$ complement of EAF or EAD. Q. E. D.

Calculation.—$BF = BC + CF$ or $CA = 20 + 80 = 100$, $BD = \frac{1}{2}BF = 50$, $CD = BD - BC = 30$, $CE = CA = 80$, $DE = \sqrt{CE^2 - CD^2} = \sqrt{80^2 - 30^2} = \sqrt{5500} = 10\sqrt{55}$. Also, $AD = AB + BD = 110$.

* III. 20. † I. 4.

PROBLEM LXIII.—In any plane triangle are given the vertical angle, the line bisecting this angle, and the *sum* of the including sides, to determine the triangle.

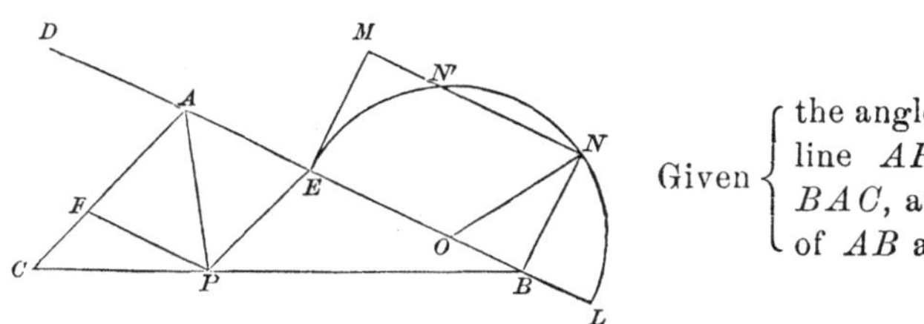

Given { the angle BAC, the line AP bisecting BAC, and the *sum* of AB and AC.

Analysis.—Let BAC represent the required triangle. Through P draw PF parallel to AB, and PE parallel to AC; then $AEPF$ is a rhombus, having all its sides equal, and known. Produce AB both ways, making $AD = AF$, and $BL = CF$; then $DL = AB + AC$, the given sum of the sides, is known; and $EL = DL - 2\,AE$ is known. Also, by similar triangles CFP, PEB, we have CF $(BL) : PF\,(EP) :: EP : EB$; wherefore $BL \times EB = EP^2$. Whence this

Construction.—Having laid $AD = AF$, and $DL =$ the given sum of the sides, on EL describe a semicircle of which the centre is O; erect the perpendicular $EM = EP$; draw MN parallel to AL, and let fall the perpendicular NB; then through B and P draw the line BPC, and BAC will be the triangle required.

Demonstration.—By similar triangles, $CF : FP\,(EP) :: EP : EB$; hence $CF \times EB = EP^2 = EM^2 = BN^2 = BL \times EB$; whence $CF = BL$, and $AB + AC = AB + AF + CF = AB + AD + BL = DL =$ given sum of sides. Q. E. D.

Calculation.—Join N and O. In triangle APE, Case 1, find $EP = EM = AE = AD = NB$; then $EL = DL - 2\,AE$. In triangle OBN, find OB; whence BE and BL are known; then $AB = AE + EB$, and $AC = AF + BL$.

Limits.—1. It will be seen that the line MN cuts the semicircle in another point N'. If the point N' had been taken instead of N, the point B would have come between E and O, and the sides would have been of the same lengths as above obtained, but AB would have been the *shorter*.

2. When MN is just *tangent* to the semicircle, the sides AB and AC will be equal, BC will be perpendicular to AP, and the *shortest possible*, or *minimum*, line that can be drawn through P to meet the lines AB and AC.

3. When MN does not meet the semicircle, the problem is impossible.

Remarks.—1. When BAC is a right angle, the rhombus $AEPF$ becomes a square.

2. This problem is the same as having a point P given in position equidistant from the lines AB and AC, to draw through P a line such that AB and AC shall have a given sum. (See Prob. XXIX. of this division.)

PROBLEM LXIV.—In a trapezium, of the six parts—viz., two opposite sides, and the angle formed where these sides meet when produced, the two diagonals, and the angle formed at their intersection—any five being given,—to construct the trapezium.

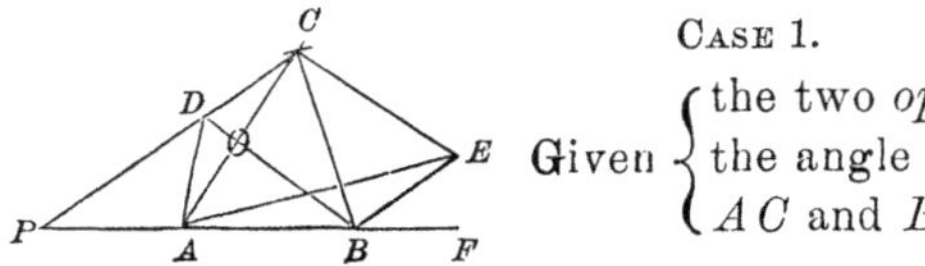

CASE 1.

Given { the two *opposite* sides AB and CD, the angle P, and the two diagonals AC and BD.

Analysis.—Suppose $ABCD$ to be the trapezium constructed as required. Produce the *given* sides CD and BA to meet in P. Draw BE parallel to DC, and CE parallel to DB, and produce AB to F; then angle $FBE =$ (I. 20, cor. 3*) the given angle P, $BE = CD$, and $CE = BD$. Whence this

Construction.—In any line, as PBF, take $AB =$ its given length, and make the angle $FBE =$ the given angle P, and $BE =$ the length of CD. Then, with A and E as centres, and radii equal to the diagonals AC and BD respectively, describe arcs intersecting in C. Join AC, EC, and AE. Draw BD parallel to CE, and CD parallel to BE, producing it to meet BA, produced, in P. Join AD, and $ABCD$ will be the trapezium required.

Demonstration.—Since PC is parallel to BE, the angle $P = FBE$ = the given angle by construction. Also, being opposite sides of a parallelogram, $CD = BE$, and $BD = CE$, the given lengths of the diagonal and side by construction. Q. E. D.

Calculation.—1. In triangle ABE, Case 3, find angle BAE, and side AE.

2. In triangle ACE, Case 4, find angle CAE; then angle $CAB = CAE + BAE$.

3. In triangle CAB, Case 3, find angle ABC, and side BC.

4. In triangle PBC, Case 1, find PC and PB; whence we have PA and PD.

* I. 29.

5. In triangle PAD, Case 3, find the angles A and D, and the side AD; we then have all the angles and all the sides of the trapezium.

NOTE.—*Either pair of opposite* sides may be given, only it must be observed to produce those sides that *are* given, and that it is the angle at *their* intersection that is given. If the *given sides*, in any case, are lettered AB and CD, and the angle formed, where they meet on their being produced, be BPC, the same letters used in the preceding analysis, construction, etc. will apply in each case.

PROBLEM LXIV., continued. (Case 2.)—In a trapezium are given the two opposite sides, the angle formed where they meet on being produced, one diagonal, and the angle formed at the intersection of the diagonals, to determine the trapezium.

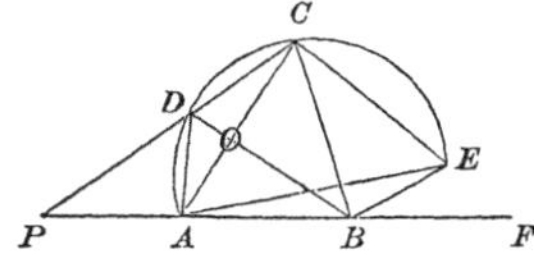

Given $\begin{cases} \text{the two opposite sides } AB \text{ and} \\ CD\text{, one diagonal, as } AC\text{, and the} \\ \text{angles } P \text{ and } O. \end{cases}$

Analysis.—As in Case 1, let $ABCD$ represent the required trapezium. Produce the *given* sides BA and CD to meet in P, and draw BE and CE parallel to DC and DB respectively; then $BE =$ the given side CD, and angle $ACE =$ (I. 20, cor. 3*) the given angle AOB. Whence this

Construction.—In any line, as PF, take $AB =$ its given length, make the angle $FBE =$ the given angle P, and $BE =$ the given length of CD. Join AE, and on it describe (III. Prob. 16†) a segment to contain an angle equal to the given angle AOB, in which apply $AC =$ the given diagonal. Join BC and EC; draw CP parallel to BE, and BD parallel to EC, and join AD; then $ABCD$ will be the required trapezium.

Demonstration.—By parallel lines, $CD = BE$, angle $P = FBE$, and angle $AOB = ACE =$ the given angles. Q. E. D.

Calculation.—1. In triangle ABE, Case 3, find angles BAE, AEB, and side AE.

2. In triangle ACE, Case 2, find angles CAE, CEA, and side $CE = DB$.

3. In triangle CAB, angle $CAB = CAE + BAE$, find, Case 3, angles ABC, ACB, and side BC.

4. In triangle PBC, Case 1, find PC and PB, thence find PA, PD, and AD, and we have all the sides and all the angles of the trapezium.

* I. 29. † III. 33.

Scholium 1.—If the diagonal BD were given instead of AC, the other given parts being the same, then, after describing the segment on AE as above, apply $EC =$ the given diagonal BD, and proceed with the construction and calculation upon the same principle as already directed.

Scholium 2.—*Either pair* of opposite sides may be given, only it must be observed to produce the *sides that are given*, and that it is the angle at *their* intersection that is given. (See note to Case 1.)

PROBLEM LXIV., continued. (Case 3.)—Having given in the trapezium the two opposite sides, the two diagonals, and the angle at their intersection, to determine the trapezium.

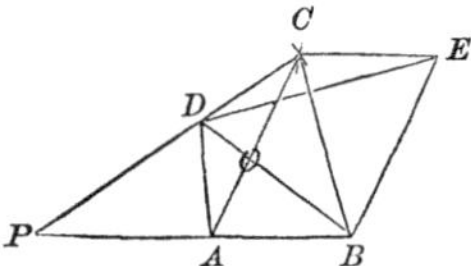

Given $\begin{cases} \text{the opposite sides } AB \text{ and } CD, \text{ the} \\ \text{diagonals } AC \text{ and } BD, \text{ and the angle} \\ O \text{ at their intersection.} \end{cases}$

Analysis.—Suppose $ABCD$ to be the required trapezium, and, as in the preceding cases, produce the *given* sides BA and CD to meet in P. Draw BE parallel to AC, and CE parallel to AB; then, by parallel lines, $BE = AC$, $EC = AB$, and the angle $DBE =$ the given angle DOC. Whence this

Construction.—Draw any line, in which take $DB =$ its given length; make the angle $DBE =$ the given angle DOC, and $BE =$ the given diagonal AC. Join DE, and with E and D as centres, and radii respectively equal to the given sides AB and CD, describe arcs intersecting in C. Join EC, CD, and CB, draw CA parallel to EB, and BAP parallel to CE, meeting CA and CD, produced, in A and P respectively. Join AD; then $ABCD$ will be the trapezium required.

Demonstration.—By parallel lines, $AB = CE$, $AC = BE$, and angle $DOC = DBE =$ the given angle. Q. E. D.

Calculation.—1. In triangle DBE, Case 3, find angles BDE and BED, and side DE.

2. In triangle DCE, Case 4, find angle CDE; then angle $CDB = CDE + BDE$.

3. In triangle CDB, Case 3, find angle DCB, and side BC.

4. In triangle CAB, Case 4, find angle ABC.

5. In triangle PBC, Case 1, find PC and PB, and thence find PD, PA, and DA; and we then have all the sides and all the angles of the trapezium. (See note to Case 1.)

PROBLEM LXIV., continued. (Case 4.)—When in the trapezium are given one of the opposite sides, both diagonals, the angle at their intersection, and the angle where the given side and the side opposite to it meet on being produced, to determine the trapezium.

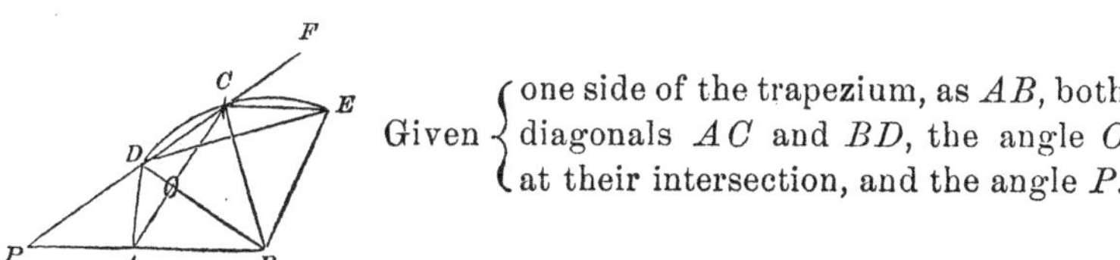

Given { one side of the trapezium, as AB, both diagonals AC and BD, the angle O at their intersection, and the angle P.

Analysis.—Suppose $ABCD$ to be the required trapezium. Draw BE parallel to AC, and CE parallel to AB; then, producing DC to F, the angle FCE, by parallel lines, will be equal to the given angle P, and the angle DBE equal to the given angle DOC. Wherefore BE, being equal to AC, is known in length and position with respect to the given diagonal DB, and angle ECD equal to the supplement of FCE or given angle P. Whence this

Construction.—Draw DB its given length; make the angle DBE = the given angle DOC, and BE = the other given diagonal AC. Join DE, and on it (III. Prob. 16*) describe a segment to contain an angle equal to the supplement of the given angle P, in which apply EC = the given length of AB. Draw CA parallel to EB, and BA parallel to EC; join AC, AD, and CD, and produce CD and BA to meet in P.

Demonstration.—By parallel lines, $AC = BE$, $AB = EC$, and angle $P = FCE$ = the supplement of DCE. Q. E. D.

Calculation.—1. In triangle DBE, Case 3, find angles BDE, BED, and side DE.

2. In triangle DCE, Case 2, find angles CDE, CED, and side CD; thence the angles CEB, CAB, and CDB are known.

3. In triangle CAB, Case 3, find angles ACB and ABC, and side BC.

4. In triangle PBC, Case 1, find PB and PC, thence PA, PD, and AD; when we have all the sides and all the angles of the trapezium.

Remark.—If the side CD were given instead of AB, then in the segment on DE apply CD = its given length, and proceed with the construction and calculation as above. This concludes all the varieties of the problem where the *opposite sides meet.*

* III. 33.

Scholium.—We now come to consider the Cases, *in the same order*, where the *opposite sides* are *parallel.*

CASE 1.—In Case 1 of the preceding problem, where the opposite sides AB and CD are given, and also the diagonals AC and BD, *if the opposite sides AB and CD are parallel,* then the angle P vanishes, and BE, in that figure, will be the prolongation of AB. Then we have this

Construction.—Produce AB (Fig. 1), making the produced part $BE=$ the length of CD; and with A and E as centres, and radii respectively equal to the given diagonals AC and BD, describe arcs intersecting in C. Join CA, CB, and CE; draw CD parallel to BE, and BD parallel to EC, and join AD; then $ABCD$ will be the required trapezoid, as is evident by parallel lines. It is manifest that in this case the angle O need not be given.

FIG. 1.

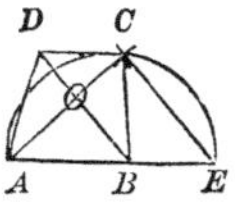

D C

A B E

Calculation.—1. In triangle ACE, Case 4, find the angles; then angle $AOB = ACE$, and angle $DCA = CAB$.

2. In triangles ACD and CAB, Case 3, find angles, and sides AD and CB.

CASE 2.—In Case 2 of the preceding problem, where AB, CD, one diagonal, as AC, and the angles O and P are given, *if the sides AB and CD* (see Fig. 1) *are parallel,* then, as in Case 1, the angle P vanishes, and on AB, produced, lay $BE = CD$, and on AE (III. Prob. 16*) describe a segment to contain an angle equal to the given angle AOB, in which apply the given diagonal AC; or, if BD is given, apply $EC = BD$, and finish the construction of $ABCD$, which will evidently be the trapezoid required.

Calculation.—1. In triangle ACE, Case 2, find angles CAE and CEA, and the unknown diagonal.

2. The angle $ABD = BDC = AEC$. In triangles ABD and BDC, Case 3, find AD and BC.

CASE 3.—In Case 3 of the preceding problem, where the two opposite sides AB and CD, the two diagonals, and the angle O at their intersection are given, *if the opposite sides AB and CD are parallel,* the angle P vanishes; then on AB, produced (see Fig. 1), take $BE = CD$, and with centres A and E, and radii respectively equal to AC and BD, describe arcs intersecting in C. Join AC,

* III. 33.

CB, and *CE;* draw *CD* and *BD* parallel to *BE* and *EC* respectively, and join *AD* and *AC;* then *ABCD* is evidently the trapezoid required.

Calculation.—The same as in Case 1.

NOTE.—It is evident the angle *O* in this case need not be given.

CASE 4.—In Case 4 of the preceding problem, where one side, both diagonals, and the angles *O* and *P* are given, *if the sides AB and CD are parallel,* the angle *P* vanishes, and we then make the angle *DBE* = (see Fig. 2) the given angle *DOC*, *DB* = its given length, and *BE* = the given length of *AC*. Join *DE;* then, if *DC* is given, lay its length from *D* to *C;* but if *AB* is given, lay its length from *E* to *C*. Join *BC*, draw *CA* parallel to *EB*, *BA* parallel to *EC*, and join *AD;* then *ABCD* is evidently the trapezoid required.

FIG. 2.

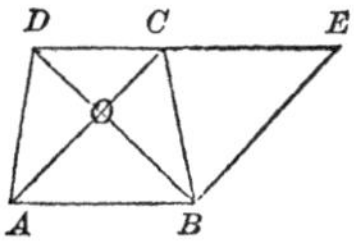

Calculation.—1. In triangle *DBE*, Case 3, find angles, and side *DE;* then the sides *AB* and *CD* are known.

2. In triangles *BAC* and *ACD*, Case 3, find *AD* and *BC*.

PROBLEMS

INVOLVING PROPERTIES OF THE CIRCLE.

PROBLEM I.—Through a given point *within* a given circle to draw a chord of a given length.

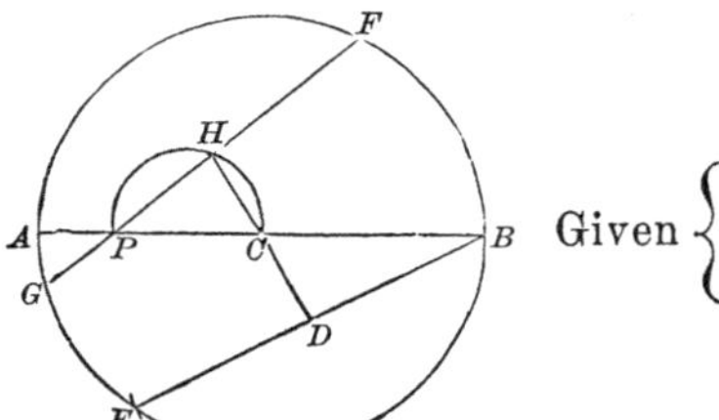

Given $\begin{cases} \text{the diameter } AB \text{ of the circle, the} \\ \text{point } P\text{, that is, } AP \text{ and } PB\text{, and} \\ \text{the length of the chord } GPF. \end{cases}$

Analysis.—Through the given point P, and the centre C, draw the diameter APB. Apply the chord $BE =$ the given chord GF; and from the centre C let fall the perpendiculars CD and CH; then, since $BE = GF$, we have (III. 8*) $CD = CH$. Whence this

Construction.—Having drawn the diameter APB, from its extremity B apply the chord $BE =$ the given chord, and from the centre C let fall on it the perpendicular CD. On CP describe a semicircle, in which apply $CH = CD$, and through P and H draw the chord $GPHF$, which will be equal to BE, and be the chord required.

Demonstration.—The angle PHC, being in a semicircle, is a right angle. Hence CH is perpendicular to the chord GPF, and, it being equal to CD by construction, we have (III. 8†) $GPF = BE$. Q. E. D.

Calculation.—1. The perpendiculars CD and CH (III. 6‡) bisect the chords BE and GF respectively; hence $BD = \frac{1}{2} BE = \frac{1}{2} FG = FH = GH$, and $CH^2 = CD^2 = BC^2 - BD^2$, $PH = \sqrt{PC^2 - CH^2}$.

2. Now, $GP = GH - PH$, and $PF = GH + PH = FH + PH$.

* I I. 14. † III. 14. ‡ III. 3.

Limits.—1. The given chord cannot be *greater* than the diameter AB.

2. The distance of the given chord from the centre cannot be *greater* than CP. When it is *equal* to CP, the required chord will be a perpendicular to the diameter through P. Hence the *shortest* chord that can be drawn through a given point is the one which is perpendicular to the diameter at that point.

PROBLEM II.—Through a given point *without* a given circle to draw a right line so that the chord intercepted by the circumference shall be of a given length.

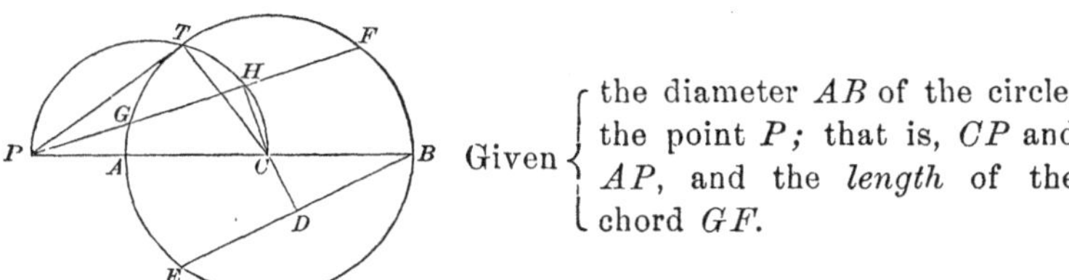

Given { the diameter AB of the circle, the point P; that is, CP and AP, and the *length* of the chord GF.

Analysis.—Through the given point P and the centre C draw the diameter PAB. Apply the chord $BE =$ the given chord GF, and from the centre C let fall the perpendiculars CD and CH; then, since $BE = GF$, we have (III. 8*) $CD = CH$. Whence this

Construction.—Having drawn $PACB$, from B, apply the chord $BE =$ the length of the given chord, on which let fall from the centre C the perpendicular CD. On CP describe a semicircle, in which apply $CH = CD$, and through the points P and H draw the line PGF; then GF will be equal to BE, and PGF be the line required.

Demonstration.—The angle PHC, being in a semicircle, is a right angle. Hence CH is perpendicular to the chord GF, and, being equal to CD by construction, $GF = BE$ (III. 8†). Q. E. D.

Calculation.—The perpendiculars CD and CH (III. 6‡) bisect the chords BE and GF respectively; hence $BD = \frac{1}{2} BE = \frac{1}{2} FG = FH = GH$. Now, $CH^2 = CD^2 = BC^2 - BD^2$, and $PH = \sqrt{CP^2 - CH^2}$. Then $GP = PH - GH$, and $PF = PH + GH = PH + FH$.

Limit.—The given chord GF must not be greater than the diameter AB.

* III. 14. † III. 14. ‡ III. 3.

Scholium.—To draw a tangent from the given point P *without* the circle, join P with the point T where the semicircle on PC intersects the semicircle AFB, and PT will be a tangent to the circle AFB, for (III. 18, cor. 2*) angle PTC is a right angle.

PROBLEM III.—With a given radius it is required to describe three equal circles which shall touch one another, and then to describe another circle which shall touch them all three.

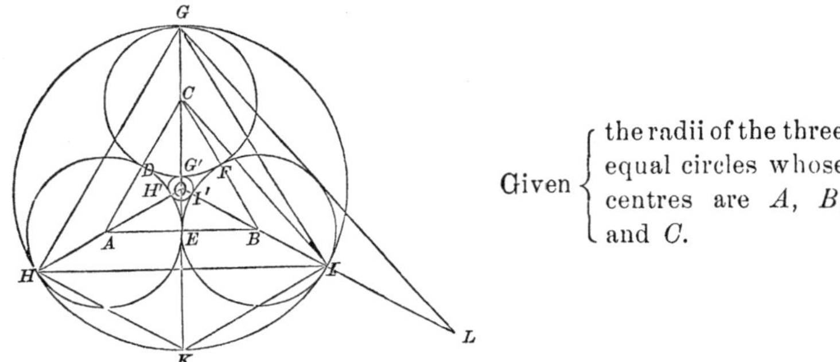

Given { the radii of the three equal circles whose centres are A, B, and C.

Analysis.—Suppose the problem constructed as in the above figure. Join the centres A, B, and C, and an equilateral triangle ABC will evidently be formed, having each of its sides equal to twice the given radius. Bisect two of the angles A and B by the lines AO and BO, and join OC, which three lines will evidently be equal. Produce these three lines; then they will intersect the equal circles in the points G, H, and I; also, before being produced, in the points G', H', and I'. Now, since the lines OA, OB, and OC are all equal, we have OH, OI, and OG all equal, as also OH', OI', and OG' all equal. Whence this

Construction.—Describe the equilateral triangle ABC, having each of its sides double the given radius. Then, with A, B, and C as centres, and the given radius, describe the three circles, which will evidently touch each other at D, E, and F, the middle points of the sides. Bisect the two angles A and B by the lines AO and BO, intersecting in O, and join OC. These three lines, and these lines produced, intersect the given circles in the points G', H', and I' and G, H, and I. Then, with centre O, and radii OH' and OH, describe the circles H', I', G' and H, I, G respectively, and the problem is constructed as required.

* III. 31.

The *demonstration* is all evident from the analysis.

Calculation.—In triangle AOB, Case 1, find OA; then $OH = OA + AH$, and $OH' = OA - AH'$, the radii required.

NOTE.—From the above is immediately deduced the following

PROBLEM IV.—Having given the radius of a circle, to inscribe therein three equal circles which shall touch one another and also touch the given circle. (See figure on opposite page.)

Analysis.—Suppose the problem constructed with OH or OI the given radius. Join GH, HI, and IG, and HIG will evidently be an equilateral triangle. Join OG, OH, OI, and CI, and draw GL parallel to CI, meeting OI, produced, in L. Then, in similar triangles CBI and GIL, since $IB = \frac{1}{2} BC$, we have $IL = \frac{1}{2} IG$. Whence this

Construction.—Having described the circle with the given radius, draw the diameter GOK, and in the circle apply the chords KI and KH, each equal to the given radius KO. Join GH, HI, and IG; then (V. 4*) HIG will be an *inscribed equilateral triangle*, having the side $GI = \sqrt{GK^2 - KI^2} = \sqrt{4\,OK^2 - OK^2} = \sqrt{3\,OK^2} = OK\sqrt{3}$. Also, join OG, OH, and OI, and produce OI, making $IL = \frac{1}{2} IG$. Join LG, draw IC parallel to LG, CB parallel to GI, and CA parallel to GH, and join AB. With the centres A, B, and C, and the radii AH, BI, and CG respectively, describe the circles HED, IFE, and GDF, which will be the equal circles required, touching each other at the points D, E, and F.

Demonstration.—Because of the parallel lines, it is evident that AH, BI, and CG are all equal, that ABC is an equilateral triangle, and that the triangles GIL and CBI are similar. Hence it only remains to be proved that the three circles touch each other.

In the similar triangles GIL, CBI, since $GI = 2\,IL$ by construction, we have $CB = 2\,BI = CF + FB = CD + DA = AE + EB$. Hence, the distances between the centres of the circles being equal to the sum of the radii (III. 13†), the circles touch each other externally in the points D, E, and F. Q. E. D.

Calculation.—By parallel lines, $OL : IL :: OG : CG = BI = AH$, and $2\,CG = AC = AB = BC$; also, $OG - CG = OC = OA = OB$.

* IV. 15. † III. 12.

PROBLEM V.—To describe a circle which shall pass through two given points and have its centre in a straight line given in position.

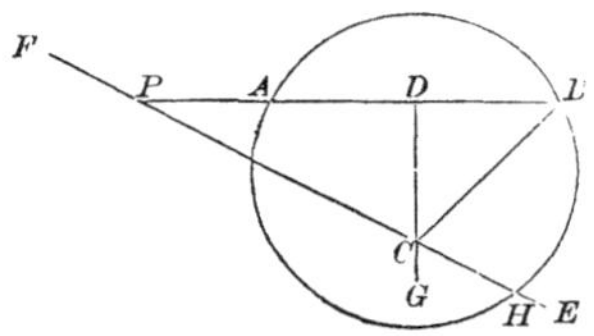

Given { the points A and B, and the line FE *in position;* that is, the angle P which it forms with the line joining A and B, or this line produced, and the distance AP or BP.

Analysis.—Suppose the figure constructed as above. Join AB, and from C, the centre of the circle, let fall on AB the perpendicular CD; then (III. 6*) $AD = DB$. Whence this

Construction.—Join the given points A and B, bisect the line AB in D, and draw DCG perpendicular to AB, and the point in which this perpendicular cuts the given line FE will be the centre of the circle required. In the figure this point is C. Join CB, and with it as a radius, and centre C, describe the circle ABH, which will be the one required.

The *demonstration* is evident from the preceding.

Calculation.—Let the line joining A and B be produced, if necessary, to meet the given line FE in P; then the angle P, and the distance PA or PB, are known from the given position of FE. Thence PA, PD, and PB are known, by adding or subtracting, as the point P is in AB produced, or between A and B. And DB or $DA = \frac{1}{2} AB$.

In triangle PDC, Case 1, find PC and CD; then radius $CB = \sqrt{CD^2 + DB^2} = CA$.

Limits.—1. When FE is *parallel* to AB, CD will be the perpendicular distance between the two parallel lines, and, as FE is given in position, the distance CD is known, and thence the radius CA or $CB = \sqrt{CD^2 + DB^2}$, as before.

2. When the given angle at P is a *right* angle, FE and DG will be parallel, and the problem is impossible; that is, in the language of mathematicians, "the *radius would be infinite*, and the arc of a circle with an infinite radius is a straight line," so that the straight line passing through A and B would fulfil the conditions of the problem mathematically.

* III. 3.

PROBLEM VI.—To divide a given circle into two segments, such that the angle contained in one shall be double of the angle contained in the other.

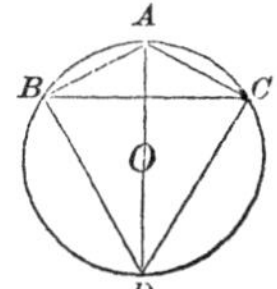

Given { the radius OA of the circle $ABDC$, to draw the chord BC, such that the angle in the segment BAC shall be double that in BDC.

Analysis.—Suppose in the annexed figure the angle BAC to be double BDC; then $BAC + BDC =$ three times BDC. But (III. 18, cor. 4*) the angles $BAC + BDC = 180°$; hence $3\ BDC = 180°$, or angle $BDC = 60°$, and consequently $BAC = 120°$. Whence this

Construction.—Draw the diameter AOD, and lay AB and AC each equal to the radius AO. Join BC, BD, and CD; then angle $BAC = 2\ BDC$.

Demonstration.—By III. 18† and V. 4,‡ it is manifest that the arcs AB and AC each contain 60°; hence the arcs BD and CD each contain 120°, the arc BDC contains 240°, or twice the arc BAC; and, as the angle is measured by *half* the arc on which it stands, the *angle* BAC in the segment $BAC =$ twice the angle BDC in the segment BDC.

Calculation.—In the right-angled triangle ABD, we have $AD = 2\ AB = 2\ AO$, and $BD = \sqrt{AD^2 - AB^2} = \sqrt{4\ AO^2 - AO^2} = \sqrt{3\ AO^2} = AO\sqrt{3} = BC = CD$.

Scholium 1.—If the angle in one segment were required to be *three* times that in the other segment, the first segment would be a quadrant, and the other *three* quadrants, and the angles 135° and 45° respectively.

Scholium 2.—If the angle in one segment were required to be *four* times that in the other segment, the circle would be divided by the side of an inscribed pentagon (V. 5§), and the contained angles would be 36° and 144°.

Scholium 3.—If the angle in one segment were required to be *five* times that in the other segment, the circle would be divided by the side of an inscribed hexagon (V. 4‖), and the contained angles would be 30° and 150°.

* III. 22. † III. 21. ‡ IV. 15. § IV. 11. ‖ IV. 15.

Problem VII.—To describe a circle which shall touch a given circle at a given point, and also touch a straight line given in position.

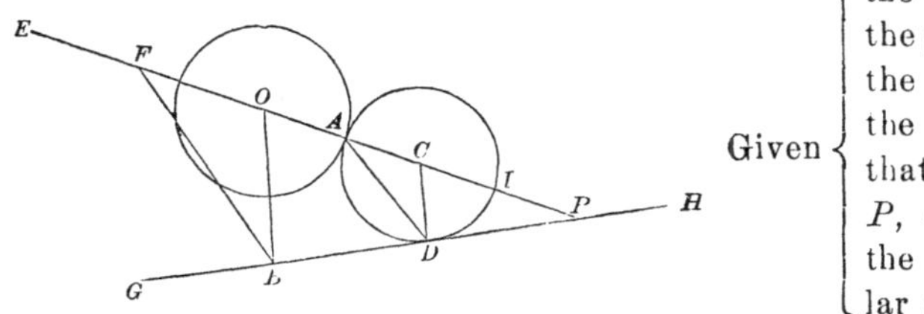

Given { the circle OA, the point A, and the *position* of the line GH; that is, the angle P, and OP or the perpendicular OB.

Analysis.—Suppose ADI to be the required circle, passing through the given point A, and touching the given line GH at the point D. Draw the radius CD, and the line OB, both perpendicular to GH. Join DA, and draw BF parallel thereto; then the triangles ACD and FOB will be similar; and, since $CA = CD$, we have $OF = OB$. Whence this

Construction.—Having let fall on GH the perpendicular OB, and produced the line joining the centre O of the given circle with the given point A, to meet the given line GH in P, produce PO, making $OF = OB$; join FB; draw AD parallel to FB, and DC parallel to BO; then, with the centre C and radius CD, describe a circle ADI, and it will pass through the given point A, and touch the line GH in D.

Demonstration.—By III. 9,* the circle touches the given line GH at the point D. By similar triangles FOB and ACD, we have $FO = OB$ by construction; hence $CA = CD$, and the circle ADI passes through the given point A, and evidently touches the given circle. Q. E. D.

Calculation.—1. In triangle POB, Case 1, find PB and PO; then $PA = PO - OA$, and $PF = PO + OB$.

2. By similar triangles PFB and PAD, we have $PF : PA :: PB : PD$; also, $PB : PD :: OB : CD = CA$.

Limits.—1. When the line GH is *parallel* to OA, then OB, the perpendicular distance between the two parallels, will be the radius of the required circle, and, laying this radius on the line OA, produced from A to C, C will be the centre of the required circle.

2. When the given line GH is *perpendicular* to OA, or OA produced, the difference between OA and OB will be the *diameter* of the required circle.

* III. 16, cor.

PROBLEM VIII.—With a given radius to describe a circle which shall pass through a given point and touch a line given *in position.*

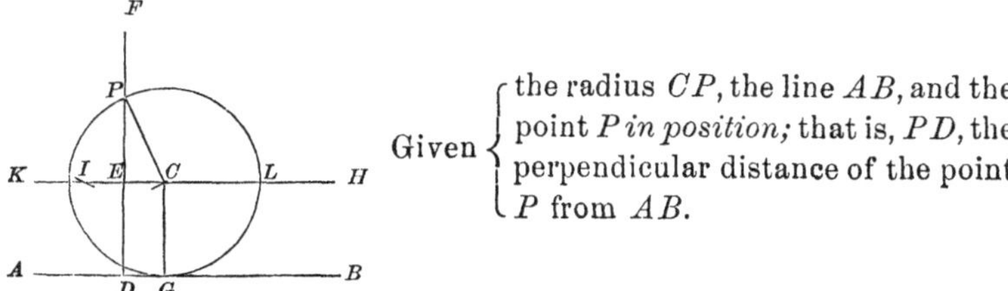

Given { the radius CP, the line AB, and the point P *in position;* that is, PD, the perpendicular distance of the point P from AB.

Analysis.—Suppose PLG to be the circle required, of which the centre is C. Through C draw KH parallel to AB, cutting the perpendicular DPF in E; then $ED = CG = CP$.

Construction.—Having drawn through the given point P the line DPF perpendicular to AB, on it lay $DE =$ the given radius, and through E, parallel to AB, draw KEH, to which apply $PC =$ the given radius, and let fall on AB the perpendicular CG. Then, with C as a centre, and radius CP, describe the circle PLG, which will be the one required.

Demonstration.—$CG = ED = CP$ by construction, and CG is perpendicular to AB, and hence the circle touches the line AB at the point G (III. 9*).

Calculation. — We have $PE =$ the difference between the given distance PD and the given radius. Then $DG = EC = \sqrt{CP^2 - PE^2}$.

Limits.—1. The given radius may be of any length not less than half DP. When the radius is *just half* DP, the required circle must be described on DP as a diameter.

2. The radius PC may be applied to KH, on *either side* of PD, at the point C or I, either of which may be taken as the centre of the circle to fulfil the conditions of the problem. We have $EI = EC$. If a perpendicular be let fall on AB from I, and IP be joined, such perpendicular will be equal to $CG = PC = PI$.

* III. 16, cor.

PROBLEM IX.—Through two given points to describe a circle which shall touch a line given in position.

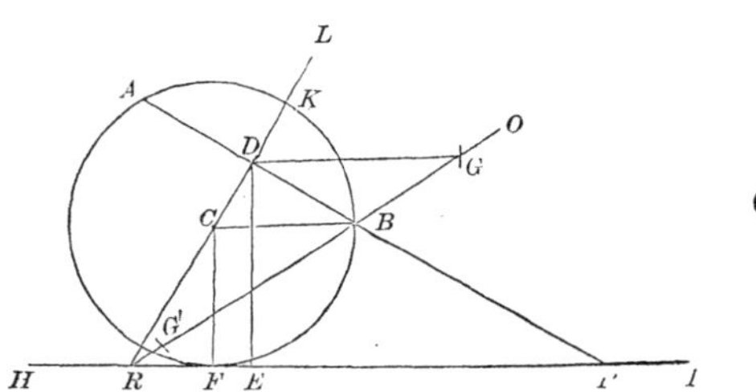

Given the points A and B, and the line HI *in position;* that is, the angle P which it forms with AB, produced, and the distance PB or PA.

Analysis.—Suppose the circle $AKBF$, whose centre is C, to be drawn to pass through the given points A and B, and to touch the given line HI at F. Join the given points A and B, and through the centre C draw a line LDC perpendicular to AB, meeting HI in R; then (III. 16*) $AD = DB$. Let fall on HI the perpendiculars DE and CF, and join CB and RB, and produce RB to O. Through D draw DG parallel to CB. Now, by parallel lines and similar triangles, since $CB = CF$, we have $DG = DE$. Whence this

Construction.—Join the given points A and B, bisect AB in D, and draw $LDCR$ perpendicular to AB, meeting HI in R. On HI let fall the perpendicular DE; draw RBO, and to it apply $DG = DE$. Draw BC parallel to DG, and CF parallel to DE; then, with the centre C and radius CF, describe the circle $AKBF$.

Demonstration.—By III. 9,† the circle $AKBF$ touches the given line HI at F. By parallel lines and similar triangles, since $DG = DE$ by construction, we have $CB = CF = CA$; hence the circle passes through the points A and B. Q. E. D.

Calculation.—1. Producing AB to meet HI in P, we have the angle P and the distance PB given, because the line HI *is given in position;* then $PD = PB + \frac{1}{2} AB$. In triangles PDR and PDE, find, Case 1, DR and $DE = DG$.

2. In triangle DRB, Case 3, find RB, and angle DBR; then angle $DBG = 180° - DBR$.

3. In triangle DBG, Case 2, find BG; then $RG = RB + BG$, and we have $RG : RB :: GD : BC = CF$. Or, after finding RB and angle DRB, we have $\sin CRF : \sin F :: CF : CR :: CB : CR ::$ $\sin CRB : \sin CBR$; then in triangle CBR, Case 1, find $BC = CF$.

* III. 3. † III. 16, cor.

Limits.—1. DE may be applied to RO at G', on the other side of AB. Then, if a line be drawn through B, parallel to $G'D$, to meet RDL, produced, it will give the centre C' of *another* circle passing through the given points A and B, and touching the given line HI.

2. When ABP is perpendicular to HI, DR and HI will be parallel, DP will be the radius of the required circle, which apply from B to LR to obtain the centre.

PROBLEM X.—To inscribe a square in a given segment of a circle.

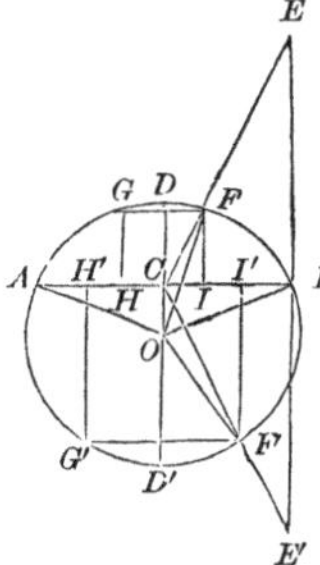

Given $\begin{cases}\text{the radius } OA \text{ or } OB,\\ \text{and the angle } AOB.\end{cases}$

Analysis.—Draw the chord AB, and perpendicular to it draw the diameter $DCOD'$; then $AC = CB$, and the arc $AD = DB$. Now, taking the *smaller* segment, let $IFGH$ represent the required square. Join CF, and produce it to meet BE, perpendicular to AB, in E. Then, by similar triangles CIF and CBE, since $IF = IH = 2\,IC$, we have $BE = 2\,BC = BA$. Whence this

Construction.—Erect the perpendicular $BE = BA$; join CE, cutting the arc BD in F; draw FG parallel to AB, and FI and GH each parallel to BE; then $IFGH$ will be the square required.

Demonstration.—$IFGH$ (I. 20, cor.*) is rectangular. Also, since, by construction, $BE = BA = 2\,BC$, we have, by similar triangles CIF and BCE, $FI = 2\,IC = IH = FG = GH$. Q. E. D.

Calculation.—1. Join OF. In triangle OCB, Case 1, find OC and CB.

2. In triangle BCE, Case 3, find CE, and angle ECB; then angle $OCF = 90° + ECB$.

3. In triangle OCF, Case 2, find CF; then, by similar triangles CIF and CBE, $CE : EB :: CF : FI = IH$.

Scholium.—When the square is to be inscribed in the *larger* segment $AD'B$, the perpendicular BE' must be made equal to BA;

* I. 29.

and, using the marked letters instead of the corresponding ones unmarked, the construction, demonstration, and calculation are *just the same* as that given for the *smaller* segment, *only* the angle OCF' in this case $= 90° - E'CB$.

If the point E' falls *on* the circumference, BA is the side of the square required, and the arc ADB is a quadrant. When the point E' falls within the circumference; a square cannot be inscribed in the *larger* segment, having one side on the chord AB. When each segment is a semicircle, BE and BE' are each equal to the diameter, and the two squares are equal.

PROBLEM XI.—To inscribe a circle in a given quadrant.

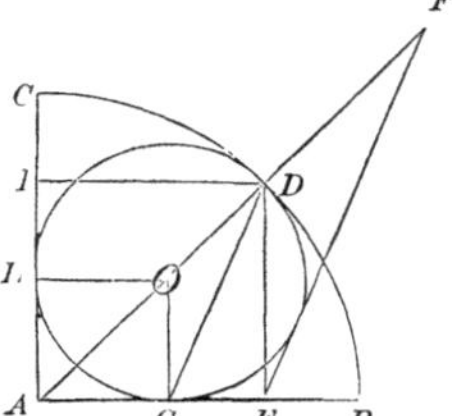

Given $\left\{\begin{array}{l}\text{in the quadrant } ABC\text{, the radius } AB \\ \text{or } AC\text{, to inscribe a circle in the} \\ \text{quadrant.}\end{array}\right.$

Analysis.—Suppose DGH to be the required inscribed circle, then D is evidently the middle point of the arc BC, and the radii OD, OG, and OH are all equal. Let fall the perpendicular DE; join GD, and parallel to it draw EF to meet AD, produced, in F; then the triangles OGD and DEF will be similar, and, since $OD = OG$, we have $DF = DE$. Whence this

Construction.—Bisect the arc BC in D; let fall the perpendicular DE; join AD, and produce it, making $DF = DE$. Join EF; draw DG parallel to EF, GO parallel to AC, and OH parallel to AB; then, with the centre O and radius OD or OG, describe the circle DGH, and it will be inscribed in the quadrant ABC.

Demonstration.—By parallel lines, the triangles ODG and DFE are similar, and, since $DF = DE$ by construction, we have $OD = OG = OH$, and AB and AC, being perpendicular to the radii OG and OH, are tangents to the circle at the points G and H (III. 9*). Q. E. D.

Calculation.—$AE = ED$. Hence $AE^2 = \frac{1}{2} AD^2$, and $AE = \sqrt{\frac{1}{2} AD^2} = DE = DF$. Also, $AF = AD + DF$; then, by similar triangles AFE and ADG, we have $AF : AD :: AE : AG = OH = OD$.

* III. 16, cor.

Remark.—If DI be drawn parallel to AB, $AEDI$ will be a square inscribed in the given quadrant ABC, whose side AE or $ED = \sqrt{\frac{1}{2}\overline{AD}^2} = \sqrt{\frac{1}{2}\overline{AB}^2} = \frac{1}{2} AB \sqrt{2}$.

PROBLEM XII.—To describe a circle which shall touch three lines given in position.

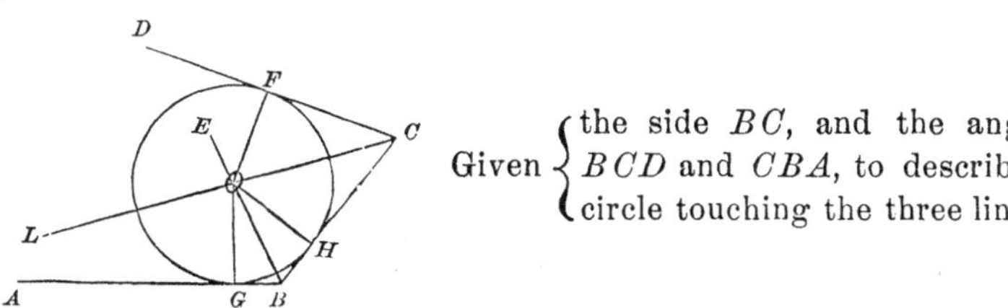

Given $\begin{cases}\text{the side } BC, \text{ and the angles} \\ BCD \text{ and } CBA, \text{ to describe a} \\ \text{circle touching the three lines.}\end{cases}$

Analysis.—Suppose the circle FGH, of which O is the centre, to touch the three given lines AB, BC, and CD in the points G, H, and F respectively. Draw the radii OG, OH, and OF, and they will be perpendicular to AB, BC, and CD respectively (III. 9*). Also, join OB and OC; then, in the right-angled triangles OGB and OHB, since $OG = OH$, we have the angle $OBG = OBH$. For a like reason, in the triangles OCF and OCH, we have the angle $OCF = OCH$. Hence the lines BO and CO bisect the angles ABC and BCD respectively. Whence this

Construction.—Bisect the angles ABC and BCD by the lines BE and CL, intersecting in O. From O, on the given lines, let fall the perpendiculars OG, OH, and OF. Then, with the centre O and any one of these lines as radius, describe a circle, and it will pass through the extremities of the others, and touch the given lines as required.

Demonstration.—Since BO bisects the angle GBH by construction, the triangles OBG and OBH are equal, and $OG = OH$. For a like reason, $OH = OF$; hence OG, OH, and OF are all equal. Also, since AB, BC, and CD are respectively perpendicular to the radii OG, OH, and OF at their extremities, they touch the circle at the points G, H, and F by reference in analysis. Q. E. D.

Calculation.—1. In triangle BOC, Case 1, find BO and OC.

2. In triangle BOH, Case 1, find $OH = OG$, and $BH = BG$; then $CF = CH = CB - BH$.

Scholium 1.—When the three lines on being produced do not meet, each one must be parallel to both the others, and the problem is impossible.

Scholium 2.—When the sum of the given angles ABC and BCD

* III. 16, cor.

is *equal to* 180°, the lines CD and BA are parallel. When the sum of the given angles is *less than* 180°, the lines CD and BA will meet, if produced, and form a triangle, when the problem becomes the same as III. Prob. 15,* to inscribe a circle in a triangle. In both cases the construction, etc. are the same as that given above.

PROBLEM XIII.—To draw a common tangent to two circles given in magnitude and position.

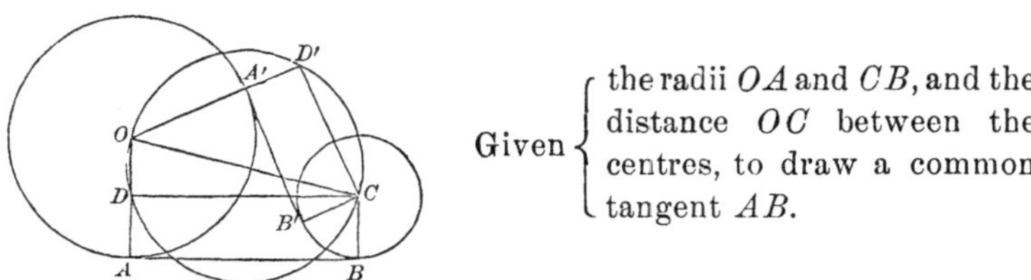

Analysis.—Let AB be the common tangent required; then (III. 9†) the radii OA and CB are perpendicular to AB, and consequently parallel to each other. Through C, the centre of the less circle, draw CD parallel to AB; then $AD = BC$, and OD is the difference between the given radii, and the triangle ODC is known. Whence this

Construction.—On the given distance OC describe a circle, in which apply $OD =$ the *difference* of the given radii. Produce OD to meet the circumference at A; draw the radius CB parallel to OA, and join AB, which will be the common tangent required.

Demonstration.—Since, by construction, OD is the difference between the radii, and OA the greater radius, we have DA equal and parallel to CB, and hence AB is equal and parallel to CD; and consequently, since ODC is a right angle, being in a semicircle, AB is perpendicular to the radii OA and CB at their extremities, and is hence a tangent to both circles (III. 9‡).

Calculation.—In triangle ODC, we have $CD = \sqrt{CO^2 - OD^2} = AB$.

Scholium.—Another common tangent may be drawn to the same circles. Suppose it done, as $A'B'$, and on it let fall the perpendiculars CB' and OA', and produce OA' to meet CD' drawn parallel to $A'B'$, in D'; then $A'D' = CB'$, the angle D' is a right angle, and consequently the point D' is in the semicircle described on the given line OC, and the line OD' is equal to the sum of the given

* IV. 4. † III. 18. ‡ III. 16, cor.

radii. In the semicircle described on OC apply OD' = the *sum* of the given radii; join CD'; draw CB' parallel to OD', and join $A'B'$, and it will be a common tangent, as is evident from the preceding demonstration. The length of $A'B' = CD' = \sqrt{OC^2 - OD'^2}$.

Remark.—AB is called the *external* common tangent, and $A'B'$ the *internal* common tangent.

Problem XIV.—On a given line to describe two semicircles touching each other such that their common tangent may be of a given length.

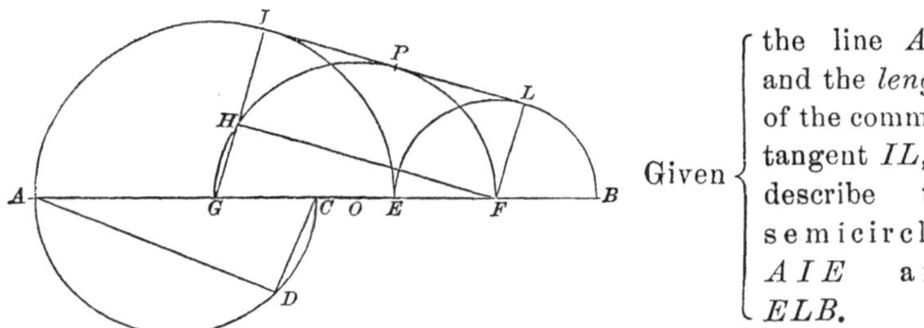

Given { the line AB, and the *length* of the common tangent IL, to describe the semicircles AIE and ELB.

Analysis.—Suppose AIE and ELB to be the required semicircles described on the given line AB, G and F, their respective centres, and IL their common tangent, I and L being the points of contact. Since $GE = GA$, and $EF = FB$, we have $GF = \frac{1}{2}AB$. On GF describe a semicircle, draw the radii GI and FL, which will be perpendicular to IL, and therefore parallel, and join F and H, the point where GI intersects the semicircle on GF. Now, since the angle GHF is a right angle, being in a semicircle, HF is parallel and hence equal to IL, and $HI = FL$, and GH is the difference between the two radii of the required semicircles. Whence this

Construction.—Bisect the given line AB in C, on AC describe a semicircle, in which apply AD = the given length of the common tangent, join CD, and make $CE = CD$. On AE and EB describe semicircles of which the centres are G and F, and on GF describe a semicircle of which the centre is O. Draw FH parallel to AD, draw GHI, and draw FL parallel to it, and join IL, and it is done.

Demonstration.—By last problem, IL is a common tangent to the two semicircles, and equal to HF. Because $AC = GF$, each being half of AB, and FH is parallel to AD by construction, the triangles ADC and FHG are equal, and we have $AD = FH = IL$; also, $GH = CD = CE$. Now, since $GE = GI$ and $CE = GH$, we have

$GC = HI = EF = FL$, and $GE - EF = GI - FL$. Hence $FL = HI$, and GH is the difference between the radii GI and FL, and is $= \frac{1}{2}$ the difference of the diameters AE and EB.

Calculation.—$CE = GH = CD = \sqrt{AC^2 - AD^2} = \sqrt{AC^2 - IL^2}$; then $AE = AC + CE$, and $EB = AC - CE$.

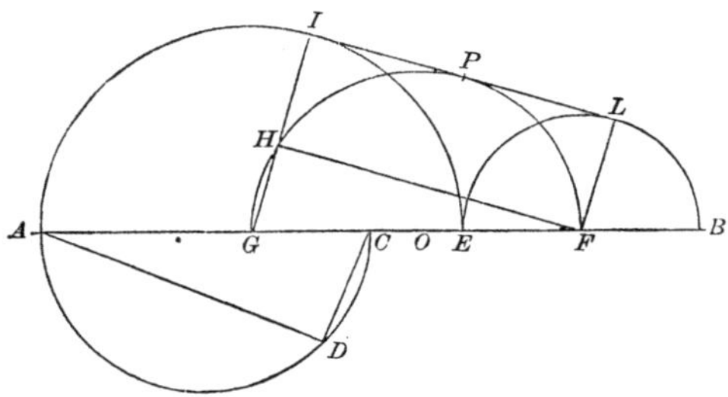

Scholium 1.—If the given common tangent is *just half* the given line AB, the semicircles will be equal. This is its superior limit.

Scholium 2.—The common tangent IL touches the semicircle GHF at P, the middle point of IL. If OP be joined, it will be perpendicular to IL, and equal to *half* $GI + FL = OG$ or $OF = \frac{1}{4} AB$; also, $GE = CF$, $GC = EF$, and $OC = OE$.

PROBLEM XV.—Of two circles which touch each other externally are given the radius of one, and the length of their common tangent, to determine the radius of the other.

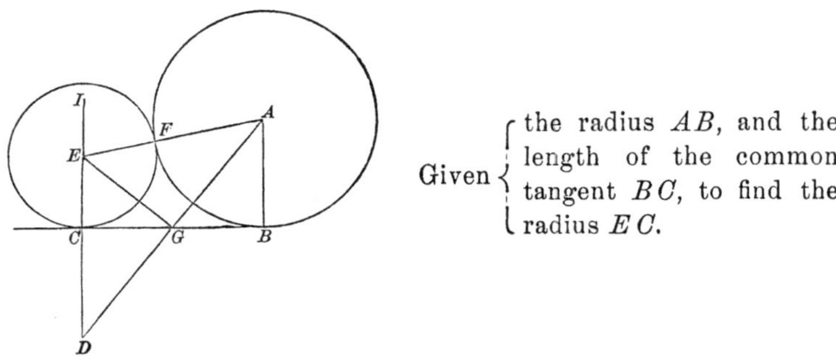

Given $\begin{cases}\text{the radius } AB, \text{ and the length of the common tangent } BC, \text{ to find the radius } EC.\end{cases}$

Analysis.—Suppose the figure constructed as above, AB the given radius, and EC the required radius. Produce the radius EC, making $ED = EA$, and join AD, intersecting BC in G, and join EG. Then, since $ED = EA$ and $EC = EF$, we have $CD = AF = AB$. By similar triangles DCG and ABG, since $CD = AB$, we have $CG = GB$, and $DG = GA$, and EG perpendicular to DA. Whence this

Construction.—Draw BC equal the given common tangent, and draw BA and DC perpendicular to BC, on opposite sides, and each equal to the given radius. Join DA, intersecting BC in G; then, evidently, by similar triangles DCG and ABG, $CG = GB$ and $DG = GA$. Draw GE perpendicular to DA, meeting DC, produced, in E; draw EFA, and with the centres E and A and radii EC and AB, respectively, describe circles, and they will touch each other at the point F, and also touch the line BC at the points B and C.

Demonstration.—BC is the common tangent by construction, and it only remains to prove that the circles are tangent to each other; that is, that $AE = AB + CE$. Since GE is perpendicular to AD at its middle point G, we have $EA = ED = CD + CE = AB + CE =$ the sum of the radii. Hence (III. 13*) the circles touch each other externally. Q. E. D.

Calculation.—$DG = \sqrt{CD^2 + CG^2}$; then, by similar triangles DGC and DEG, we have $DC : DG :: DG : DE = EA$, and $DE - DC = EC = EF$.

Limit.—The common tangent can be taken *any* length.

Scholium.—When two circles touch each other externally, half the common tangent is a mean proportional between the radii; that is, in the figure, $EC : CG$ or $GB :: CG : AB$.

PROBLEM XVI.—Given, the radii of two concentric circles, to find the radius of a circle which shall touch one of these circles externally and have a common tangent of a given length with the other.

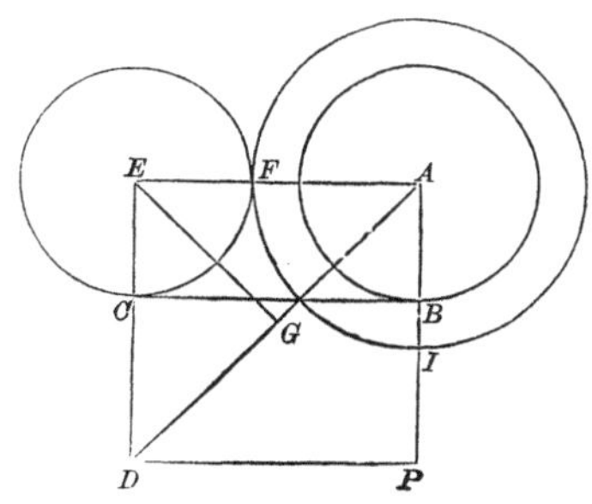

Given the radii AB and AI, and the *length* of the common tangent BC, to find the radius EC.

Analysis.—Suppose the figure constructed as above. Produce EC, making $CD = AI$, join AD and AE, and let fall the perpendicular EG; then $ED = EA$, $DG = GA$, and BC, perpendicular to the radii EC and AB at their extremities, is the given common tangent. Whence this

Construction.—With the given radii AB and AI having described

* III. 12.

the concentric circles, draw BC perpendicular to ABI, and equal to the given length of the common tangent. Draw CD perpendicular to CB and equal to AI. Join AD; bisect it in G; and draw GE perpendicular to DA, meeting DC, produced, in E; and draw EFA. With centre E and radius EC describe a circle, and it will touch one of the given circles at F, and have the common tangent BC with the other circle of the given length.

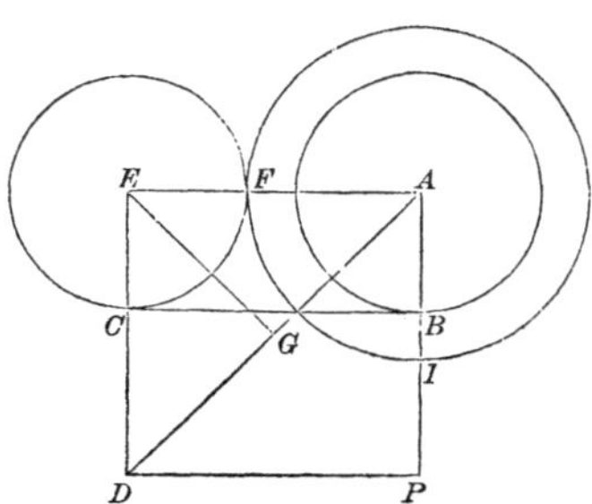

Demonstration. — BC is evidently the common tangent to the circles AB and EC, and it was made the given length by construction. Since GE is perpendicular to DA at its middle point G, we have $EA = ED = EC + CD = EC + AI =$ the sum of the radii. Hence (III. 13*) the circles touch each other at F. Q. E. D.

Calculation.—Through D, parallel to CB, draw a line to meet AI, produced, in P; then $DP = CB$, and $AP = AB + CD = AB + AI$; also, $AD = \sqrt{AP^2 + DP^2}$, and $DG = \frac{1}{2} AD$. By similar triangles APD, DGE, we have $AP : PD :: DG : DE$; that is, $AI + AB : BC :: \frac{1}{2} AD : DE$; then $DE - CD = DE - AI = EC = EF$.

Scholium.—If $AB = AI$, it reduces to the preceding problem. If CA and DI be joined, they will be parallel.

PROBLEM XVII.—Of two circles which intersect each other there are given the radius of one, their common chord, and their common tangent, to determine the radius of the other.

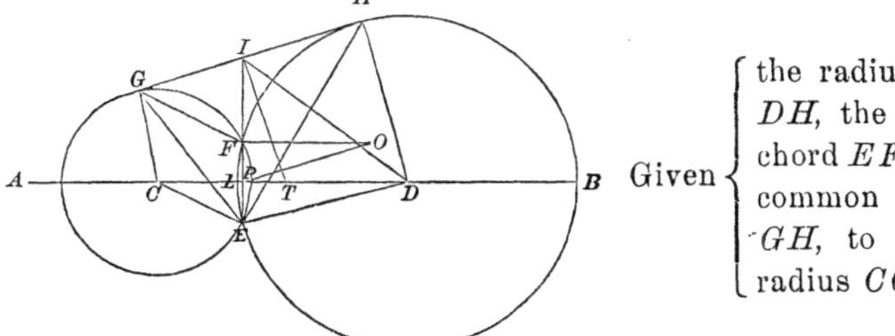

Given { the radius DB or DH, the common chord EF, and the common tangent GH, to find the radius CG or CE.

Analysis.—Suppose the problem constructed, D being the centre of the given circle, and C the centre of the required circle. From

* III. 12.

the extremities of the common tangent GH, draw GC, GE, and GF; HD, HE, and HF; and produce the common chord EF to meet the common tangent in I; then (IV. 30*) $IH^2 = IE \times IF = IG^2$; hence $IH = IG$, and the *common chord, produced, bisects the common tangent.* Whence this

Construction.—In the given circle, whose centre is D, apply the given chord EF, through the middle point P of which draw the diameter BDL. Draw FO perpendicular to FE, or parallel to LB, and equal to *half* the length of the given common tangent. Join PO, and produce EF until $PI = PO$, and apply $IH = FO$, half the given common tangent. Produce HI, making $IG = IH$; draw GC parallel to DH, meeting BDL, produced, in C; and draw CE, GC, GE, and GF; HD, HE, and HF; and with centre C and radius CG, describe the circle $AGFE$, and it will be the one required.

Demonstration.—$IE \times IF = (IP + PF) \times (IP - PF) = IP^2 - PF^2 = OP^2 - PF^2 = FO^2 = IH^2 = IG^2$. Hence (IV. 30†) IG is a tangent to the circle GFE passing through F and E, and IH is a tangent to the circle HFE passing through the same points; consequently, GH is a *common tangent to both circles*, and it is of the given length. Q. E. D.

Calculation.—Draw IT parallel to HD or GC; then, since $GI = IH$ by construction, we have $CT = TD$ and $IT = \frac{1}{2}(DH + CH)$. Now, $PI = PO = \sqrt{FO^2 + FP^2}$, and $PD = \sqrt{DE^2 - EP^2}$. Join ID; then in the right-angled triangle IHD, Case 3, find angle IDH; and in triangle IDP, Case 3, find angle IDP. Their sum = angle $PDH = PTI$. Then, in triangle PTI, Case 1, find $IT = \frac{1}{2} HD + \frac{1}{2} CG$. Whence we have $CG = 2\,IT - HD$.

* III. 36. † III. 36.

PROBLEM XVIII.—From one extremity of the diameter of a given semicircle, to draw a right line such that the part intercepted between the *circumference* and a *tangent* at the other extremity shall be of a given length.

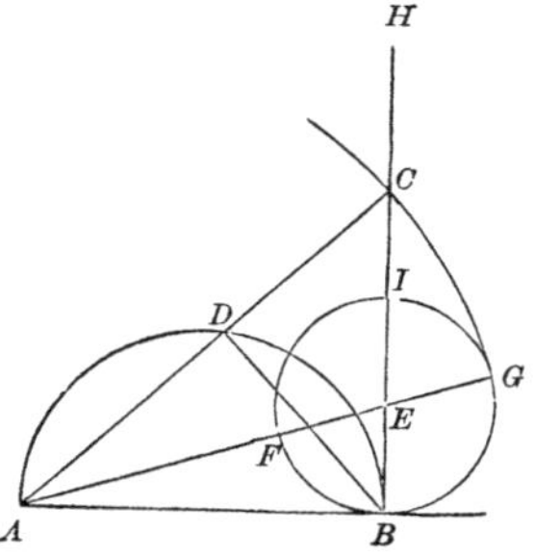

Given $\begin{cases} \text{the diameter } AB \text{ of the semicircle } ADB, \text{ and the intercepted} \\ \text{distance } DC \text{ of the line } ADC. \end{cases}$

Analysis.—Let ADB represent the given semicircle, BH a tangent, and DC the given intercepted distance. Join BD; then the triangles ABC and ADB are similar, and we have (IV. 18*) $AC : AB :: AB : AD$. Whence this

Construction.—On the tangent BH lay $BI =$ the given intercepted line, and on BI describe a semicircle of which E is the centre. Draw the line $AFEG$, and to BH apply $AC = AG$, cutting the semicircle in D; then AD will equal AF, and $DC = FG = BI$.

Demonstration.—By similar triangles and IV. 30,† we have $AC : AB :: AB : AD$, and $AG\,(AC) : AB :: AB : AF$; hence $AD = AF$; and we have $DC = AC - AD = AG - AF = FG = BI$. Q. E. D.

Calculation.—$AE = \sqrt{AB^2 + BE^2}$; then $AG = AE + EB = AC$, and $AF = AE - EF = AD$. Also, $BC = \sqrt{AC^2 - AB^2}$.

Limit.—DC may be of *any* length.

* VI. 4.

† III. 36.

PROBLEM XIX.—Having a circle given, it is required to inscribe in it an equilateral triangle, and also to circumscribe an equilateral triangle about the circle.

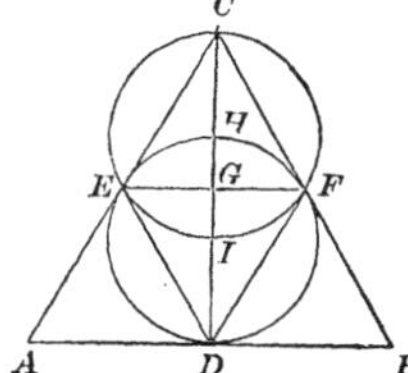

Given $\begin{cases} \text{the diameter } DH = 2R, \text{ or the radius} \\ ID = R. \end{cases}$

Analysis.—Let $EDFH$ represent the given circle, of which the centre is I, EDF the inscribed triangle, and ABC the circumscribed triangle. Join CD, which will be perpendicular to AB; and, since AB is a tangent to the circle at D, CD will pass through the centre I, and through G and H, the middle points of the chord EF and of the arc EHF, respectively. Also, since AC and BC are tangents to the circle at the points E and F, respectively, a circle described about H as a centre, with the radius HI, will pass through the points E, C, and F. Whence this

Construction.—Having the circle, perpendicular to its diameter DIH draw an indefinite line ADB. With the centre H and radius HI describe the circle FIE, meeting DH, produced, in C. Join CE and CF, and produce them to meet the perpendicular through D, in A and B. Also join EF, ED, and DF; then EDF will be the inscribed triangle, and ABC the circumscribed one, required, as is evident from the analysis and V. 4.*

Calculation.—CD is evidently equal to $3R$, and $DG = \frac{1}{2} CD = \frac{3R}{2}$, and $DG^2 = \frac{9R^2}{4}$. Now, in triangle DGF, $DG^2 = DF^2 - FG^2 = DF^2 - (\frac{1}{2} DF)^2 = DF^2 - \frac{1}{4} DF^2 = \frac{3}{4} DF^2$. Hence $\frac{3}{4} DF^2 = \frac{9R^2}{4}$. Whence $DF^2 = 3R^2$, or $DF = R\sqrt{3} = FB$. Also, $AB = BC = 2BF = 2DF = 2R\sqrt{3}$.

Corollary.—The side AB being equal to twice EF, the area of the circumscribed triangle ABC will be four times the area of the inscribed triangle DFE. The four partial equilateral triangles composing ABC are all equal, and each one-fourth the area of ABC; and each of the twelve sides of these four triangles is equal to half of each side of the triangle ABC.

* IV. 15.

Problem XX.—Given, the lengths of two chords cutting each other at right angles in a circle, and the distance of their point of intersection from the centre of the circle, to determine the radius of the circle.

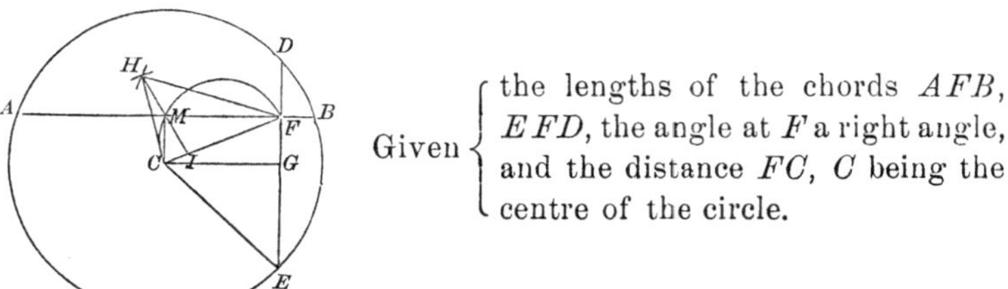

Given $\left\{ \begin{array}{l} \text{the lengths of the chords } AFB, \\ EFD, \text{ the angle at } F \text{ a right angle,} \\ \text{and the distance } FC,\ C \text{ being the} \\ \text{centre of the circle.} \end{array} \right.$

Analysis.—Suppose $ADBE$ to be the circle constructed as required. Then (IV. 28, cor.*) $AF \times FB = EF \times FD$. Bisect the given chords AB and DE by the perpendiculars CM and CG, which will be parallel to DE and AB respectively. Then $AF \times FB = (AM + MF) \times (AM - MF) = AM^2 - MF^2$. Also, $EF \times FD = (EG + GF) \times (EG - GF) = EG^2 - GF^2$. Hence we have $AM^2 - MF^2 = EG^2 - GF^2$, or $AM^2 - EG^2 = MF^2 - GF^2 = MF^2 - CM^2$. Whence this

Construction.—Draw $CF =$ the given distance from the centre to the intersection of the chords, and on it describe a semicircle. From C and F as centres, with radii equal to *half* the given chords DE and AB respectively, describe arcs intersecting in H. Join CH and HF, and draw HMI perpendicular to CF. Join CM and FM, and produce FM both ways, making MB and MA each equal to FH, half the given chord AB; then $AB =$ the given length. With the centre C and radius equal to CA or CB describe a circle, and through F, parallel to CM, draw a line to meet the circumference in D and E; then DE will equal 2 $CH =$ the given length.

Demonstration.—Draw CG parallel to AB. The angle CMF, being in a semicircle, is a right angle, and hence all the angles in the parallelogram $CMFG$ are right angles, and CG is perpendicular to DE, and bisects it. Now, by the analysis, we have $EG^2 - GF^2 = AM^2 - MF^2$, or $AM^2 - EG^2 = MF^2 - GF^2 = MF^2 - CM^2 =$ (Prob. 36, cor., in "Triangles," etc.) $FI^2 - IC^2 = FH^2 - CH^2 = AM^2 - CH^2$; hence $EG^2 = CH^2$, and $EG = CH$, and $DE = 2\ EG = 2\ CH =$ the given length.

Calculation.—We have $MF^2 + CM^2 = CF^2$, and $MF^2 - CM^2 =$

* III. 35.

$(MF^2 - FG^2) = AM^2 - EG^2$ by analysis. Their sum gives $2\,MF^2 = CF^2 + AM^2 - EG^2$, or $MF = \sqrt{\dfrac{CF^2 + AM^2 - EG^2}{2}} = CG$. Their difference gives $2\,CM^2 = CF^2 - AM^2 + EG^2$, or $CM = \sqrt{\dfrac{CF^2 - AM^2 + EG^2}{2}} = FG$; whence EF, FD, and AF, FB are known. Also, radius $CE = \sqrt{CG^2 + EG^2}$.

PROBLEM XXI.—Given, the three distances from a point either *within* or *without* a circle to the three nearest extremities of two diameters which intersect each other at right angles, to find the diameter of the circle. (See Problems XL. and XLI., "Triangles," etc.)

FIG. 1. FIG. 2.

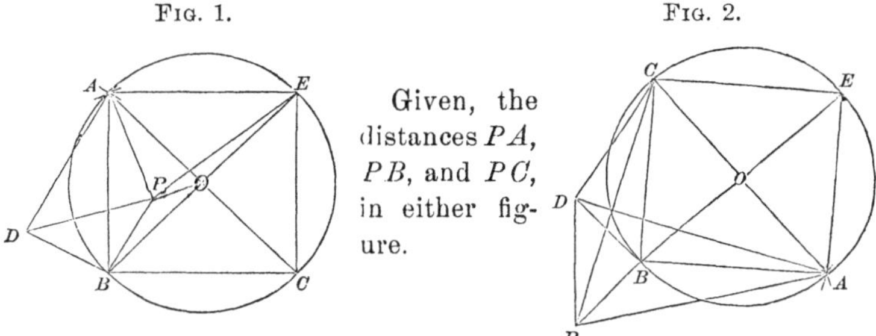

Analysis.—Suppose AC and BE (referring to either figure) to be the two diameters crossing each other at right angles at the centre O, and that we have the distances PA, PB, and PC given. Join the points AB, BC, CE, and EA, and $ABCE$ will be a *square*, and ABC an *isosceles triangle*, having the angle ABC included between the equal sides given, and also the three distances from the point P, PA, PB, and PC, to the three angular points, to construct the triangle and find the sides, which is done in Problem XL. of "Triangles, Quadrilaterals, and Parallels." Then AC will be the diameter required. Complete the square $ABCE$, and describe the circle with the radius OA or OC. The same letters used in Problem XL. apply to these figures in the construction, demonstration, etc.

Property of the circle.—If from any point within or without a circle lines be drawn to the extremities of diameters of the circle, the sum of the squares of the two lines drawn to the extremities of *one* diameter, will be equal to the sum of the squares of the two lines drawn to the extremities of *any other* diameter.

Demonstration.—Join PE and PO (Fig. 1); then (in either figure) (IV. 14*) $PA^2 + PC^2 = 2\,PO^2 + 2\,AO^2 = 2\,PO^2 + 2\,OB^2 =$

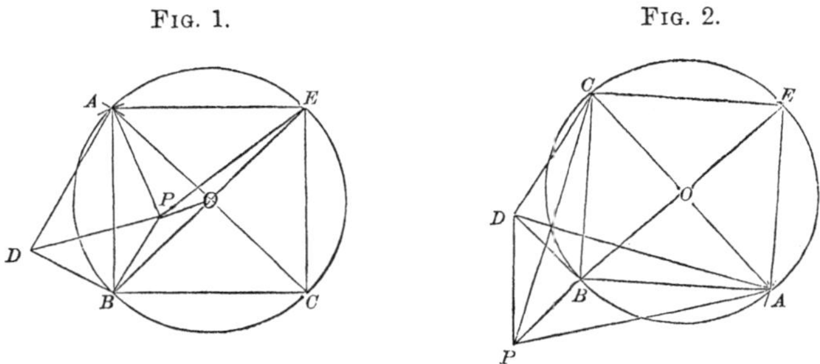

FIG. 1. FIG. 2.

$PB^2 + PE^2$. And if two lines be drawn from P to the extremities of *any other* diameter, they together with such diameter regarded as the base will form a triangle of which PO will always be a line drawn from the vertex to the middle of the base, and hence the sum of the squares of these two lines (IV. 14†) will be equal to twice $PO^2 +$ twice the square of the radius $= 2\,PO^2 + 2\,OA^2 = PA^2 + AC^2$. Q. E. D.

PROBLEM XXII.—Having given the positions of two lines, and of a point between them, it is required to describe a circle which shall pass through the given point and touch each of the lines given in position.

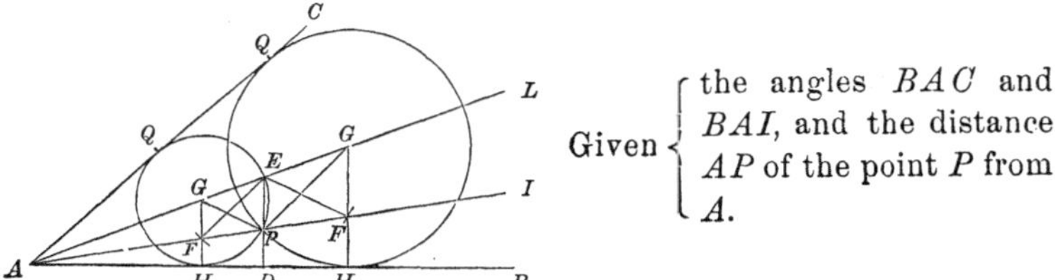

Given { the angles BAC and BAI, and the distance AP of the point P from A.

Analysis.—Suppose G‡ to be the centre of the required circle passing through the given point P, and touching the lines AB and AC given in position in the points H and Q respectively. Through the centre G draw the line AGL indefinitely; then (III. Prob. 15§) AL bisects the given angle BAC. Join GP, let fall the perpen-

* II. A. † II. A.

‡ The student may give attention to either circle, but he should confine it to one at a time. The analysis, construction, etc. apply to both circles alike, taken separately.

§ IV. 4.

dicular GH, and through the given point P draw ED parallel to GH, and draw EF parallel to GP. Then, by parallel lines, we have $AG : GH :: AE : ED$, and $AG : GP\,(GH) :: AE : EF$; wherefore $EF = ED$; and we have this

Construction.—Through the given point P draw ED perpendicular to AB, the line AEL bisecting the given angle BAC. Apply $EF = ED$; draw PG parallel to EF; let fall the perpendicular GH; and with the centre G and radius GH describe the circle HQ, which will pass through the given point P and touch the lines AB and AC.

Demonstration.—By the principles and reasoning in the analysis, $GP = GH$, and hence the circle passes through P, and by III. Prob. 15* it touches AB and AC. Q. E. D.

Calculation.—In triangle APD, Case 1, find AD. Then, in triangle ADE, Case 1, find AE and $ED = EF$. Now, angle $EAF = \frac{1}{2} BAC - BAI$. In triangle AEF, Case 2, find AF, which has two values. Taking the *less* value of AF, we have $AF : FE\,(ED) :: AP : GP$, the radius of the *greater* circle. And taking the *greater* value of AF, we have $AF : FE\,(ED) :: AP : GP$, the radius of the *less* circle.

Limit.—If P coincide with D, G will coincide with E, ED will be the required radius, and there will be but one circle.

* IV. 4.

PROBLEM XXIII.—Having given the positions of two right lines, and also a circle whose centre is between them is given in position and magnitude, it is required to describe another circle which shall touch each of the said lines and also the given circle.

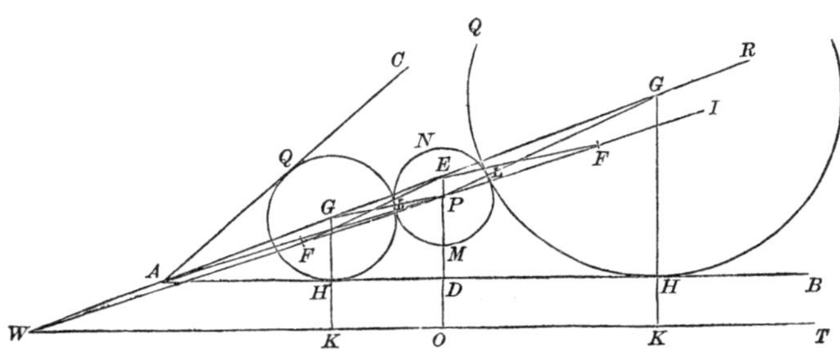

Given $\begin{cases}\text{the angles } BAC \text{ and } BAP, \text{ the perpendicular } DP, \text{ and} \\ \text{radius } PM \text{ of the circle } MLN.\end{cases}$

Analysis.—Suppose G* to be the centre of the required circle, touching the lines AB and AC in the points H and Q, and the given circle, whose centre is P, in L. Through the centre G draw the line AGR indefinitely, and (III. Prob. 15†) the line AGR will bisect the given angle BAC. Draw GLP; let fall on AB the perpendicular GH; through P draw EPD parallel to GH, and draw EF parallel to GP. Also, produce ED, making $DO =$ the radius PM, and through O, parallel to AB, draw a line, meeting RA, produced, in W, and GH, produced, in K; then $GK = GH + HK = GL + PM$ or $PL = GP$. Also, by parallel lines, $WG : GK :: WE : EO$, and $WG : GP\,(GK) :: WE : EF$; hence $EF = EO$; and we have this

Construction.—Draw AR bisecting the given angle BAC; through the centre P of the given circle draw EPD perpendicular to AB, and produce it, making $DO = PM$. Through O, parallel to AB, draw a line WT, meeting RA, produced, in W. Through P draw WPI, to which apply $EF = EO$. Draw PG parallel to EF; let fall the perpendicular GHK; and with the centre G and radius GH describe the circle HQL, which will be the one required.

Demonstration.—By parallel lines, since $EF = EO$ by construction, we have $GP = GK = GH + HK = GH + PL = GL + PL$. Hence the circles touch externally at the point L. And by III.

* The student may give attention to either of the required circles, but he should confine it to one at a time. The analysis, demonstration, etc. apply to both circles (QLH) alike, taken separately.

† IV. 4.

Prob. 15* HQL touches the lines AB and AC, the angle BAC being bisected by the line AR. Q. E. D.

Calculation.—In the triangle PAD, Case 1, find AD. Then, in the triangle AED, Case 1, find AE and ED; then $EO = ED + PM$; also, $PO = PD + PM$. By similar triangles EDA, EOW, we have $ED : EO :: AD : WO$, and $ED : EO :: AE : WE$. In triangle PWO, Case 3, find angle PWO and side WP; then angle $EWF = EWO - PWO =$ (I. 20, cor. 3†) $EAB - PWO$. In triangle EWF, Case 2, find WF, which will have two values. Taking the *less* value of WF, we have, by parallel lines, $WF : WP :: EF (EO) : PG$, from which take PL or PM, and we have $GL = GH$, the radius of the *greater* circle HLQ, the point Q being on the line AC. Also, taking the *greater* value of WF, we have $WF : WP :: EF (EO) : PG$; then $GL = GP - PL = GP - PM =$ the radius of the *less* required circle HLQ.

PROBLEM XXIV.—To divide the arc of a given semicircle into two segments such that the radius shall be a mean proportional between the chords of the segments.

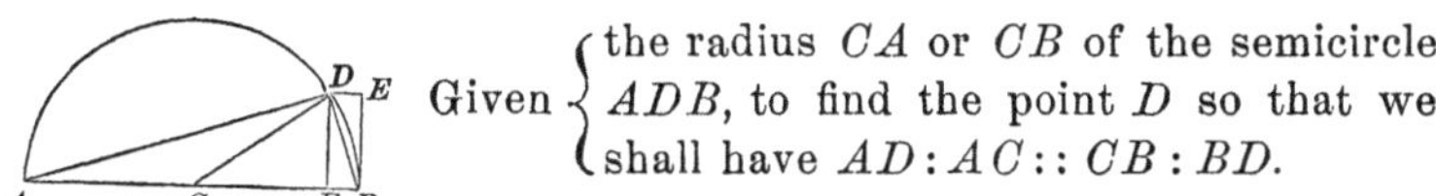

Given $\begin{cases} \text{the radius } CA \text{ or } CB \text{ of the semicircle} \\ ADB, \text{ to find the point } D \text{ so that we} \\ \text{shall have } AD : AC :: CB : BD. \end{cases}$

Analysis.—Suppose the figure constructed as above. Draw DE parallel to AB to meet the perpendicular BE at E; then the angles DAB and DBE, being each the complement of DBA, are equal, and the triangles DAB and DBE are consequently similar, and we have $AB : AD :: BD : BE$. Whence this

Construction.—Having the semicircle described on the diameter ACB, since, in the above proportion, the first term AB is *twice* the radius, we must erect the perpendicular $BE =$ *half* the radius, draw ED parallel to AB, and join AD and DB, and they will be the chords of the required segments.

Demonstration.—In similar triangles DAB and DBE, we have $AB (2\,BC) : AD :: BD : BE (\frac{1}{2} BC)$. Hence $AD \times DB = 2\,BC \times \frac{1}{2} BC = BC^2 = AC \times BC$. Wherefore we have $AD : AC :: BC : BD$. Q. E. D.

Calculation.—Join CD, and on AB let fall the perpendicular DF; then $CF = \sqrt{CD^2 - DF^2} = \sqrt{CD^2 - (\frac{1}{2}\,CD)^2} = \sqrt{\frac{3}{4}\,CD^2} =$

* IV. 4. † I. 29.

$\frac{1}{2} CD \sqrt{3}$. Also $AF = AC + CF$, and $BF = AC - CF$; then we have $AD = \sqrt{AF^2 + DF^2}$, and $BD = \sqrt{BF^2 + DF^2}$.

Problem XXV.—Through two given points to describe a circle which shall touch another circle given in position and magnitude.

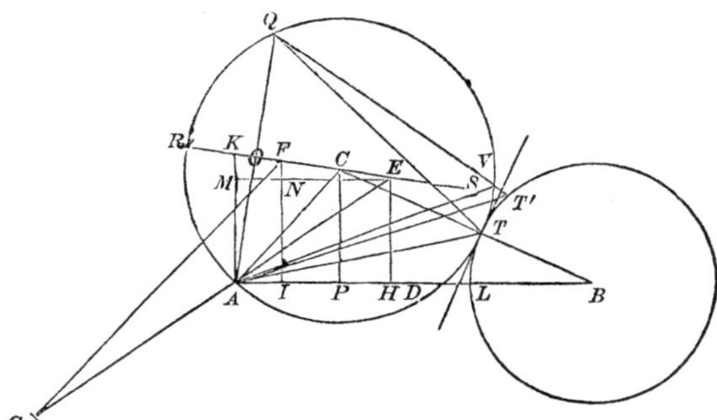

Given, the points A and Q, the angle QAB, the distance AB, and the radius BL or BT, to describe the circle AQT.

Analysis.—Suppose AQT to be the required circle, of which C is the centre, touching the given circle BL in T. Through the centre C draw ROS perpendicular to AQ, and draw CA and CTB. Then on the base AB we have to construct a triangle ABC, whose vertex C shall be in the line ROS, and the difference of whose sides BC and CA shall be equal to the radius BL, which is done by Problem VIII. in "Analysis by Algebra."

Calculation.—See calculation in Problem VIII. "Analysis by Algebra." By construction, $QAB : BL :: BL : DH$; then $AH = AD - DH$; also $HI = BL$ by construction. Draw ENM parallel to HI; then $EN = HI$. Also draw AK perpendicular to AB; then angle $MEK = OAK =$ the complement of the given angle QAB.

Now, 1. In triangle ENF, Case 1, find EF; and in triangle EMK, having $EM = AH$, find KM, Case 1.

2. In triangle AOK, Case 1, find AK; then $EH = AM = AK - KM$.

3. In triangle AHE, Case 3, find angle $HAE = AEM$; then angle $AEF = AEM + MEK = GEF$.

4. In triangle EGF, Case 2, having $FG = AB$, find angle $FGE = CAE$; then angle $CAP = CAE + HAE$, and angle $OAC = OAB - CAP$. In triangle OAC, Case 1, find $AC = CT =$ the required radius.

Problem XXVI.—From two given points without a circle which is given in position and magnitude, it is required to draw two lines to meet in the circumference of the given circle such that the angle formed by these lines at the circumference shall be the *greatest possible.*

Given { the same things as in the last problem (see figure to Problem XXV.), to draw the lines AT and QT such that the angle ATQ shall be the *greatest possible.*

Analysis.—Join AT and QT, T being the point in which the circle AQT touches the circle LT, in the figure on the opposite page; then ATQ is the angle required.

Demonstration.—In the circumference of the given circle LT, take *any other* point, as T', and draw AT' and QT'. One of these lines must cut the circumference ATQ in a point, as V. Draw a line from V to the other extremity of the base, as VA; then (III. 18, cor. 1*) the angle $ATQ = AVQ$. But (I. 25, cor. 6†) the angle AVQ is *greater* than $AT'Q$. Hence the angle ATQ is greater than $AT'Q$, and, as T' is *any point* in the circumference of the given circle, the angle ATQ is the *greatest possible* angle that can be formed by two lines drawn from the points A and Q to meet in the circumference of the given circle LTT'. Wherefore, as in the last problem, we must describe a circle to pass through the points A and Q, and touch the given circle in the point T, and draw the lines AT and QT, which will be the lines required.

Problem XXVII.—Having given the chord and tangent of a circle, to determine the radius.

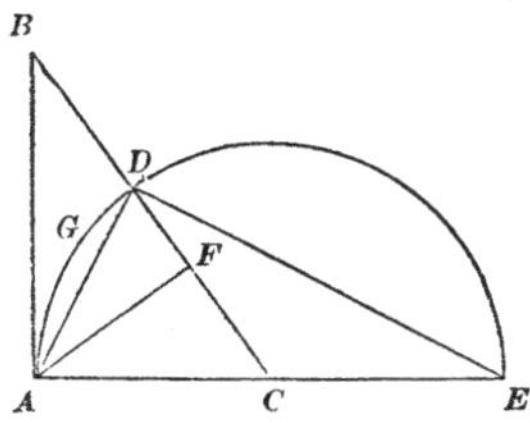

Given { the chord AD, and the tangent AB, to find the radius AC of the arc AGD.

Analysis.—Let AD be the chord and AB the tangent of the arc AGD, and CDB the secant. Complete the semicircle ADE; join DE; and on BC let fall the perpendicular AF; then (IV. 23‡) angle $BAF = BCA = DCA = 2\,DEA =$ (III. 21§) $2\,DAB$; hence

* III. 21. † I. 16. ‡ IV. 8. § III. 32.

AD bisects the angle BAF. Wherefore, in the right-angled triangle BFA, we have given the hypothenuse AB, and the line AD bisecting the acute angle BAF, to determine the triangle BFA by Problem XII. in "Analysis by Algebra."

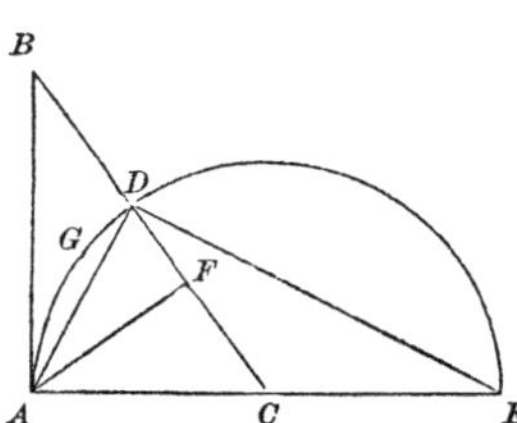

Then draw AC perpendicular to AB to meet BF, produced, in C, and AC will be the required radius.

Calculation.—After finding BF and BA, as in the problem referred to, we have $BF : FA :: BA : AC =$ the radius required.

PROBLEM XXVIII.—Having given the angle formed by two lines, the length of a line bisecting this angle, and the radius of a circle whose centre is the remote extremity of this bisecting line, to draw a tangent to the circle such that the part intercepted between the two lines, whose inclination to each other is given, may be of a given length.

FIG. 1.

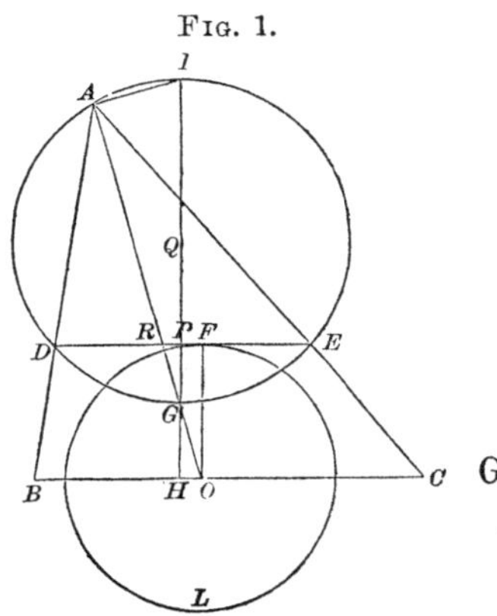

FIG. 2.

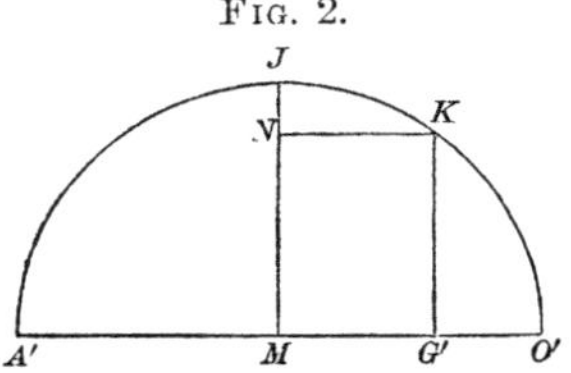

Given $\begin{cases} \text{the angle } BAC, \text{ the line } AO = \\ A'O', \text{ the radius } OF, \text{ and the} \\ \text{length of the tangent } DE, \text{ in-} \\ \text{tercepted between } AB \text{ and } AC. \end{cases}$

Analysis.—Suppose the problem constructed as in figure 1. About the triangle ADE describe the circle $AIEGD$. Through O draw BC parallel to DE, and through G, where the circumference cuts AO, draw $HGPI$ perpendicular to BC or DE; then GI is the diameter of $AIEGD$; $DP = PE$, arc $DG = GE$, and $OG \times GA = HG \times GI$. Whence this

Construction.—On the given tangent DE (III. Prob. 16*) describe a circle such that the segment DAE shall contain the given

* III. 33.

angle. Bisect DE by the diameter IPG, which produce until $PH =$ the given radius. Draw BHC parallel to DE. On $A'O'$ (Fig. 2), equal to the given line AO, describe a semicircle of which M is the centre; draw the radius MJ perpendicular to $A'O'$; lay MN a mean proportional between HG and GI. Draw NK parallel to $A'O'$, and KG' parallel to MJ; then $A'G' \times G'O' = G'K^2 = MN^2 = HG \times GI$. Apply $G'O'$ from G to O in BC, and produce OG to meet the circumference in A. Draw ADB, AEC; let fall the perpendicular OF on DE, and with centre O and radius OF describe the circle O,FL, to which the line DFE will be tangent.

Demonstration.—It remains to show only that $AG = A'G'$. Join AI; then, in similar triangles GAI and GHO, $GH : GO :: GA : GI$; hence $GO \times GA$ (or $G'O' \times GA$) $= GI \times GH =$ (by construction) $G'O' \times G'A'$. Hence $GA = G'A'$ and $OA = O'A'$.

Calculation.—$OH = \sqrt{OG^2 - GH^2} = PF$. By similar triangles GHO and OFR, $GH : GO :: OF : OR$, and $GH : HO :: OF : FR$. Whence RP, RE, DR, and AR become known; and by similar triangles, BO, OC, and BC also. Then in triangles ABC and ACO we have, Case 2, AB and AC; and thence, by similar triangles, we have AD and AE.

Scholium.—If the point O comes between DE and the arc DGE, we must *produce* $A'O'$ to G'' (IV. Prob. 4*) so that the rectangle $A'G'' \times G''O' = GH \times HI = MN^2$, and proceed in every other respect as already directed. (See Problem XXXII. of "The Circle.")

* II. Prob. 11.

PROBLEM XXIX.—From a given point within a given semicircle, it is required to draw a line to a point in the diameter of the semicircle such that that line may be equal to the ordinate of the semicircle at the point at which the line meets it.

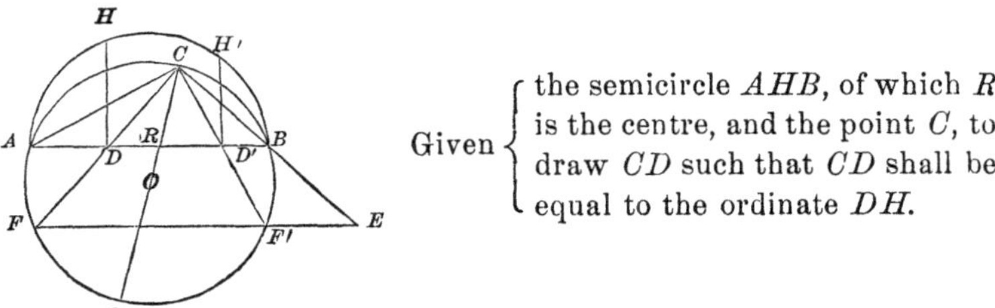

Given $\left\{\right.$ the semicircle AHB, of which R is the centre, and the point C, to draw CD such that CD shall be equal to the ordinate DH.

Analysis.—Since AB and the point C are given, the triangle ACB is given. About ACB describe a circle of which the centre is O, and suppose CD drawn so that $CD = DH$; then $CD^2 = DH^2 =$ (by the property of the semicircle) $AD \times DB$; so that this problem resolves into Problem L. of "Triangles, Quadrilaterals, and Parallels," and the *construction, demonstration,* and *calculation* are the same as in that problem.

Then at D and D' erect the perpendiculars, or draw the ordinates DH and $D'H'$, which will be equal to DC and $D'C$ respectively, as is evident from the demonstration of that problem.

Scholium.—If it were desired that CD should have to DH any given ratio, as m to n, then produce CB, so that $CB : BE :: m^2 : n^2$ (IV. Prob. 11*); draw $EF'F$ parallel to AB; then draw CDF and $CD'F'$, and CD or CD' will be the line required.

For, since $m^2 : n^2 :: CB : BE :: CD : DF$, we have $DF = \frac{n^2}{m^2} \cdot CD$.

Now, $DH^2 = AD \times DB = CD \times DF = \frac{n^2}{m^2} \cdot CD^2$. Hence $DH = \frac{n}{m} \cdot CD$; that is, $m : n :: CD : DH$. In the same way $m : n :: CD' : D'H'$. Q. E. D.

* See Problem XXXIII. of "The Circle."

Theorem and Problem XXX. (Proposed by Francis Miller.) —If from *any point*, either within or without a circle, lines be drawn to the extremities of diameters to the circle, the sum of the squares of the two lines drawn to the extremities of *one* diameter will be equal to the sum of the squares of the two lines drawn to the extremities of *any other* diameter. And having given three of these lines (and consequently the fourth) drawn from the given point to the extremities of *two diameters making a given angle with each other*, it is required to find the diameter of the circle.

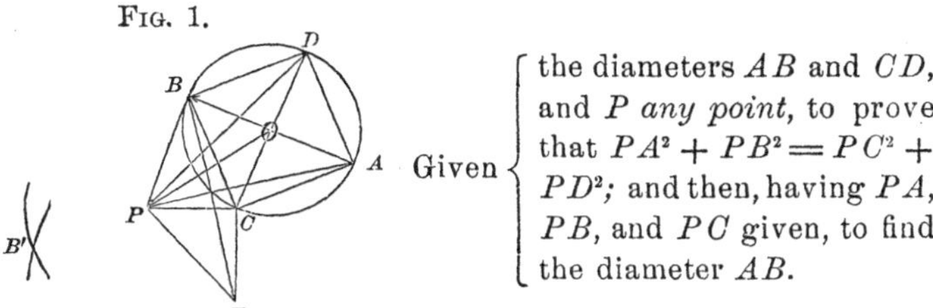

Fig. 1.

Given { the diameters AB and CD, and P *any point*, to prove that $PA^2 + PB^2 = PC^2 + PD^2$; and then, having PA, PB, and PC given, to find the diameter AB.

Demonstration of the Theorem.—Join PO; then, in each of the three figures, we have $PA^2 + PB^2 =$ (IV. 14*) $2\,PO^2 + 2\,OB^2 = 2\,PO^2 + 2\,OD^2 = PC^2 + PD^2$. Q. E. D. *Corollary.*—$PD = \sqrt{PA^2 + PB^2 - PC^2}$.

Analysis of the Problem. (Case 1.)—When the diameters cross at right angles (Figure 1), draw CE perpendicular and equal to CP, and join EB; then, in triangles ECB and PCA, we have $EC = PC$, $CB = CA$, and angle $ECB = ECP + PCB = ACB + PCB = PCA$; hence $EB = PA$; and we have this

Construction.—Draw EC and CP at right angles, and each equal to the shortest of the given lines. With P and E as centres, and radii respectively equal to the given lines PB and PA, describe arcs intersecting in B. Join EB, PB, and BC. On BC describe the square $BCAD$; draw the diagonals AOB, COD; and with centre O and radius OB or OC, describe the circle $BCAD$, which will be the one required.

Demonstration.—In the triangles PCA, ECP (I. 5†), $PA = EB$; and PB and PC were made of the given lengths. Q. E. D.

Calculation.—Join EP; then $EP = \sqrt{PC^2 + CE^2} = 2\,PC^2$. In triangle EPB, Case 4, find angle EPB; then angle $CPB = EPB - 45°$. In triangle CPB, Case 3, find BC; then AB or $CD = \sqrt{2\,BC^2}$.

* II. A. † I. 4.

Scholium 1.—When the given point is *within* the circle, let the two arcs described about P and E meet on the *other side* of PE, as at B'; join $B'C$, and on it describe the square, the diagonal of which will be the diameter required. In the calculation, the angle CPB will sometimes equal $45° - EPB$.

Scholium 2.—This problem solves the one for constructing a *square* when the three distances from a given point, either within or without the square, are given to the three nearest corners of the square. The fourth distance is then also given. (See Problem XLI., "Triangles," etc.)

CASE 2.—When the diameters cross at *any* angle (which of course includes Case 1).

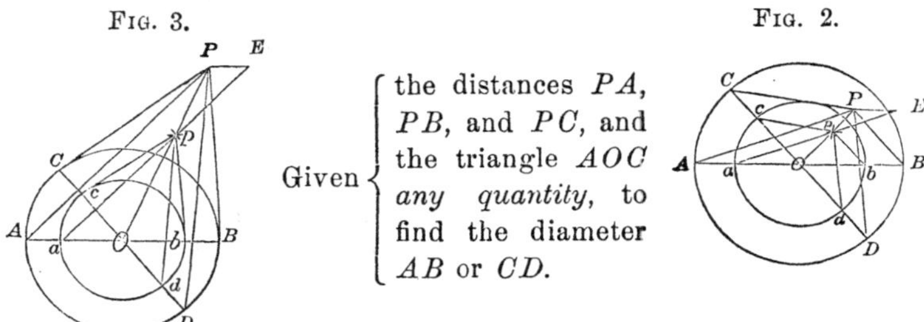

FIG. 3. FIG. 2.

Given $\begin{cases} \text{the distances } PA, \\ PB, \text{ and } PC, \text{ and} \\ \text{the triangle } AOC \\ \textit{any quantity}, \text{ to} \\ \text{find the diameter} \\ AB \text{ or } CD. \end{cases}$

Analysis.—Join OP; and in OP, or OP produced, take *any* point p. Parallel to PA, PB, PC, and PD, respectively, draw the lines pa, pb, pc, and pd; then, since, by similar triangles, $PO : pO :: \begin{cases} OA \\ OB \\ OC \\ OD \end{cases} : \begin{cases} Oa \\ Ob \\ Oc \\ Od \end{cases}$ we have Oa, Ob, Oc, and Od all equal, and the points a, b, c, d lie in the circumference of a circle of which O is the centre. Also, by similar triangles, we have $PA : pa :: PB : pb$; $PC : pc :: PD : pd$; $PA : pa :: PC : pc$, etc. Whence this

Construction.—With centre O and *any* radius, describe the circle $adbc$, and draw the diameters aOb and cOd, making the angle $aOc =$ the given angle. Then, as in Problem XXIV. of "Triangles, Quadrilaterals, and Parallels," describe an arc such that two lines drawn from a and b to any point in this arc shall have the ratio of the two given lines PA and PB. In the same manner describe another arc such that two lines drawn from the points c and d to any point in this arc shall have the ratio of the two given lines PC

and PD, and let these arcs intersect in a point p. Join pa, pb, pc, pd, and pO. On ap produced, if necessary, lay aE equal to the given length of AP, and parallel to ab draw EP, meeting Op produced, if necessary. Then draw PA, PB, PC, and PD, respectively, parallel to pa, pb, pc, and pd, meeting the diameters ab and cd, produced, if necessary, in the points A, B, C, and D respectively. With centre O and radius OA describe a circle, and it will pass through the points C, B, and D, and be the circle required, as is evident from the analysis.

Calculation.—As in Problem XXIV. of "Triangles, Quadrilaterals," etc., find the lines ap and cp; then $ap : AP :: aO : AO$, the double of which is AB or CD, the diameter of the circle required.

Scholium.—The two circles or arcs which intersect at p would intersect also in another point on the *opposite side* of ab, showing that there are two points that will fulfil the conditions of the problem. The *construction*, *demonstration*, etc. are the same in both instances.

PROBLEM XXXI. (To illustrate the mode of "discussing a problem.")—(Two circles being given in magnitude and position, it is required to apply a line which shall have an extremity in the circumference of each circle and be parallel to a given line.

FIG. 1.

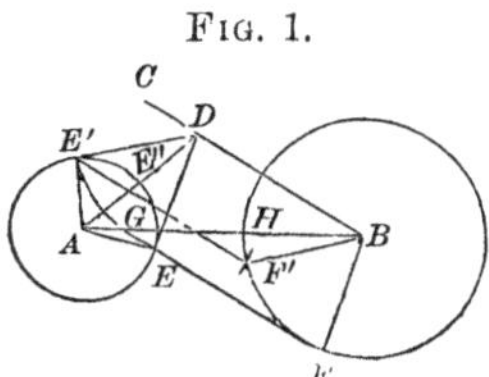

Given, AB, the angle ABC, the radii AG and BH, and the line BD, to apply EF parallel and equal to BD.

FIG. 2.

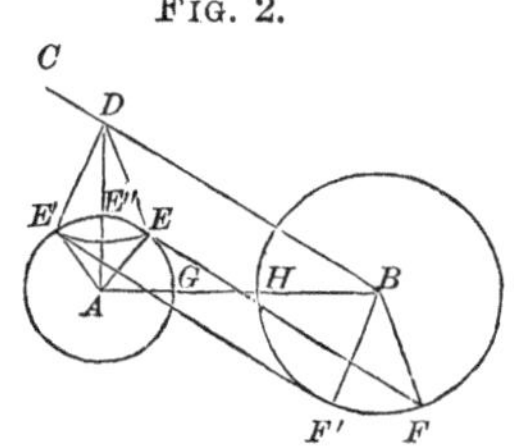

Analysis.—Suppose the line EF (either figure) applied parallel and equal to BD. Join BF and DE; then (I. 30*) $BDEF$ will be a parallelogram, and we will have $DE = BF = BH$. Whence this

Construction.—From D apply DE, DE' to the circumference of AG, equal to the radius BH, then apply EF or $E'F'$ equal to the given line BD, either of which will be the applied line required, equal and parallel to BD.

Demonstration.—Join BF, BF', DE and DE', and DA; then, since $DE = BF$ and $EF = BD$ by construction, EF is parallel to

* I. 33.

BD (I. 30*). Also, since $DE' = BF'$ and $E'F' = BD$, $E'F'$ is parallel to BD. Q. E. D.

Calculation.—What is required is the angle BDE or DEF, and the angle BDE' or $DE'F'$. Join AE and AE'; then in triangle

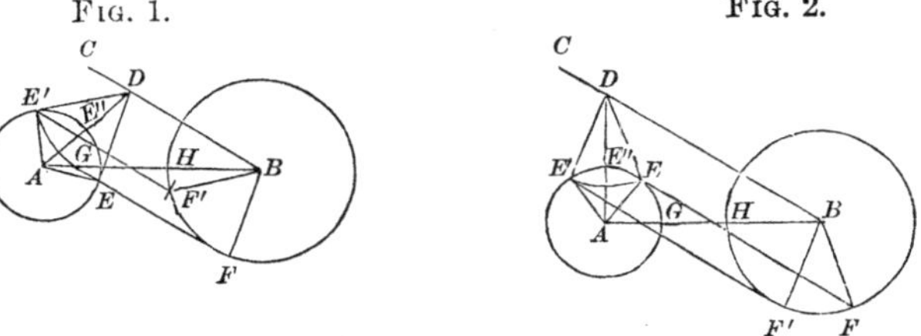

FIG. 1. FIG. 2.

ABD, Case 3, find angle ADB, and side AD. In triangle ADE, Case 4, find angle $ADE = ADE'$; then angle $BDE = ADB - ADE$, and angle $DEF = 180° - BDE$. Also, angle $BDE' = ADB + ADE'$, and angle $DE'F' = 180° - BDE'$.

Discussion of the problem. 1.—If the radius BH, applied from D, does not reach the circumference of AG, the problem is impossible, as is evident.

Discussion 2.—Regarding Figure 1, it will be seen that, as BD must always be of the length of the given applied line, and DE equal to the radius BH, as the applied line is taken shorter and shorter, D approaches B. But as D approaches B it gets more remote from A, and E and E' approach each other, till the angles ADE and ADE' vanish, the points E and E' meet at E'', and $DE'' = DE' = DE = BH$, in which case the applied line will be a *minimum*, or the *shortest line possible*, to have an end in the circumference of each circle and be parallel to the line BC.

FIG. 3.

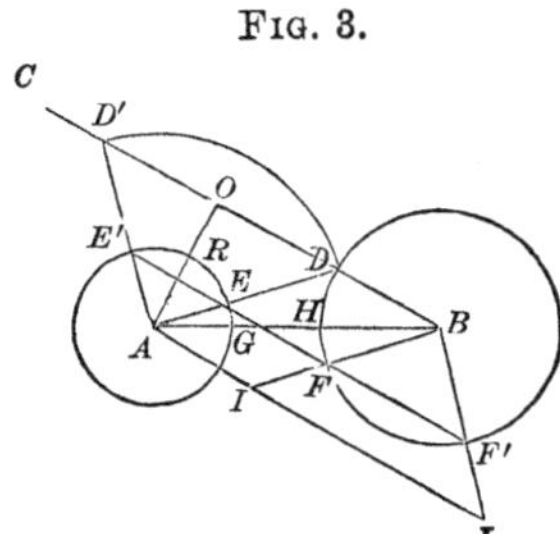

To obtain this minimum line by construction, with centre A (Fig. 3) and radius AD equal to the *sum* of the radii AG and BH, describe an arc cutting BC in a point D *nearer* B than the foot of a perpendicular AO let fall from A on BC. Draw AED; also draw EF parallel to BC, and join BF; then will EF evidently be equal and parallel to BD, and be the minimum line required. To find its length, in triangle ABD, Case 2, find $BD = EF$.

* I. 33.

Discussion 3.—In like manner, regarding Figure 2, as BD must always be of the length of the given applied line, and DE equal to the radius BH, it will be seen that, as the applied line is taken longer and longer, D recedes from B, AD increases, E and E' approach each other, till the angles ADE and ADE' vanish, E and E' meet at E'', when $DE''=DE=DE'=BH$, and then the applied line will be the *longest possible*, or a maximum.

To obtain this maximum line by construction, with centre A (Fig. 3) and radius equal to the *sum* of the two radii AG and BH, describe an arc, cutting BC in D', *further* from B than the foot of the perpendicular AO let fall from A on BC. Draw $AE'D'$, and draw $E'F'$ parallel to BC. Join BF'; then will $BD'E'F'$ be a parallelogram, and $E'F'$ will evidently be equal and parallel to BD', and be the required maximum line. To determine its length by calculation, in triangle ABD', Case 2, find $BD'=E'F'$.

NOTE.—Angle $AD'B=180°-ADB$.

Discussion 4.—Through A (Fig. 3), parallel to BC, draw a line meeting BF and BF', produced, in I and L respectively; then, since $AD=AD'$, and $AE=AE'$, EE' is parallel to BC. But EF is parallel to BC. Hence $E'EF$ is a straight line. Again, since $BL=AD'$ and $BF'=D'E'$, we have $F'L=AE'$. In like manner $BI=AD$ and $BF=ED$. Therefore $FI=AE=AE'=F'L$. Hence FF' is parallel to IL or BC. But EF is parallel to BC; wherefore the three lines $E'E$, EF, and FF' are in one and the same straight line, parallel to BC. Therefore, the *shortest* line possible (EF), and the longest line possible ($E'F'$), that can be applied between the circles, parallel to a line given in length and position, *are in the same straight line*, the minimum (EF) being part of the maximum ($E'F'$).

Discussion 5.—In case of the minimum and maximum lines EF and $E'F'$ (Fig. 3), the angle $EAG=HBF$, and arc EG is similar to arc HF. Also, the angle $E'AG=HBF'$, and arc $E'G$ is similar to arc HF'. Also, $ADBI$, $AEFI$, $AD'BL$, $AE'F'L$ are parallelograms.

Discussion 6.—The *maximum angle* which the line BC, given in position, can make with the line AB, is that made by the internal *common tangent* to the two given circles, as EPF, $E'PF'$ (Fig. 4). For it is evident that a line drawn from any point in the circle BH, making with AB an angle *greater* than BPF or BPF', will not meet the circle AG; and that a line drawn from any point in the circle AG, making with AB an angle *greater* than APE or APE', will not meet the circle BH.

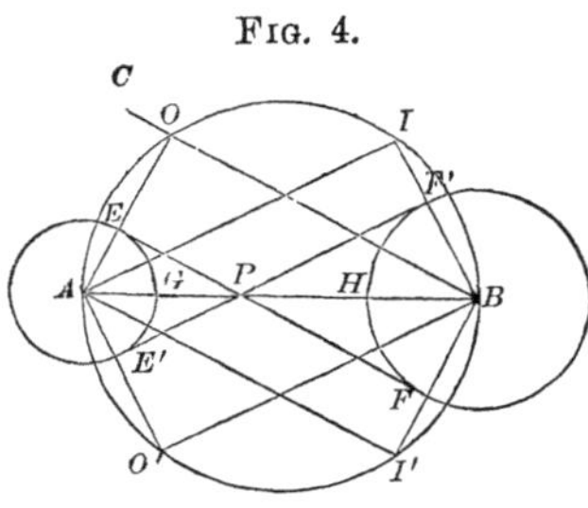

FIG. 4.

These two lines EF and $E'F'$ are equal; they intersect each other in the same point P on AB; and are the only lines which can be applied between the circles, making with AB an angle as great as APE or APE'.

To find these lines by construction.—On AB as a diameter describe a circle, in which apply AO and AO' equal to the *sum* of the radii AG and BH. Draw BOC, and join BO', parallel to which lines, respectively, draw EF and $E'F'$, and join BF and BF'; then (Problem XIII. of "The Circle") EF and $E'F'$ will be tangents to both circles.

To find the lengths of these common tangents by calculation.—We have $EF = E'F' = BO = \sqrt{AB^2 - AO^2}$.

Corollary.—If BF and BF' be produced to meet the circle described on AB in I' and I, then $FI' = AE = AE' = F'I$, and EF is parallel to AI' as well as to BO, also $E'F'$ is parallel to AI and to BO'.

Discussion 7.—The *minimum angle* which the line BC, given in position, can make with AB is 0°. In this case BC (Fig. 5) will coincide with BA, and, making BD = the length of the given line, and applying DE, DE' each equal to BH, and drawing EF and $E'F'$ each parallel to BD, we have the construction just as in Figure 1.

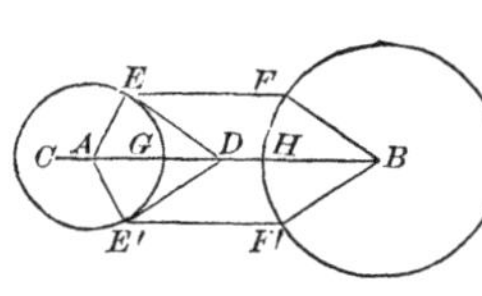

FIG. 5

Calculation.—In triangle AED, Case 4, find angle $ADE = DEF = DE'F' = ADE'$; then $EDB = 180° - ADE = E'DB$.

Discussion 8.—In Figure 3, when the perpendicular AO is *just equal* to the sum of the radii AG and BH, then D and D' coincide at O, OR (Fig. 3) becomes equal to DE or BH, and the maximum and minimum lines $E'F'$ and EF coincide in the common tangent EF, as in Figure 6, which makes with AB the maximum angle $APE = ABC$, as shown in Discussion 6.

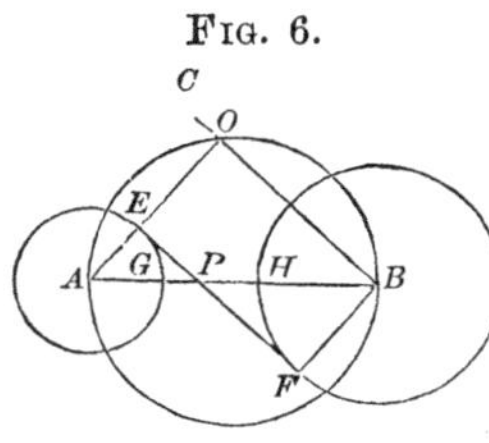

FIG. 6.

PROBLEM XXXII.—To produce a line of a given length so that the rectangle of the *whole line thus produced*, and the *part produced*, shall be equal to the square of a given line. (See IV. Prob. 4.*)

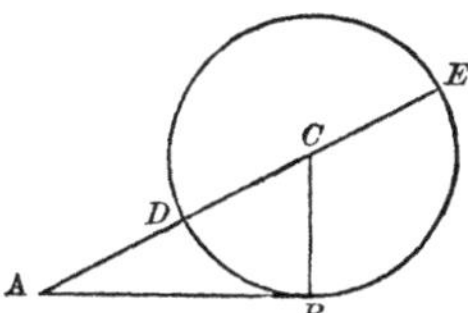

Given $\begin{cases}\text{the } length \text{ of } ED, \text{ to produce it so} \\ \text{that } EA \times AD = AB^2.\end{cases}$

Analysis.—It is evident that AB is tangent to the circle whose radius CB is half the given line ED, to be produced so that $EA \times AD = AB^2$. Whence this

Construction.—With a radius equal to half the length of the given line to be produced, describe a circle, as BDE. Draw *any* radius, as CB, perpendicular to which draw the tangent $BA =$ the given line to whose square the rectangle is to be equal. Through A and C draw $ADCE$, the line required.

Demonstration.—By IV. 30,† $EA \times AD = AB^2$. Q. E. D.

Calculation.—$AC = \sqrt{AB^2 + BC^2}$; then $AE = AC + CB$, and $AD = AC - CB$.

PROBLEM XXXIII.—To find a square which shall be to a given square in a given ratio, as m to n. (See IV. Prob. 11.)

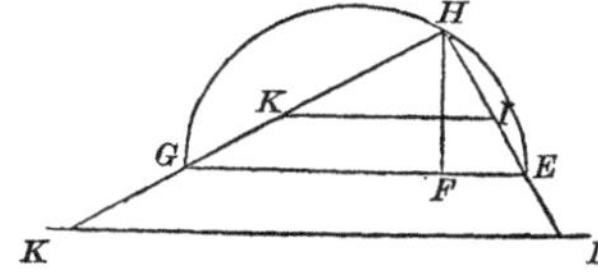

Given $\begin{cases}\text{the two segments } FE \text{ and} \\ FG, \text{ to find two lines whose} \\ squares \text{ shall have the same} \\ \text{ratio as } EF : FG \text{ as } m : n.\end{cases}$

Analysis.—Having laid EF and FG, or m and n, contiguous on the same line EG, on EG describe a semicircle; erect the perpendicular FH to meet the semicircle in H. Join HG and HE, and produce them indefinitely. Then, drawing KI, parallel to GE, *anywhere* above or below GE, we always have (IV. 11, cor.‡) $EF : FG :: HE^2 : HG^2 :: HI^2 : HK^2 :: HI^2 : HK^2$ in the given ratio.

Construction and *demonstration* are manifest.

Calculation.—If HK is the side of the given square, then $GF : FE$ or $n : m :: HK^2 : HI^2 = \frac{m}{n} HK^2$.

If HI is given, then $EF : FG$ or $m : n :: HI^2 : HK^2 = \frac{n}{m} HI^2$.

* II. 11. † III. 36. ‡ I. 47.

ANALYSIS BY ALGEBRA.

PART I.

CONSTRUCTION OF ALGEBRAIC EQUATIONS OF ONE UNKNOWN QUANTITY OF THE FIRST DEGREE, IN WHICH EACH LETTER REPRESENTS A RIGHT LINE.

NOTE.—In a fraction, when the sum of the indices in the numerator exceeds the sum of the indices in the denominator *by one*, the fraction will represent *a line;* if *by two*, it will represent *a square* or *surface;* if *by three*, *a cube* or *solid.*

PROBLEM I.—Given, the equation $x = a + b + c + d$, to find x by construction.

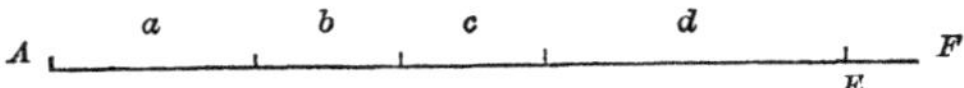

Draw an indefinite line AF, and on it lay the lines a, b, c, and d from A to E; then $x = AE$.

Limit.—The given lines a, b, c, d, etc. may be of *any* number.

PROBLEM II.—Given, the equation $x = a + b - c + d + e - f - g$, to find x by construction.

Draw an indefinite line AF, and on it lay all the *affirmative* quantities a, b, d, e successively from A to B, their extremities being marked by short lines *above* the line AF.

Then, beginning at B, lay all the *negative* quantities back from B to E, their extremities being marked by short lines *below* the line AF. Whence we have $x = a + b + d + e - (c + f + g) = AB - BE = AE$.

Limit.—The lines a, b, c, d, e, etc. may be continued to any number. When the sum of the *negative* quantities *exceeds* the sum of the *affirmative*, E will be on the line to the *left* of A, and AE will be *negative*, or a *minus* quantity.

PROBLEM III.—Given, the equation $x = \frac{abc}{de}$, to find x by construction. Here, the sum of the indices in the numerator being *one more* than the sum in the denominator, the result will be a line.

Lay the lines represented by the letters, as in the annexed figure, $AD = d$, $DF = a$, $AE = b$. Draw EO parallel to AB, and lay $EH = e$, $HI = c$. Join DE, and draw FG parallel to DE. Join HG, and draw IL parallel to HG. Then GL will equal $\frac{abc}{de} = x$. For we have, by parallel lines, $d : a :: b : x'$, and $e : c :: x' : x$ or GL. By multiplying the corresponding terms, we have $de : ac :: b : x = \frac{abc}{de} = GL$.

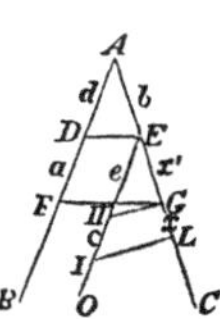

NOTE.—The figure looks more symmetrical when EO is drawn *parallel* to AB, as above, but it may be drawn making *any angle* with EC.

PROBLEM IV.—Given, the equation $x = \frac{a^2bcd^2efg}{h^2ik^2lmn}$, to find x by construction.

Lay the lines represented by the given letters, as in the annexed figure, and draw the corresponding parallel lines, observing to take the first term, or *antecedent*, in the successive proportions, from the *denominator*, and the *consequents* from the numerator; and, also, that the *second consequent* or *last term* in each proportion becomes the *second antecedent* or *third term* in the next proportion.

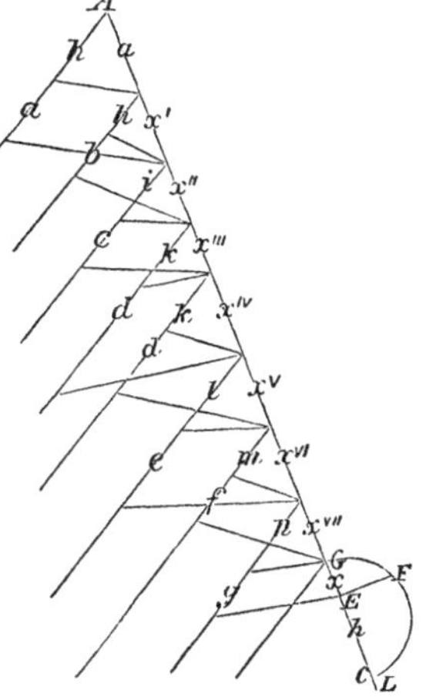

Then we have the proportions $h : a :: a : x'$, hence $x' = \frac{a^2}{h}$;

$h : b :: x' : x''$, hence $x'' = \frac{a^2b}{h^2}$, etc.;

$i : c :: x'' : x'''$

$k : d :: x''' : x^{iv}$

$k : d :: x^{iv} : x^{v}$

$l : e :: x^{v} : x^{vi}$

$m : f :: x^{vi} : x^{vii}$

$n : g :: x^{vii} : x$; then, by multiplying the corresponding terms of these proportions, and observing that all the third terms but the first, and all the fourth terms but the last, cancel each other, we have $h^2ik^2lmn : abcd^2efg :: a : x = \frac{a^2bcd^2efg}{h^2ik^2lmn} = GE$.

Limits.—The terms may be continued to *any* number, always observing that if the sum of the indices in the numerator is *one greater* than the sum in the denominator, the result will be *a line;* if *two greater, a square;* if *equal,* it *will be unity;* if *one less,* the result will represent *unity divided by a line;* if *two less, unity divided by a square,* imaginary quantities.

PART II.

CONSTRUCTION OF ALGEBRAIC EQUATIONS OF THE SECOND DEGREE.

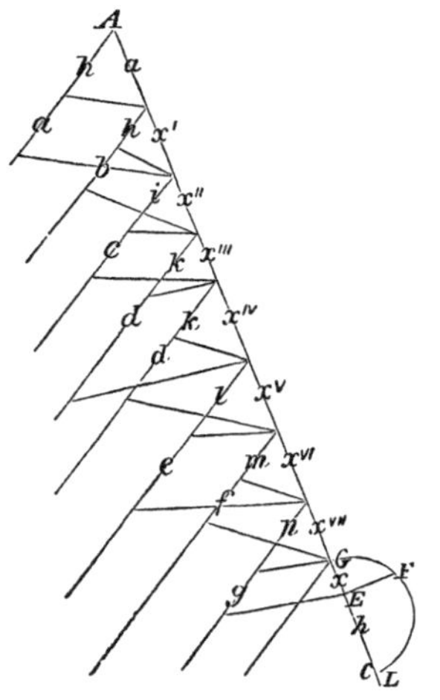

PROBLEM V.—Given, the equation $x^2 = \frac{a^2bcd^2efgp}{h^2ik^2lmn}$, to find x by construction.

As in last problem, construct the *line* $\frac{a^2bcd^2efg}{h^2ik^2lmn}$, and thus obtain GE. Produce GE to L, and on GL lay $EC = p$. On GC describe a semicircle, and erect the perpendicular EF. Then (IV. Prob. 3*) we have $EF^2 = GE \times EC = GE \times p = \frac{a^2bcd^2efgp}{h^2ik^2lmn} = x^2$, or $GE = x$.

* VI. 13.

PROBLEM VI.—Given, the equation $x^2 = a^2 + b^2 + c^2 + d^2 + e^2 + f^2 + g^2$, to find x by construction.

Draw two lines AB and BC at right angles; make $AB = a$; $BC = b$; CD perpendicular to $AC = c$; DE perpendicular to $AD = d$; EF perpendicular to $AE = e$; FG perpendicular to $AF = f$; GH perpendicular to $AG = g$, and join AH; then will AH be the line required $= x$. For $AC^2 = AB^2 + BC^2 = a^2 + b^2$; $AD^2 = AC^2 + CD^2 = a^2 + b^2 + c^2$; $AE^2 = AD^2 + DE^2 = a^2 + b^2 + c^2 + d^2$; $AF^2 = AE^2 + EF^2 = a^2 + b^2 + c^2 + d^2 + e^2$; $AG^2 = AF^2 + FG^2 = a^2 + b^2 + c^2 + d^2 + e^2 + f^2$; and $AH^2 = AG^2 + GH^2 = a^2 + b^2 + c^2 + d^2 + e^2 + f^2 + g^2 = x^2$; whence $x = AH$.

Scholium.—These squares to whose sum x^2 is to be equal may be of *any* number. Also, we have $AC^2 = a^2 + b^2$, and hence $AC = \sqrt{a^2 + b^2}$; $AD^2 = a^2 + b^2 + c^2$, and hence $AD = \sqrt{a^2 + b^2 + c^2}$; $AE^2 = a^2 + b^2 + c^2 + d^2$, and hence $AE = \sqrt{a^2 + b^2 + c^2 + d^2}$, etc., etc., to any number, greater or less than the number given in the problem.

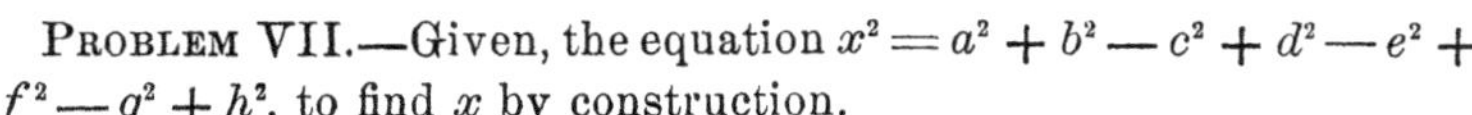

PROBLEM VII.—Given, the equation $x^2 = a^2 + b^2 - c^2 + d^2 - e^2 + f^2 - g^2 + h^2$, to find x by construction.

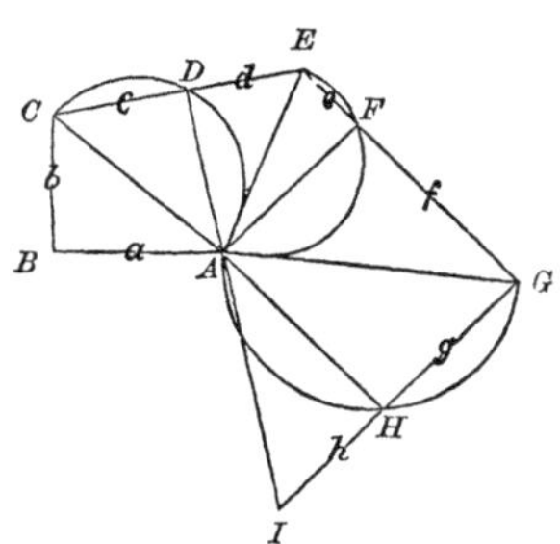

Proceed in the same manner as in Problem VI., and when we come to a quantity that is *negative* (as c^2) on the line (as AC) whose square represents the value of all the preceding quantities, describe a semicircle, in which apply the line whose square is negative (as c); join A with this point (as AD), and then proceed as in the last problem. It will be observed that producing the line applied in the semicircle will be erecting a perpendicular on the line drawn from A to the remote end of the applied line. Thus, CD, produced, gives DE perpendicular to AD. Now, we have, as in last problem, $AC^2 = AB^2 + BC^2 = a^2 + b^2$; $AD^2 = AC^2 - CD^2 = a^2 + b^2 - c^2$; $AE^2 = AD^2 + DE^2 = a^2 + b^2 - c^2 + d^2$; $AF^2 = AE^2 - EF^2 = a^2 + b^2 - c^2 + d^2 - e^2$, etc., etc.; and, finally, we have $AI^2 = a^2 + b^2 - c^2 + d^2 - e^2 + f^2 - g^2 + h^2 = x^2$; whence $x = AI$.

Scholium.—The squares to whose collected value x^2 is to be equal may be of *any* number.

PROBLEM VIII.—Given, the equations $x^2 + ax = b^2$, and $x^2 - ax = b^2$, to find the value of x in each.

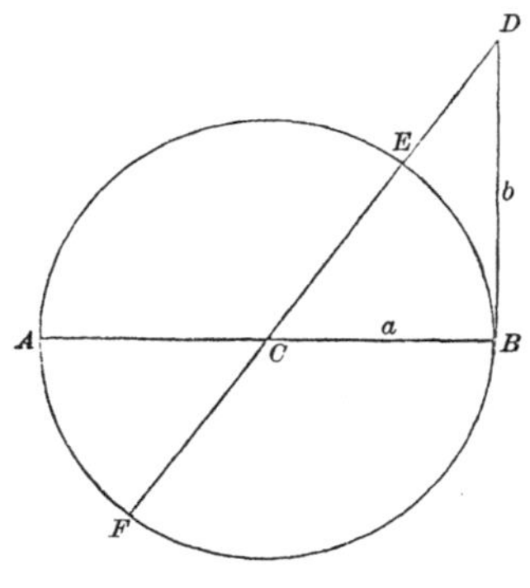

Construction.—Draw two lines AB and BD at right angles, and make $AB = a$ and $BD = b$. On AB describe a circle whose centre is C, and draw the secant line $DECF$; then $EF = AB = a$; and, in the equation $x^2 + ax = b^2$, $DE = x$; while in the equation $x^2 - ax = b^2$, $DF = x$.

Demonstration.—$DE \times DF = DB^2$ (IV. 30*). Now, if $DE = x$, $DF = x + a$, and $DE \times DF = x \times (x + a) = x^2 + ax = DB^2 = b^2$. Also, if $DF = x$, $DE = x - a$, and $DE \times DF = x \times (x - a) = x^2 - ax = DB^2 = b^2$.

Limit.—The given lines a and b may be of any length whatever.

Examples.—1. Construct the equation $x^2 + 12x = 8^2$, and find $x = 4$, measured from the scale.

2. Construct the equation $x^2 - 12x = 8^2$, and find $x = 16$, measured from the scale.

3. Construct the equation $x^2 + 21x = 10^2$, and find $x = 4$, measured from the scale.

4. Construct the equation $x^2 - 21x = 10^2$, and find $x = 25$, measured from the scale.

PROBLEM IX.—Given, the equation $x^2 - ax = -b^2$, or $ax - x^2 = b^2$, to find the two values of x.

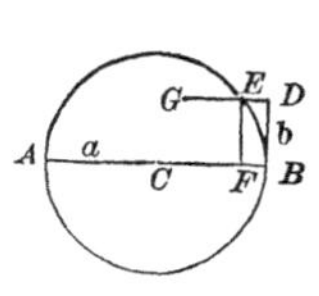

Construction.—Draw two lines AB and BD at right angles, making $AB = a$, and $BD = b$; and on AB describe a semicircle of which C is the centre. Parallel to AB draw DG, meeting the circumference in E, and let fall the perpendicular EF on the diameter AB; then $EF = BD = b$, and either BF or $AF = x$.

Demonstration.—By IV. 23, cor.,† $AF \times FB = FE^2 = BD^2 = b^2$. Now, if either BF or AF is x, the other will be $= a - x$, and we always have $AF \times FB = x \times (a - x) = ax - x^2 = b^2$. In an equation of the form $ax - x^2 = b^2$, there are always two *positive* values of x.

* III. 36. † VI. 8, cor.

Limit.—BD must always be *less* than the radius BC; that is, b must be less than $\frac{1}{2}a$.

Examples.—1. Construct the equation $x^2 - 20x = -8^2$, or $20x - x^2 = 8^2$, and find $x = 4$ or 16.

2. Construct the equation $x^2 - 29x = -10^2$, or $29x - x^2 = 10^2$, and find $x = 4$ or 25.

PROBLEM X.—Given, the equations $x^2 + ax = d$, and $x^2 - ax = d$, to find the value of x.

CASE 1.—When d can be divided into two factors b and c, whose difference is *less* than a, then the equations become $x^2 + ax = bc$, or $x^2 - ax = bc$.

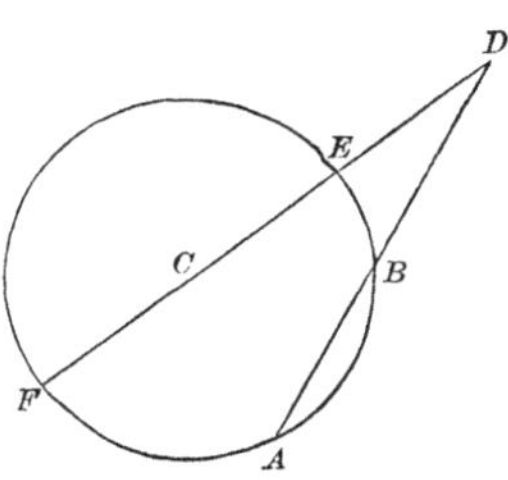

Construction.—With a radius equal to $\frac{1}{2}a$, describe a circle, the centre of which is C. In this circle apply the chord AB = the *difference* between b and c. Produce AB, making the produced part BD = the *less* factor, then AD = the *greater* factor. Through C draw $DECF$; then DE will be the value of x in the equation $x^2 + ax = d = bc$, and DF will be the value of x in the equation $x^2 - ax = d = bc$.

Demonstration.—For (IV. 29*) $DE \times DF = DB \times DA = bc$. Now, if $DE = x$, we have $DF = x + a$. Whence $DE \times DF = x(x + a) = x^2 + ax = bc = d$.

Also, if $DF = x$, we have $DE = x - a$, and $DE \times DF = (x - a)x = x^2 - ax = bc = d$.

Examples.—1. Construct the equations $x^2 + 8x = 48 = 8 \times 6$, and $x^2 - 8x = 48 = 8 \times 6$, and find $x = 4$ and 12.

2. Construct the equations $x^2 + 10x = 24 = 6 \times 4$, and $x^2 - 10x = 24 = 6 \times 4$, and find x.

CASE 2.—When d cannot be divided into two factors whose difference is less than a, then take the square root of a to three figures, and, putting this $\sqrt{d} = b$, we have $d = b^2$, and the given equations become $x^2 + ax = b^2$, and $x^2 - ax = b^2$, whence x is found as in Problem VIII.

Examples.—1. Construct the equations $x^2 + 6x = 112$, and $x^2 - 6x = 112$. Now, $112 = 56 \times 2 = 28 \times 4 = 16 \times 7 = 14 \times 8$, and the difference of no two factors is less than 6 or a. Hence take

* III. 36, cor.

$\sqrt{112} = 10.9 = b$; then $d = 10.9^2 = b^2 = 112$, and we have $x^2 + 6x = 10.9^2$, and $x^2 - 6x = 10.9^2$.

2. Construct the equations $x^2 + 10x = 24$, and $x^2 - 10x = 24$. Here $\sqrt{24} = 4.90$, and the equations become $x^2 \pm 10x = 4.90^2$, to find x as in Problem VIII.

PROBLEM XI.—Given, the equation $x^2 - ax = -d$, or $ax - x^2 = d$, to find x.

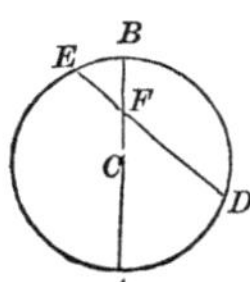

CASE 1.—When d can be divided into two factors b and c, whose *sum* is less than a, then we have $x^2 - ax = -bc$, or $ax - x^2 = bc$.

Construction.—With a radius equal to $\frac{1}{2}$ a, and centre C, describe a circle, in which apply the chord $DE =$ the sum of the given lines b and c. On DE lay $DF =$ to one of these lines, then EF will be equal to the other. Through C and F draw the diameter $ACFB$; then x will be either AF or FB.

Demonstration.—If either AF or FB is taken as x, the other will be $a - x$. Now (IV. 28, cor.*), we always have $AF \times FB = DF \times FE$; that is, $x(a - x) = bc$, or $ax - x^2 = bc = d$.

Examples.—1. Construct the equation $19x - x^2 = 60 = 10 \times 6$, and find $x = 15$ or 4.

2. Construct the equation $19x - x^2 = 48 = 6 \times 8$, and find $x = 16$ or 3.

CASE 2.—When d cannot be divided into two factors whose sum is less than a.

Construction.—Take the square root of d to three figures, and put the root $= b$; then $\sqrt{d} = b$, or $d = b^2$. Then $x^2 - ax = -b^2$, or $ax - x^2 = b^2$, which construct as in Problem IX.

Examples.—1. Construct the equation $x^2 - 22x = -112$, or $22x - x^2 = 112 = 10.9^2$, and find $x = 14$ or 8.

2. Construct the equation $22x - x^2 = 85 = 9.2^2$, and find $x = 17$ or 5.

NOTE.—The results of all these numerical equations should be verified by solving the equations algebraically.

Remark.—The preceding problems under the head of "Construction of Algebraic Equations" contain *all the forms* of equations of one unknown quantity of the first and second degrees, which are all that can be constructed by Plane Geometry. The *results* may be *combined* in various ways in a problem.

* III. 35.

PROBLEMS

ILLUSTRATING THE USE OF ALGEBRA

IN GEOMETRICAL ANALYSIS.

PROBLEM I.—In a plane triangle are given the base, the sum of the other two sides, and the line drawn from the vertex to the middle of the base, to determine the triangle.

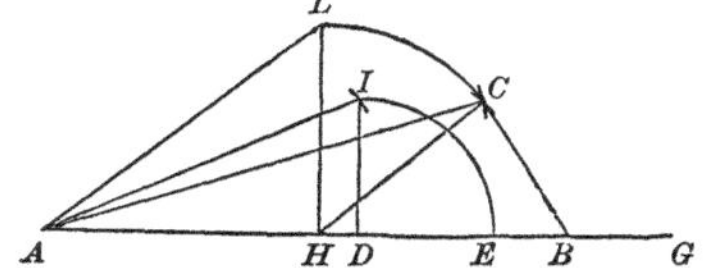

Given the base AB, the *sum* of AC and CB, and the line CH, H being the middle of AB.

Analysis by algebra.—Let ABC represent the required triangle, having $AH = HB$. Put $b = AH$ or HB, $d = CH$, $s = \frac{1}{2}$ the sum of AC and CB, and $x = \frac{1}{2}$ the difference of AC and CB; then $AC = s + x$, and $CB = s - x$. Now (IV. 14*), $AC^2 + CB^2 = 2\,AH^2 + 2\,CH^2$; that is, $(s+x)^2 + (s-x)^2 = 2b^2 + 2d^2$; or, by expanding the two quantities in the first member of the equation, $2s^2 + 2x^2 = 2b^2 + 2d^2$, whence $x^2 = b^2 + d^2 - s^2$; and we have this

Construction.—In any straight line take $AG =$ the given sum of the sides, and $AB =$ the given base, and bisect them in the points D and H respectively; then $AD = s$ and $AH = b$. At H erect the perpendicular $HL = d$, the given line to bisect the base; join AL, and to a perpendicular at D apply $AI = AL$; then $DI^2 = AI^2 - AD^2 = AL^2 - AD^2 = AH^2 + HL^2 - AD^2 = b^2 + d^2 - s^2 = x^2$; hence $x = DI$. Make $DE = DI$; then $AE = s + x$, and $EG = s - x$, the two required sides. From H as a centre and a radius HL describe an arc LC, to which apply $AC = AE$; join CB and CH, and ACB will be the triangle required.

Demonstration geometrically.—We have to prove only that $BC = EG$. Now, $AE^2 + EG^2 = \overline{AD + DE}^2 + \overline{AD - DE}^2 = 2\,AD^2 + 2\,DE^2 = 2\,AD^2 + 2\,DI^2 = 2\,AI^2 = 2\,AL^2 = 2\,AH^2 + 2\,HL^2 = 2\,AH^2 + 2\,CH^2 =$ (IV. 14†) $AC^2 + BC^2 = AE^2 + BC^2$. Hence $BC^2 = EG^2$, or $BC = EG$. Q. E. D.

* II. A. † II. A.

Calculation.—By construction, $DE = DI = \sqrt{AH^2 + HL^2 - AD^2}$. Then $AC = AE = AD + DE$, and $BC = EG = DG - DE = AD - DE$.

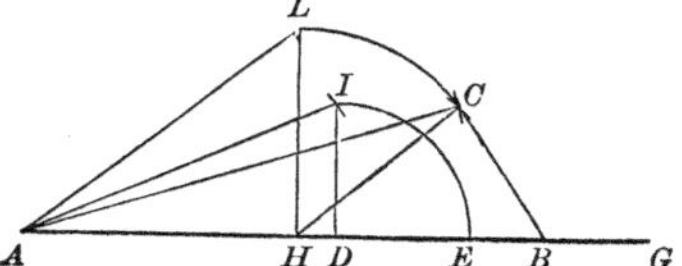

Limit.—The sum of the squares of AH and CH can never be *less* than the square of AD, half the sum of the sides. If $AH^2 + CH^2 = AD^2$, then $DI = 0$ and $DE = 0$, and the sides AC and BC will be equal, HC will coincide with HL, and AL will be half the sum of the sides.

PROBLEM II.—In a plane triangle are given the ratio of the two sides, and the segments of the base made by a perpendicular let fall from the opposite angle on the base, to determine the triangle.

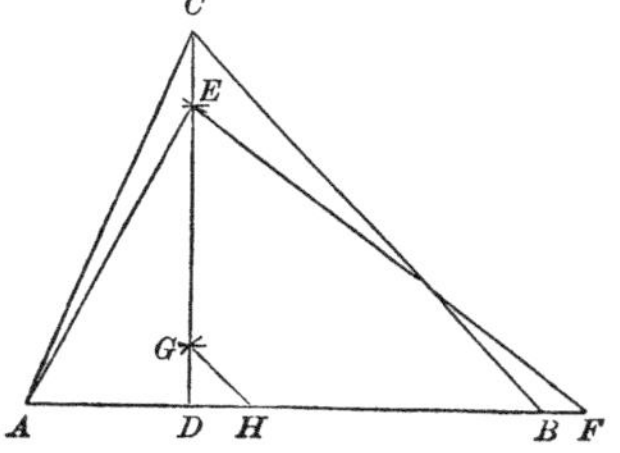

Given { the ratio of BC to AC as m to n, and the segments AD and DB made by the perpendicular CD on the base AB.

Analysis by algebra.—Let ACB represent the required triangle, having CD perpendicular to AB. Put $BD = a$, $AD = b$, $mx = BC$, and $nx = AC$; then $BC : AC :: mx : nx :: m : n$; and, since $BC^2 - BD^2 = CD^2 = AC^2 - AD^2$, we have $BC^2 - AC^2 = BD^2 - AD^2$; that is, $m^2x^2 - n^2x^2 = a^2 - b^2$, whence $x^2 = \frac{a^2 - b^2}{m^2 - n^2}$, $x = \frac{\sqrt{a^2 - b^2}}{\sqrt{m^2 - n^2}}$, $mx = \frac{m\sqrt{a^2 - b^2}}{\sqrt{m^2 - n^2}} = BC$, and $nx = \frac{n\sqrt{a^2 - b^2}}{\sqrt{m^2 - n^2}} = AC$. Whence this

Construction.—Take $DH = n$, and apply $HG = m$; then $DG = \sqrt{m^2 - n^2}$. Apply $AE = BD = a$; then $DE = \sqrt{a^2 - b^2}$. Draw EF parallel to GH. Now, by similar triangles DGH, DEF, DG $(\sqrt{m^2 - n^2}) : GH (m) :: DE (\sqrt{a^2 - b^2}) : EF = \frac{m\sqrt{a^2 - b^2}}{\sqrt{m^2 - n^2}} =$ the value of BC as found in the analysis; also $DG (\sqrt{m^2 - n^2}) : DH (n) :: DE (\sqrt{a^2 - b^2}) : DF = \frac{n\sqrt{a^2 - b^2}}{\sqrt{m^2 - n^2}} =$ the value of AC, as found in the analysis. Apply $BC = FE$, and join AC, then ABC will be the required triangle.

Demonstration by geometry.—By construction and parallel lines, we have EF or $BC : FD :: GH : HD :: m : n$; hence we have only to prove that $FD = AC$. By the analysis, $BC^2 - AC^2 = BD^2 - AD^2 = AE^2 - AD^2 = DE^2 = EF^2 - FD^2 = BC^2 - FD^2$. Hence $AC^2 = FD^2$ and $AC = FD$. Q. E. D.

Calculation.—We have $DG = \sqrt{m^2 - n^2}$, and $DE = \sqrt{a^2 - b^2}$; then, by similar triangles DGH and DEF, we have $DG : DE :: GH : EF = \frac{GH \times DE}{DG} = BC$, and $DG : DE :: DH : DF$ or $AC = \frac{DH \times DE}{DG}$.

Example.—If $BD = 17$, $AD = 8$, $m = GH = 5$, $n = DH = 4$, then $DG = 3$, $DE = 15$, $BC = 25$, and $AC = 20$.

NOTE.—The same problem can be analyzed geometrically. (See Problem XXIII. in "Triangles, Quadrilaterals, and Parallels.")

PROBLEM III.—It is required to draw a line from *one angle* of a given square so that the part intercepted between the side and the side produced, *that meet at the opposite angle*, may be of a given length.

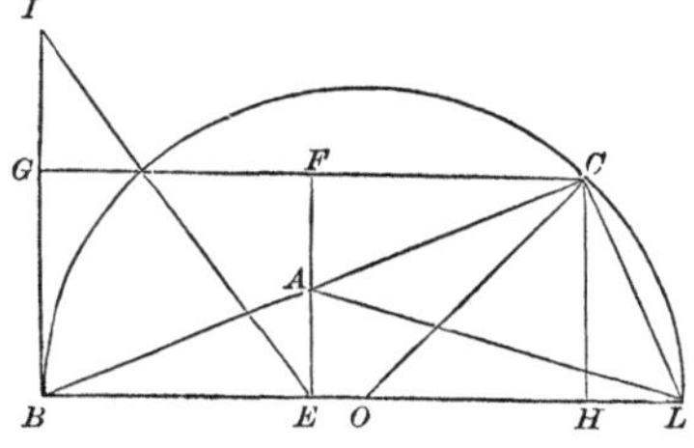

Given $\begin{cases} \text{the square } BEFG\text{, and} \\ \text{the length of the part} \\ AC \text{ of the line } BAC \\ \text{intercepted between the} \\ \text{sides } EF \text{ and } GF\text{, pro-} \\ \text{duced, that meet at } F. \end{cases}$

Analysis by algebra.—Let BAC represent the required line of which the length of AC is given. Draw CL perpendicular to AC to meet BE, produced, in L, and on BL let fall the perpendicular CH; then $CH = EF = BE$, and (IV. 21*) the triangles CHL and BAE are similar and equal, having $CL = AB$ and $HL = AE$. Now, put $AC = b$, $EF = a$, $EL = x$, and $CL = y = AB$. In similar triangles EBA and CBL, we have $AB\,(y) : BE\,(a) :: BL\,(a + x) : BC\,(b + y)$; wherefore $by + y^2 = a^2 + ax$, or $2by + 2y^2 = 2a^2 + 2ax$. Also, $BL^2 = BC^2 + CL^2$; that is, $(a + x)^2 = (b + y)^2 + y^2$, or $a^2 + 2ax + x^2 = b^2 + 2by + 2y^2 =$ (from above) $b^2 + 2ax + 2a^2$. From which $x^2 = a^2 + b^2$. Whence this

* VI. 8.

Construction.—On BG, the side of the given square (produced, if necessary), lay $BI =$ the given intercepted line b. Join EI, and make $EL = EI = \sqrt{EB^2 + BI^2} = \sqrt{a^2 + b^2} = x$. On BL describe a semicircle of which the centre is O, meeting GF, produced, in C. Draw BAC; then will $AC = BI =$ the given length.

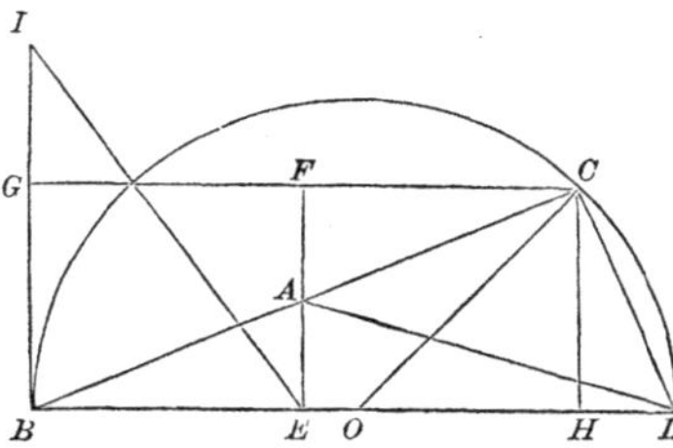

Demonstration by geometry.—Join CL, AL, and OC, and let fall on BL the perpendicular CH; then $AC^2 + CL^2 = AL^2 = AE^2 + EL^2 = AE^2 + EI^2 = AE^2 + EB^2 + BI^2 = AB^2 + BI^2 = CL^2 + BI^2$. Taking CL^2 from the first and last of these equals, we have $AC^2 = BI^2$, or $AC = BI$. Q. E. D.

Calculation.—We have $EL = EI = \sqrt{EB^2 + BI^2} = \sqrt{EB^2 + AC^2}$, $BL = BE + EL$, $OC = OL = \frac{1}{2} BL$, $OH = \sqrt{OC^2 - CH^2}$, $OL - OH = HL = AE$, $EO = OB - BE = OL - BE$, $FC = EH = EO + OH$, and $GC = GF + FC$, $CL = AB = \sqrt{BE^2 + AE^2}$.

Limit.—The line AC may be taken *any* length.

PROBLEM IV.—In a right-angled triangle are given the hypothenuse, and the side of the inscribed square, to determine the triangle.

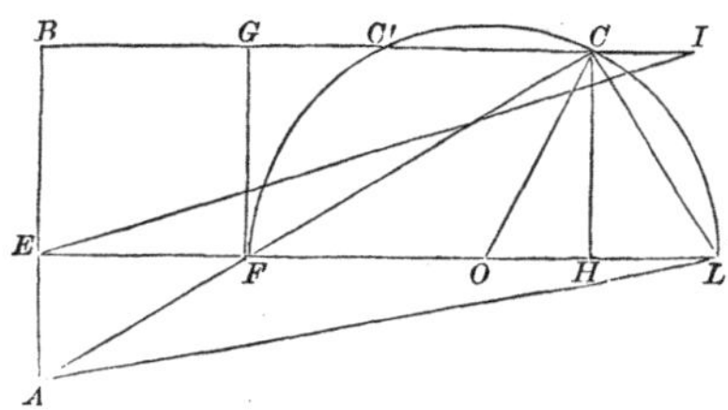

Given $\begin{cases} \text{the hypothenuse } AC \text{ of the triangle } ABC, \\ \text{and the side } BG \text{ of the inscribed square } BGFE. \end{cases}$

Analysis by algebra.—Let ABC represent the required triangle of which the hypothenuse AC and the inscribed square $BGFE$ are given. Draw CL perpendicular to AC, meeting EF, produced, in L, and let fall the perpendicular CH, which will be equal to GF or EF; then (IV. 21*) the triangles CHL and EAF are similar and equal, having $CL = AF$ and $HL = AE$. Now, put $AC = b$, $EF = a$, $EL = x$, and $CL = y = AF$. In similar triangles AFE and CFL, we have $AF\,(y) : FE\,(a) :: FL\,(x - a) : FC\,(b - y)$. Where-

* VI. 8.

fore $by - y^2 = ax - a^2$, or $-2by + 2y^2 = -2ax + 2a^2$. Also, $FL^2 = FC^2 + CL^2$; that is, $(x-a)^2 = (b-y)^2 + y^2$, or $x^2 - 2ax + a^2 = b^2 - 2by + 2y^2 =$ (from above) $b^2 - 2ax + 2a^2$. From which we have $x^2 = a^2 + b^2$, and this

Construction.—On BG, the side of the given square, produced, lay $BI =$ the given hypothenuse. Join EI, and make $EL = EI = \sqrt{EB^2 + BI^2} = \sqrt{a^2 + b^2} = x$. On FL describe a semicircle of which the centre is O, cutting BI in C; draw CFA; then will $AC = BI$, the given length of the hypothenuse.

Demonstration by geometry.—Join CL, AL, and OC, and on EL let fall the perpendicular CH; then $AC^2 + CL^2 = AL^2 = AE^2 + EL^2 = AE^2 + EI^2 = AE^2 + EB^2 + BI^2 = AE^2 + EF^2 + BI^2 = AF^2 + BI^2 = CL^2 + BI^2$. Whence $AC^2 = BI^2$, or $AC = BI$. Q. E. D.

Calculation.—$EL = EI = \sqrt{EB^2 + BI^2} = \sqrt{EB^2 + AC^2}$, $FL = EL - EF$, OF or $OC = \frac{1}{2} FL$, $OH = \sqrt{OC^2 - CH^2}$; then $FH = FO + OH = GC$, and $HL = FO - OH = AE$. Also, $BC = BG + GC$, and $AB = BE + AE$.

Limit.—When the semicircle described on FL does not meet BI, the problem is impossible. If the semicircle *just touches* GI, that is, when the radius $OC = FG$, the triangle ABC will be isosceles, having $BA = BC$, and AC will be the *shortest possible* line which can be drawn through F to meet BE and BG produced.

Scholium.—As the semicircle on FL cuts GI in the point C', as well as C, if through C' and F a line had been drawn to meet BE, produced, in a point A', the triangle $A'BC'$ thus formed would have been equal in all its parts to ABC, having $A'C' = AC$, $A'B = BC$, and $BC' = AB$.

PROBLEM VI.—In a plane triangle are given the base, the perpendicular height, and the *sum* of the other two sides, to determine the triangle.

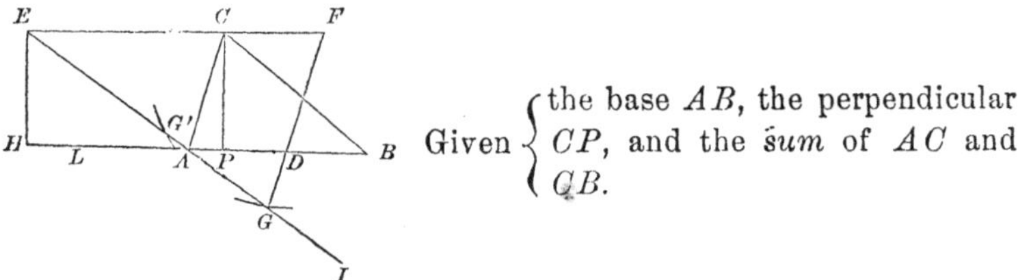

Given $\begin{cases} \text{the base } AB, \text{ the perpendicular} \\ CP, \text{ and the } sum \text{ of } AC \text{ and} \\ CB. \end{cases}$

Analysis by algebra.—Let ACB represent the required triangle. Bisect the given base AB in D. Produce BA indefinitely, and lay

$BL =$ the given sum of AC and CB. Let fall the perpendicular CP. Now, put $BL = a$, $AB = b$, $CP = p$, $AC = x$, and $AP = y$; then we have $BC = a - x$, and $BP = b - y$. And, since (cor. to Prob. XXXVII., "Triangles," etc.) $BC^2 - AC^2 = BP^2 - AP^2$, we have $(a-x)^2 - x^2 = (b-y)^2 - y^2$, or $a^2 - 2ax = b^2 - 2by$; whence $y = \frac{ax}{b} - \left(\frac{a^2}{2b} - \frac{b}{2}\right)$. Now, construct this known quantity $\left(\frac{a^2}{2b} - \frac{b}{2}\right)$.

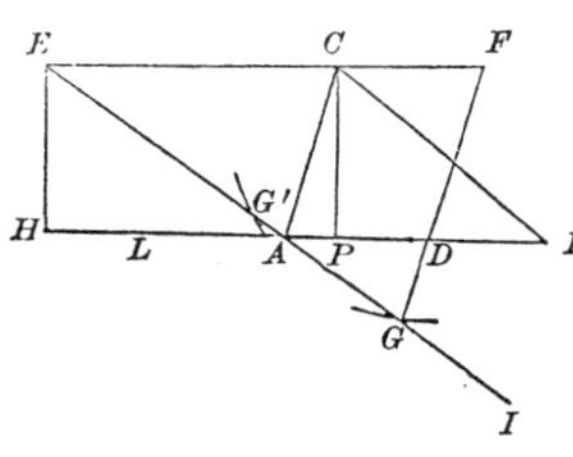

Take a third proportional (IV. Prob. 2, cor.*) to $2b$ and a, and lay it from D to H; then $2\,AB\,(2b) : BL\,(a) :: BL\,(a) : DH = \frac{a^2}{2b}$. And $\frac{a^2}{2b} - \frac{b}{2} = DH - AD = AH$. Then, AH being known, erect the perpendicular HE to meet a line through C, parallel to AB, and we have $HE = CP$. Now, the above equation $y = \frac{ax}{b} - \left(\frac{a^2}{2b} - \frac{b}{2}\right)$ becomes $y = \frac{ax}{b} - AH$, or $\frac{ax}{b} = y + AH = AP + AH = PH = EC$; that is, we have $\frac{ax}{b} = EC$, or $b : a :: x : EC$; or, by inversion, $a\,(BL) : b\,(AB) :: EC : x\,(CA)$. Hence, if EF is made equal to BL (a), and FG is drawn parallel to AC, we shall have $FG = AB$. Whence this

Construction.—Take DH a third proportional to $2\,AB$ and BL. Erect the perpendicular $HE =$ the given perpendicular. Draw EF parallel and equal to BL. To EA, joined and produced, apply $FG = AB$. Draw AC parallel to FG, join CB, and let fall the perpendicular CP; then ACB is the required triangle.

Demonstration by geometry.—We have to prove only that $AC + BC = BL$. By parallel lines we have $FG\,(AB) : EF\,(BL) :: AC : EC\,(PH)$. Hence $AB \times PH = BL \times AC$, or $2\,BL \times AC = 2\,AB \times PH$. By construction, $2\,AB : BL :: BL : DH$; hence $BL^2 = 2\,AB \times DH$. Taking the first of these equations from the second, we have $BL^2 - 2\,BL \times AC = 2\,AB \times DH - 2\,AB \times PH = 2\,AB\,(DH - PH) = 2\,AB \times DP = AB \times 2\,DP = AB\,(BP - AP) = (BP + AP) \times (BP - AP) = BP^2 - AP^2 = BC^2 - AC^2$. To the first and last of these equals add AC^2, and we have $BL^2 - 2\,BL \times AC + AC^2 = BC^2$; that is, $(BL - AC)^2 = BC^2$, or $BL - AC = BC$. Whence $AC + BC = BL$. Q. E. D.

* VI. 11.

Calculation.—By construction, $2AB : BL :: BL : DH$; then $AH = DH - AD$.

1. In triangle HAE, Case 3, find $\angle EAH = \angle GEF$.
2. In triangle EFG, Case 2, find $\angle EGF = \angle EAC$; then $\angle CAP = 180° - (EAH + EAC)$.
3. In triangle CAP, Case 1, find AC and AP; then $BP = AB - AP$, and $BC = BL - AC$.

Scholium 1.—In applying $FG = AB$ to the line EAI, unless it be perpendicular to EI, which would be the shortest line possible, it will meet EI in another point, as G'. Either point may be used in the construction, but in the latter case the line drawn from A will be the longer of the two lines, and equal to the present BC.

Scholium 2.—PC may be any quantity *less than* $\sqrt{\left[\left(\frac{BL}{2}\right)^2 - \left(\frac{AB}{2}\right)^2\right]}$. When $PC = \sqrt{\left(\frac{BL}{2}\right)^2 - AD^2}$, AC and BC will be equal, and each half of BL.

Scholium 3.—When $PC^2 = (BL - CP)^2 - AB^2$, the angle BAC will be a right angle.

Scholium 4.—When PC^2 is *less* than $(BL - CP)^2 - AB^2$, the angle BAC is an *acute* angle, as in the present problem. When PC^2 is *greater* than $(BL - CP)^2 - AB^2$, the angle BAC is an *obtuse* angle, as in the case to which the following problem reduces.

PROBLEM VII.—In any plane triangle are given the *difference* of *the segments* of the base, the perpendicular height, and the *sum* of the two sides, to determine the triangle.

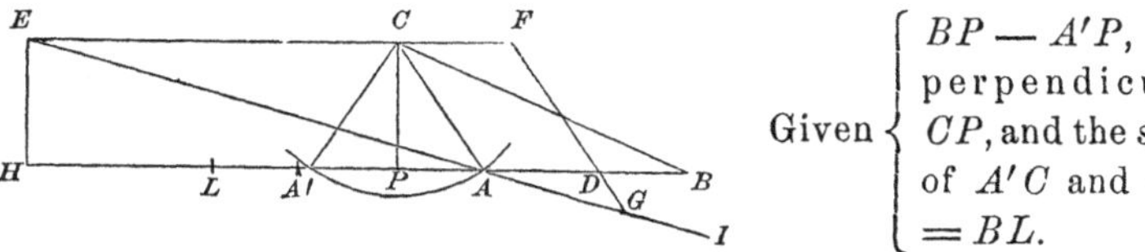

Given $\begin{cases} BP - A'P, \text{ the perpendicular} \\ CP, \text{ and the } sum \\ \text{of } A'C \text{ and } CB \\ = BL. \end{cases}$

Analysis.—Let $A'CB$ represent the required triangle, of which CP is the perpendicular, and BP and $A'P$ the segments of the base whose difference is given. Take $PA = PA'$, and join CA; then AB will be the given difference of the segments, and $AC = A'C$ (I. 5*).

Now, in the triangle ACB, we have given the base AB, the perpendicular CP, and the *sum* of the sides AC and CB, to con-

* I. 4.

struct the triangle ACB, as in the last problem. Then make $CA' = CA$, when PA' will equal PA, and $A'CB$ will be the required triangle.

The *construction, demonstration, calculation*, and *limits* are precisely as given in the last problem.

PROBLEM VIII.—In a plane triangle are given the base, the perpendicular height, and the difference between the two sides, to determine the triangle.

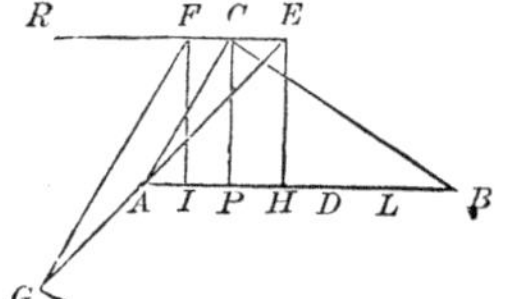

Given $\begin{cases}\text{the base } AB, \text{ the perpendicular} \\ \text{height } CP \text{ or } EH, \text{and the } \textit{difference} \\ \text{between } BC \text{ and } CA.\end{cases}$

Analysis by algebra.—Let ABC represent the required triangle. Bisect the base AB in D, and lay $BL =$ the given difference between the sides BC and CA. Let fall the perpendicular CP. Put $BL = a$, $AB = b$, $CP = p$, $AC = x$, and $AP = y$; then $BC = x + a$, and $BP = b - y$. Since $BC^2 - AC^2 = BP^2 - AP^2$, we have $(a + x)^2 - x^2 = (b - y)^2 - y^2$. That gives $a^2 + 2ax = b^2 - 2by$. Whence $y = \left(\frac{b}{2} - \frac{a^2}{2b}\right) - \frac{ax}{b}$. Now, construct the quantity $\frac{b}{2} - \frac{a^2}{2b}$ in this equation. We have $\frac{b}{2} = AD$. Take (IV. Prob. 2, cor.*) a third proportional to $2b$ and a, and lay it from D to H. Then $2b : a :: a : DH = \frac{a^2}{2b}$. Hence $AH = AD - DH = \frac{b}{2} - \frac{a^2}{2b}$. Then, AH being known, erect the perpendicular HE to meet a line through C parallel to AB, and we have $HE = CP$, whence the point E is determined.

The above equation $y = \frac{b}{2} - \frac{a^2}{2b} - \frac{ax}{b}$ now becomes $y = AH - \frac{ax}{b}$, or $\frac{ax}{b} = AH - y = AH - AP = PH = EC$; that is, we have $b : a :: x : EC$. Or, by inversion, $a\ (BL) : b\ (AB) :: EC : x\ (AC)$, or $EC : AC\ (x) :: a\ (BL) : b\ (AB)$. Hence, if EF is made equal to BL, and FG be drawn parallel to AC, then $EC : AC : EF (BL) : FG$, and FG will equal AB. Whence this

Construction.—Take DH a third proportional to $2\,AB$ and BL.

* VI. 11.

Erect the perpendicular HE = the given perpendicular CP. Draw EF parallel and equal to BL. To EA, produced, apply $FG = AB$. Draw AC parallel to FG; join CB, and let fall the perpendicular CP; then ACB is the required triangle.

Demonstration by geometry.—We have to prove only that $BC - AC = BL$. By parallel lines, $FG\,(AB) : EF\,(BL) :: AC : EC\,(PH)$; hence $BL \times AC = AB \times PH$. By construction, $2\,AB : BL :: BL : DH$; hence $BL^2 = 2\,AB \times DH$. To this equation add *twice* the preceding one, and we have $BL^2 + 2\,BL \times AC = 2\,AB \times DH + 2\,AB \times PH = 2\,AB \times (DH + PH) = 2\,AB \times DP = AB \times 2\,DP = (BP + AP) \times (BP - AP) = BP^2 - AP^2 = BC^2 - AC^2$. Add AC^2 to the first and last of these equations, and we have $BL^2 + 2\,BL \times AC + AC^2 = BC^2$, or $(BL + AC)^2 = BC^2$; whence $BL + AC = BC$, or $BC - AC = BL$. Q. E. D.

Calculation.—By construction, $2\,AB : BL :: BL : DH$; then $AH = AD - DH$.

1. In triangle AHE, Case 3, find angle $EAH = FEG$.
2. In triangle GEF, Case 2, find angle $EGF = CAE$; then angle $CAP = EAH + CAE$.
3. In triangle CAP, Case 1, find AC and AP; then $BC = AC + BL$; also, $BP = AB - AP$.

Limits.—1. The given difference may be any quantity less than the base.

2. The perpendicular PC may be any quantity.
3. When $PC^2 = (BL + CP)^2 - AB^2$, the angle BAC will be a *right* angle.
4. When PC^2 is *less* than $(BL + CP)^2 - AB^2$, the angle BAC is an *acute* angle.
5. When PC^2 is *greater* than $(BL + CP)^2 - AB^2$, the angle BAC is an *obtuse* angle.

NOTE 1.—If we have given the *difference of the segments of the base*, the perpendicular height, and the difference of the sides of the triangle, then (see the figure to the preceding problem) lay $PA = PA'$, and join CA; then AB is the given difference of the segments. Wherefore we have to construct the triangle ABC as just shown in the problem, let fall the perpendicular CP, make $PA' = PA$, join $A'C$, and $A'CB$ will be the triangle required.

NOTE 2.—The line EFR, in which the vertex of the required triangle is to be, may be *any* straight line given in position. We must, if ER is not parallel to AB, lay $HI = BL$ (see figure to the present problem), and erect the perpendiculars HE and IF to meet the line ER, given in position, in the points E and F. Then, in the demonstration, we have, by parallel lines, $FG\,(AB) : AC :: FE : CE :: IH\,(BL) : PH$. Whence $BL \times AC = AB \times PH$, as before. (See Problem XXV. of "The Circle.")

PROBLEM IX.—Given, the perimeter of a rectangle, and its area, to find the sides.

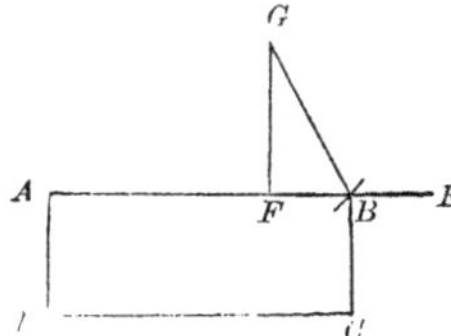

Given $\begin{cases} \text{the sum of } AB,\ BC,\ CD, \text{ and } DA, \\ \text{and the area of the rectangle } ABCD \\ = L^2. \end{cases}$

L

Analysis by algebra.—Put the given perimeter $= 2a$; then $AB + BC = a$. Let $x = AB$; then $a - x = BC$. Now, the area of $ABCD = AB \times BE = x(a - x) = ax - x^2 = L^2$, which can be constructed by Problem IX., "Algebraic Equations," or thus:

Construction.—Make $AE = a =$ half the given perimeter, and bisect it in F. Erect the perpendicular $FG = L$. Apply $GB = AF$ or FE. Draw BC perpendicular to AE and equal to BE, and complete the rectangle $ABCD$, which will be the one required, having $AB = x$.

Demonstration.—The area of $ABCD = AB \times BC = (AF + FB) \times (AF - FB) = AF^2 - FB^2 = GB^2 - FB^2 = FG^2 = L^2 =$ the given area. Also, $AB + BC + CD + DA = 2(AB + BC) = 2(AB + BE) = 2AE$. Q. E. D.

Calculation.—$FB = \sqrt{GB^2 - FG^2}$; then $AB = AF + FB$, and $BC = BE = AF - FB$.

Limit.—L may be any quantity not greater than half AE, or half a. When $L = \frac{1}{2} AE$, the required rectangle will be a square, having L for one of its sides.

PROBLEM X.—To construct a square such that the *difference* between the diagonal and the side shall be a given quantity.

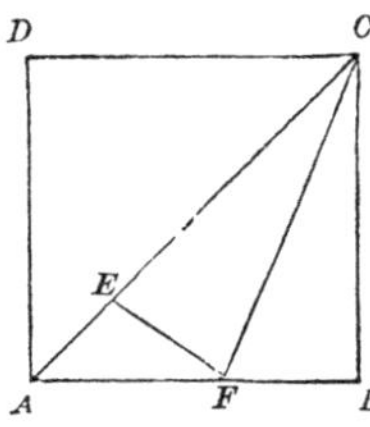

Given $\begin{cases} \text{in the square } ABCD, \text{ the } \textit{difference} \\ \text{between the diagonal } AC \text{ and side } BC. \end{cases}$

Analysis by algebra.—Let $ABCD$ represent the square. Put $d =$ the given difference, and $x =$ the side of the square; then the diagonal $AC = x + d$. Now, $AB^2 + BC^2 = AC^2$; that is, $x^2 + x^2$

$= (x + d)^2 = x^2 + 2dx + d^2$. Whence $x^2 - 2dx = d^2$, or $x^2 - 2dx + d^2 = 2d^2$. Hence $(x - d)^2 = 2d^2$, or $x - d = d\sqrt{2}$, or $x = d\sqrt{2} + d$. Whence this

Construction.—Draw two lines AB and AD at right angles. Bisect the angle BAD by the line AC, on which lay $AE = d$, the given difference. Draw EF perpendicular to AC, meeting AB in F; then $EF = AE = d$, and $AF = \sqrt{AE^2 + EF^2} = \sqrt{2\,d^2} = d\sqrt{2}$. Now, make $FB = FE = d$; then $AB = d\sqrt{2} + d = x$. Draw BC perpendicular to AB; then, since the angle $BAC =$ half a right angle, $BC = AB$. Complete the square $ABCD$, and it will be the one required.

Demonstration by geometry.—Join CF; then the right-angled triangles CEF and CBF are equal in all their parts, and we have $CE = CB$. Hence $AC - CB = AC - CE = AE =$ the given difference. Q. E. D.

Calculation.—$AF = \sqrt{2\,AE^2} = AE\sqrt{2}$; $AB = AF + FB = AF + AE = AE\sqrt{2} + AE =$ the side of the square. Also $AC = AE + CE = AE + AB = AE\sqrt{2} + 2\,AE = AE\,(\sqrt{2} + 2)$.

PROBLEM X.—In a right-angled triangle are given one leg, and the *difference of the segments* made by a perpendicular let fall from the right angle on the hypothenuse, to determine the triangle.

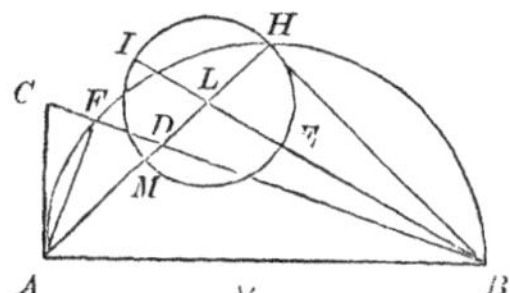

Given $\begin{cases} \text{in the triangle } BAC,\ AB \text{ and} \\ BF - CF = BD,\ FD \text{ being equal} \\ \text{to } FC. \end{cases}$

Analysis by algebra.—Let BAC be the right-angled triangle, and AF perpendicular to BC. Take $FD = FC$; then AB, and $BD = BF - CF$, the difference of the segments, are given. Now, put $AB = a$, $BD = b$, and $BF = x$; then $CF = x - b$, and $BC = BF + CF = x + x - b = 2x - b$. By IV. 23,* $BF : BA :: BA : BC$; that is, $x : a :: a : 2x - b$. Hence $2x^2 - bx = a^2$, or $x^2 - \frac{b}{2}x = \frac{a^2}{2}$. Whence this

Construction.—On the given leg AB describe a semicircle, which bisect in H, and join AH and HB; then $AH^2 = HB^2 = \frac{1}{2}AB^2 = \frac{1}{2}a^2$. (See Problem VIII., "Algebraic Equations.") Lay $HM = \frac{b}{2}$, the coefficient of x; on it describe a circle of which the centre is

* VI. 8, cor.

L, and draw $BELI$; then $BI = x$. For (IV. 30*) $BE\ (BI - HM) : BH :: BH : BI$. Whence $BI \times (BI - HM) = BH^2$; that is, taking $x = BI$, $x\left(x - \frac{b}{2}\right) = BH^2 = \frac{a^2}{2}$, or $x^2 - \frac{b}{2}x = \frac{a^2}{2}$. Now, apply $BF = BI$; draw BFC to meet the perpendicular AC, and join AF; then will BAC be the required triangle, having AF perpendicular to BC.

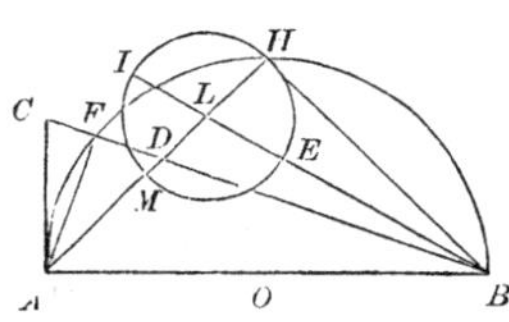

Demonstration by geometry.—We have to prove that $BD = 2\,HM = 2\,IE$. We have $AB^2 = 2\,BH^2 =$ (IV. 30†) $2\,BI \times BE = 2\,BI\,(BI - IE) = 2\,BI^2 - BI \times 2\,IE = 2\,BF^2 - BF \times 2\,IE$. Also, $AB^2 =$ (IV. 23‡) $BF \times BC = BF \times (BF + FC) = BF \times (BF + FD) = BF \times (BF + BF - BD) = BF\,(2\,BF - BD) = 2\,BF^2 - BF \times BD$. Hence $2\,BF^2 - BF \times BD = 2\,BF^2 - BF \times 2\,IE$. Wherefore $BD = 2\,IE = 2\,HM =$ the given difference.

Calculation.—We have $BL = \sqrt{BH^2 + HL^2} = \sqrt{\frac{AB^2}{2} + \frac{BD^2}{16}}$. Also, $BF = BI = BL + LH$; then $BF : BA :: BA : BC$, and $AC = \sqrt{BC^2 - AB^2}$. (See next problem.)

Limit.—The given difference must be less than AB, and consequently HM must be less than the radius AO.

PROBLEM XII.—Given, the hypothenuse of a right-angled triangle, and the line bisecting one of the acute angles and terminating in the opposite leg, to determine the triangle.

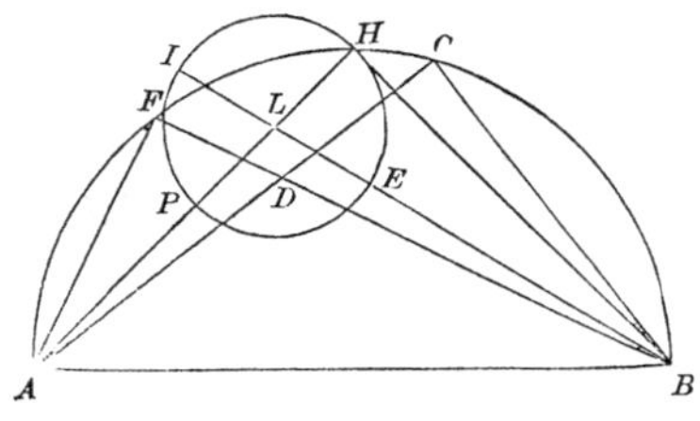

Given { in triangle ABC, AB and BD, ACB a right angle, and angle ABD to be equal to DBC.

Analysis by algebra.—Let ABC represent the required triangle, in which the hypothenuse AB, and the line BD bisecting the angle ABC, are given. On AB describe the semicircle ACB; produce BD to meet it in F; and join AF; then (III. 18, cor. 1§) angle $FAC = FBC = FBA$. Hence arcs AF and FC are equal, and the triangles AFD, AFB, and BCD are similar.

* III. 36. † III. 36. ‡ VI. 8, cor. § III. 21.

Now, put $AB = a$, $BD = b$, $BF = x$, and $AF = y$. Then, in similar triangles AFD and AFB, we have $BF(x) : AF(y) :: AF(y) : FD(x - b)$; hence $y^2 = x^2 - bx$. Also, $a^2 = x^2 + y^2 = x^2 + x^2 - bx = 2x^2 - bx$; hence $x^2 - \frac{b}{2}x = \frac{a^2}{2}$, and we have from this equation (see Problem VIII., "Algebraic Equations") the following

Construction.—Take H, the middle point of the semicircle AHB, and join AH and HB; then AH^2 or $HB^2 = \frac{1}{2}AB^2 = \frac{a^2}{2}$. On AH lay $HP = \frac{b}{2} = \frac{1}{2}$ the given bisecting line BD. On HP describe a circle $PEHI$, of which the centre is L. Draw $BELI$, and apply $BF = BI$. Make the arc $FC = FA$, and join ADC and BC, and ABC will be the required triangle. For (IV. 30*) $BI(x) : BH :: BH : BE(BI - HP)$; that is, $BI \times BE = BH^2$, or $x \times (x - \frac{b}{2}) = \frac{a^2}{2}$, which gives $x^2 - \frac{b}{2}x = \frac{a^2}{2}$. Hence BF or BI equals x in this equation.

Demonstration geometrically.—Since the arcs AF and FC are equal by construction, the angles FBA and FBC are equal, and BD bisects the angle ABC. It remains to prove only that $BD = 2\,PH = 2\,IE$.

In the similar triangles FAD and FBA, we have $BF : AF :: AF : FD$. Hence $AF^2 = BF \times FD = BF \times (BF - BD) = BF^2 - BD \times BF$; then $AB^2 = BF^2 + AF^2 = BF^2 + BF^2 - BD \times BF = 2\,BF^2 - BD \times BF$. Also, $AB^2 = 2\,BH^2 =$ (IV. 30†) $2\,BI \times BE = 2\,BI \times (BI - IE) = 2\,BI^2 - BI \times 2\,IE =$ (by construction) $2\,BF - BF \times 2\,IE$.

Wherefore we have $2\,BF^2 - BD \times BF = 2\,BF^2 - BF \times 2\,IE$. Whence $BD = 2\,IE = 2\,PH =$ the given bisecting line. Q. E. D.

Calculation.—In triangle BHL, we have $BL = \sqrt{BH^2 + HL^2} = \sqrt{\frac{AB^2}{2} + \frac{BD^2}{16}}$. Then $BF = BI = BL + LH$.

Also, $AF = \sqrt{AB^2 - BF^2}$; and $FD = BF - BD$. Then, by similar triangles BAF, FAD, and DBC, we have $BF : BA :: AF : AD$, and $AD : AF :: DB : BC$, and $AD : DF :: DB : DC$; then we have $AC = AD + DC$.

NOTE.—This problem and the preceding one depend upon the same principles, and result in like equations, as may be seen.

* III. 36. † III. 36.

ANALYSIS BY THE DIFFERENTIAL CALCULUS.

PROBLEMS ANALYZED BY THE DIFFERENTIAL CALCULUS AND CONSTRUCTED AND DEMONSTRATED BY PLANE GEOMETRY.

PRELIMINARY REMARKS.—The differential of x is written $d.x$, or simply dx; of y, dy; of z, dz; the letter d being merely a symbol for *differential.*

To find the differential of *any power* of an unknown quantity, multiply by the index denoting the power, diminish this index by unity, and multiply by the differential of the root, and the product of these three quantities will be the differential required. A constant or known quantity has no differential; and when a quantity is a maximum or a minimum, its differential is 0.

Examples.—1. x^0 being equal to 1, a constant quantity, it has no differential; that is, $d.x^0 = 0$.

2. The differential of x^1, or $d.x^1 = 1 \times x^{1-1} \times dx = x^0 \times dx = 1 \times dx = dx$.*

3. $d.x^2 = 2 \times x^1 \times dx = 2xdx$; and $d.xy = xdy + ydx$.

4. $d.x^3 = 3 \times x^2 \times dx = 3x^2dx$; and $d.x^2y^2 = x^2 \times d.y^2 + y^2 \times d.x^2 = 2x^2ydy + 2y^2xdx$.

5. $d.ax^4 = a \times d.x^4 = a \times 4 \times x^3 \times dx = 4ax^3dx$.

6. $d.3ax^5 = 15ax^4dx$.

7. $d.(ax^2 + 4x^3 + 5x^4) = 2axdx + 12x^2dx + 20x^3dx = (2ax + 12x^2 + 20x^3)\,dx$.

8. $d.(by^2 + 3y + a - 4) = 2bydy + 3dy = (2by + 3)\,dy$. The *constants*, 3 and —4, have no differential.

* The division of powers of the same quantity or root, as is shown in algebra, is effected by subtracting the index of the *divisor* from the index of the *dividend* to obtain the index of the *quotient;* that is, $\frac{x5}{x3} = x5—3 = x^2$. Also, $\frac{x5}{x5} (= 1) = x5—5 = x^0$. And $1 = \frac{x}{x} = x^{1-1} = x^0$.

9. $d.\frac{a}{x^3} = d.ax^{-3} = -3 \times a \times x^{-4} \times dx = -3ax^{-4}dx = \frac{-3adx}{x^4}$.

10. $d.\sqrt{a^2 + x^2} = d.(a^2 + x^2)^{\frac{1}{2}} = \frac{1}{2} \times (a^2 + x^2)^{-\frac{1}{2}} \times 2xdx = \frac{xdx}{(a^2 + x^2)^{\frac{1}{2}}} = \frac{xdx}{\sqrt{a^2 + x^2}}$.

11. $d.\frac{a}{\sqrt{a^2 + x^2}} = d.a \times (a^2 + x^2)^{-\frac{1}{2}} = -\frac{1}{2} \times a \times (a^2 + x^2)^{-\frac{3}{2}} \times 2xdx = -\frac{axdx}{(a^2 + x^2)^{\frac{3}{2}}} = -\frac{axdx}{\sqrt{(a^2 + x^2)^3}}$.

NOTE 1.—The term employed for this process is to *differentiate*, *differentiating*, or *differentiation*, according to the connection in which the term is used.

NOTE 2.—In problems involving the calculus, the letter d is not usually employed by mathematicians to denote a *quantity*.

NOTE 3.—The quantity which dx is multiplied by is called the *differential co-efficient*, as $2by + 3$ in Ex. 8.

PROBLEM I.—In a plane triangle are given the base and the vertical angle, to construct it such that the sum of the other two sides shall be a *maximum;* that is, the *greatest amount possible.*

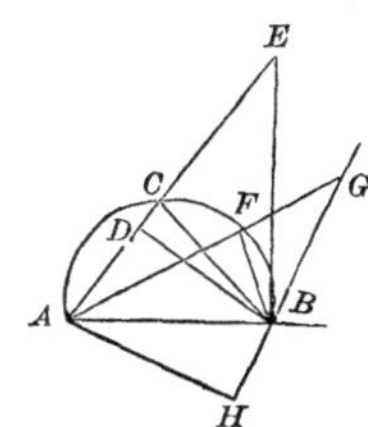

Given $\begin{cases} \text{the base } AB\text{, the angle } ACB\text{, and } AC + \\ CB \text{ to be a } maximum. \end{cases}$

*Analysis by the differential calculus.**—Let ABC represent the required triangle. On AC, produced if necessary, let fall the perpendicular BD. Put the given angle $ACB = \theta$, the given side $AB = a$, the side $AC = x$, and $BC = y$. Then $CD = y \cos \theta$.

By IV. 12,† $AC^2 + BC^2 - 2AC \times DC = AB^2$; that is, $x^2 + y^2 - 2xy \cos \theta = a^2$. By the problem, $x + y = \text{max}$. By differentiating,

* A student who has not yet studied the calculus may *omit the analysis* by that method, and just employ the results for the *construction*, and observe how beautifully the geometrical demonstration verifies the result obtained by the calculus, and how entirely the different processes of mathematical investigation harmonize with each other. Then when he studies the calculus he will feel increased confidence in the results of its determinations.

† II. 13.

$2xdx + 2ydy - 2y \cos\theta\, dx - 2x \cos\theta\, dy = 0$, and $dx + dy = 0$, or $dy = -dx$. For dy put its equal $-dx$ in the other differential equation, and we have $2xdx - 2ydx - 2y \cos\theta\, dx + 2x \cos\theta\, dx = 0$. Or $x(1 + \cos\theta) = y(1 + \cos\theta)$. Whence $x = y$. That is, the sum of AC and CB will be *a maximum* when *they are equal to each other*. Whence this

Construction.—On AB describe an arc to contain the given angle, which arc bisect in C. Join AC and BC, which will evidently be equal, and ABC will be the required triangle.

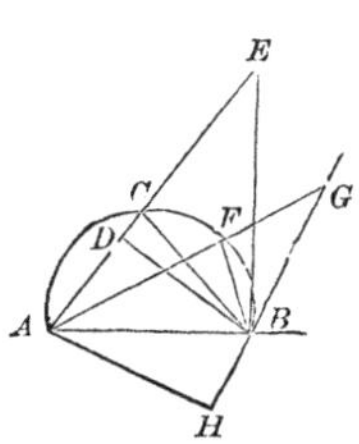

Demonstration geometrically.—Produce AC, making $CE = CB$, and join EB; then $AE = AC + CB$, angle AEB (I. 11 and 25, cor. 6*) $= \frac{1}{2} ACB$; and since C is the centre of a semicircle that would pass through the points A, B, and E, the angle ABE is a right angle. Now, take *any other* point, as F, in the arc ACB; join AF and FB, and produce AF till $FG = FB$. Join GB, and on it, produced, let fall the perpendicular AH. Then $AG = AF + FB$, and the angle AGH, being half AFB or ACB, is equal to AEB. Hence the triangles AEB and AGH are similar. But the hypothenuse AB, which is the base of ABE, is *greater* than the leg AH, which is the base of AHG. Therefore AE, which is the sum of $AC + CB$, *is greater* than AG, the sum of $AF + FB$, the sides of *any other* triangle. Q. E. D.

Calculation.—In the isosceles triangle ABC, find AC and BC, by Case 1.

* I. 5 and 29.

PROBLEM II.—In a plane right-angled triangle the hypothenuse is given, to determine the triangle such that the *base* added to *twice the perpendicular* shall be a *maximum.*

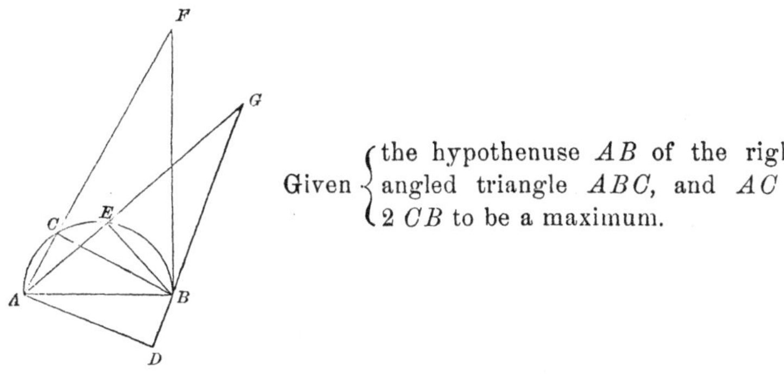

Given $\begin{cases} \text{the hypothenuse } AB \text{ of the right-} \\ \text{angled triangle } ABC, \text{ and } AC + \\ 2\,CB \text{ to be a maximum.} \end{cases}$

Analysis by the differential calculus.—Let ACB represent the required triangle. Put $AB = a$, $AC = x$, and $BC = y$. Then $x^2 + y^2 = a^2$, and $x + 2y = \text{max.}$, the differential of each of which is 0. Hence $2xdx + 2ydy = 0$, and $dx + 2dy = 0$. Or $dx = -2dy$. Putting $-2dy$ for dx in the other equation, we have $-4xdy + 2ydy = 0$, or $y = 2x$. That is, in order that the sum of $AC + 2\,CB$ may be a maximum, BC must equal *twice* AC. Whence this

Construction.—On AB describe a semicircle, and erect the perpendicular $BF = 2\,AB$. Draw ACF, and join BC. Then, in the three similar triangles ABF, ACB, and BCF (IV. 23*), since $BF = 2\,AB$, we have $BC = 2\,AC$, and $CF = 2\,BC = 4\,AC$.

Demonstration by plane geometry.—Since $CF = 2\,BC$, we have $AF = AC + 2\,BC$, which we have to prove is greater than the sum of one side and double the other side of *any other right-angled triangle* formed on AB.

In the semicircle ACB take *any other* point, as E, and draw AE and EB; then we have to prove that the sum $AC + 2\,BC$ is greater than $AE + 2\,BE$. Produce AE so that $EG = 2\,EB$; join GB, and on it, produced, let fall the perpendicular AD. Then, since $EG = 2\,EB$ and $BF = 2\,AB$, the triangles GEB and ABF, and consequently GAD, are all similar. But AB is greater than AD. Hence AF, which is the sum of $AC + 2\,BC$, is greater than AG, which is the sum of $AE + 2\,BE$. Q. E. D.

Calculation.—$AF = \sqrt{AB^2 + BF^2} = \sqrt{5\,AB^2} = AB\sqrt{5}$. Also, $AC = \frac{1}{5}\,AF = \frac{1}{5}\,AB\sqrt{5}$, and $BC = 2\,AC = \frac{2}{5}\,AB\sqrt{5}$.

* VI. 4.

Scholium.—If the sum of the base added to n times the perpendicular were to be a maximum, the solution would be precisely similar to the above.

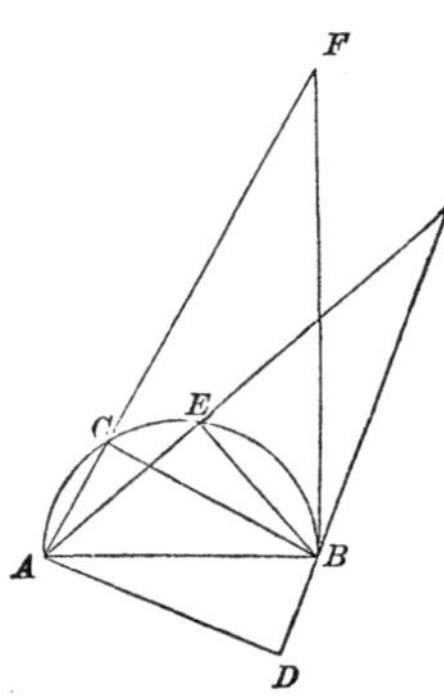

Put $AB = a$, $AC = x$, and $BC = y$; then $x^2 + y^2 = a^2$, and $x + ny = \text{max.}$ By differentiating, we have $2xdx + 2ydy = 0$, and $dx + ndy = 0$, or $dx = -ndy$. By substitution, $-2nxdy + 2ydy = 0$, or $y = nx$. Whence $BC = n$ times AC, and we have this

Construction.—Make $BF = n$ times AB; draw ACF, and join BC; then ABC is the required triangle. For, since $BF = n \,.\, AB$, we have $BC = n \,.\, AC$, and $CF = n \,.\, BC = n^2 \,.\, AC$.

The *demonstration* and *calculation* are precisely as in the problem, using n instead of 2.

PROBLEM III.—To determine the arc of a given circle, the *sum* of whose sine and versed sine, or the *difference* of whose sine and versed sine, shall be a maximum.

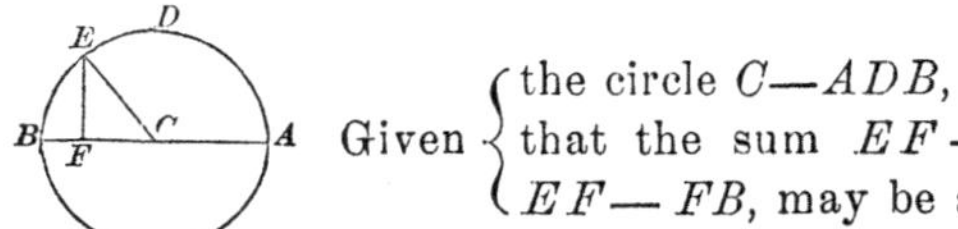

Given $\begin{cases} \text{the circle } C—ADB, \text{ to find the point } E \text{ such} \\ \text{that the sum } EF + FA, \text{ or the difference} \\ EF — FB, \text{ may be a } \textit{maximum}. \end{cases}$

Analysis.—Suppose E to be the required point; join EC, and let fall the perpendicular EF. Then, since the radius AC is a constant quantity, $AF + FE$ will be the greatest when $CF + FE$ is the greatest; that is, by Problem I. of this chapter, when $CF = FE$. Hence, if the quadrant BD be bisected in E, AE will be the arc required, having $EF + FA$, the sum of the sine and cosine, *a maximum*.

Again, $FB = BC - CF$. Hence $EF - FB = EF - (BC - CF) = EF + CF - BC$. Wherefore, since BC is a constant quantity, $EF - FB$ will be the greatest when $EF + CF$ is the greatest; that is, by Problem I., when EF and CF are equal. Hence, if the quadrant BD be bisected in E, BE will be the arc required, having $EF - BF$, the difference between the sine and cosine, *a maximum*.

Calculation.—Since the quadrant BD is bisected in E, the arc $BE =$ half a quadrant $= 45°$, and the arc $AE =$ three-fourths of a semicircle $= 135°$.

Also, $CF = FE = \sqrt{\frac{1}{2}\,CE^2} = \sqrt{\frac{1}{2}\,CB^2}$. Then $AF = AC + CF = CB + \sqrt{\frac{1}{2}\,CB^2}$, and $BF = CB - CF = CB - \sqrt{\frac{1}{2}\,CB^2}$.

PROBLEM IV.—From a given point within a given circle, to draw a straight line which shall make the *least possible* angle with the circumference.

NOTE.—The angle which a straight line makes with the *circumference of a circle* at any point is the same as that made with a *tangent to the circle* at that point.

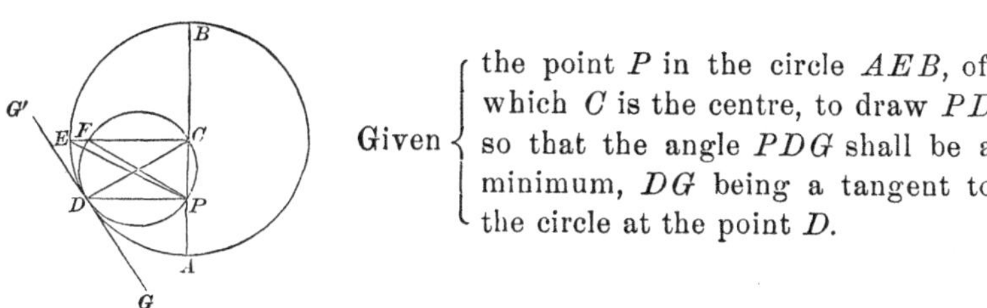

Given the point P in the circle AEB, of which C is the centre, to draw PD so that the angle PDG shall be a minimum, DG being a tangent to the circle at the point D.

Analysis by the differential calculus.—Through the given point draw the diameter $APCB$. Suppose PD to represent the required line; draw the radius CD and tangent DG. Then angles $CDP + PDG = 90°$; consequently, when the angle PDG is the *least possible*, or a *minimum*, the angle PDC will be the *greatest possible*, or a *maximum*.

Now, put $CD = r$, $CP = a$, and the angle $CPD = \theta$. Then $r : a :: \sin\theta : \sin PDC = \frac{a}{r}\cdot\sin\theta =$ max. Hence, since $\frac{a}{r}$ is a constant quantity, $\sin\theta =$ max. By differentiating, we have $\cos\theta . d\theta = 0$, or $\cos\theta = 0$; whence $\theta = 90°$. That is, PD must be perpendicular to AB. Whence this

Construction.—Draw PD perpendicular to AB, draw the radius CD and tangent DG; then will the angle PDC be the *greatest possible*, and consequently its complement PDG, which PD makes with the circumference, the *least possible*.

Demonstration geometrically.—On the radius CD describe the circle CPD, which will touch the given circle at the point D. From the point P draw *any other* line, as PE, to the circumference; draw CFE, and join PF; then we have to show that the angle PDC is greater than PEC. By III. 18, cor. 1,* angle $PDC = PFC$, which is *greater* than PEC (I. 25, cor. 6†).

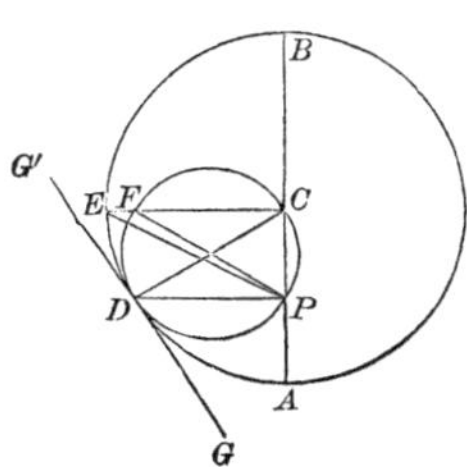

Calculation.—In triangle PDC, Case 2, find angle PDC; then angle $PDG = 90° - PDC$.

Scholium 1.—If the tangent GD be produced to G', PDG' will be the *maximum* angle made by a line from P with the circumference.

Scholium 2.—At the points A and B the angle CDP *vanishes;* that is, becomes 0; hence at these points PA and PB make an angle of 90° with the circumference.

Scholium 3.—PD may be drawn to the right or left of AB.

PROBLEM V.—Through the points of intersection of two given circles, to draw the longest possible lines terminated by the circumferences of the two circles.

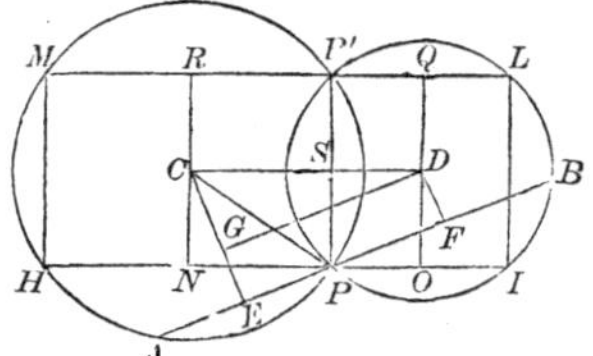

Given $\left\{ \begin{array}{l} CD, \text{ and the radii } CP \text{ and } PD, \\ \text{to draw through } P \text{ or } P' \text{ the} \\ \text{longest possible line } HPI \text{ or} \\ MP'L. \end{array} \right.$

Analysis.—Let C and D be the centres of the two given circles which intersect in P and P'. Join CD and PP'. Through P draw *any line*, as APB, terminated by the circumferences, and on it let fall the perpendiculars CE and DF, which will respectively bisect the chords AP and BP; hence $EF = \frac{1}{2} AB$. Draw DG parallel to AB; then $DG = EF = \frac{1}{2} AB$. Now, in order for AB to be the longest *possible*, DG, *its half*, must be the *longest possible*, which will be the case when DG coincides with DC, and the line APB becomes parallel to CD. Whence this

Construction.—Parallel to CD, through the points P and P', draw HPI and $MP'L$, and they will be the lines required, each being the *longest line possible* through the point through which it passes.

Demonstration.—Through the centres C and D draw the perpendiculars RCN and QDO, and they will respectively bisect the equal

* III. 21. † I. 16.

chords HP and MP', and PI and $P'L$; then $HI = ML = 2\,RQ = 2\,NO = 2\,CD$, which is *greater* than $2\,DG$, and hence *greater* than $2\,EF$ or APB, which is *any other* line than HPI drawn through the point P. Hence HPI is the *longest line possible*, or *maximum line*. Q. E. D.

Calculation.—$HI = ML = 2\,CD$. In triangle CPD, Case 4, find CS and SD; then $PS = \sqrt{CP^2 - CS^2}$. And $PP' = 2\,PS$, $HP = 2\,CS$, $PI = 2\,SD$; and MH, RN, PP', QO, and LI are all parallel and equal.

Scholium.—If a line be drawn through P', parallel to APB, and terminated by the circumferences, it will be equal to twice DG or twice EF, and will, consequently, be equal to APB. If the ends of these equal lines in each circle be joined, there will be a parallelogram formed. Now, of all the parallelograms which can be thus formed by lines drawn through the points P and P', the rectangle $HILM$ is the *greatest possible*, because its area is equal to $HI \times PP'$, and HI is longer than *any other* line through P, and PP' is longer than a perpendicular between *any other* two parallels through P and P'.

PROBLEM VI.—About a given triangle to circumscribe the *greatest possible* triangle, which shall have its angles respectively equal to the angles of a given triangle.

NOTE.—The angles of the required triangle may be given in degrees.

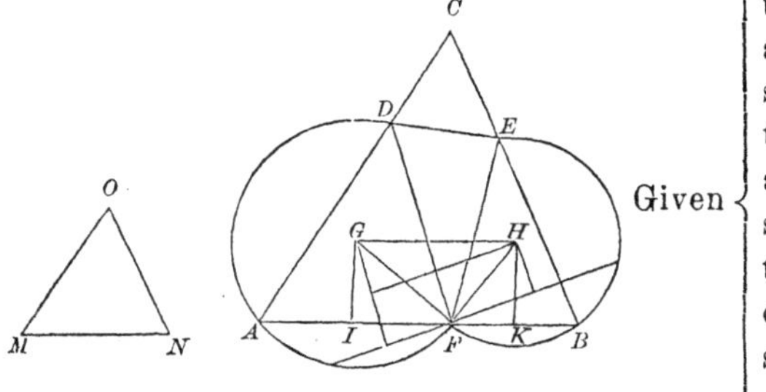

Given the triangles DEF and MNO, to circumscribe about DEF the *maximum* triangle ABC, which shall be equiangular to the triangle MNO, or have its angles respectively of given quantities.

Analysis.—Suppose ABC to be the required maximum triangle circumscribed about DEF, having the angle A equal to the angle M, the angle B equal to the angle N, and, consequently, the angle C equal to the angle O. Through the points F, A, D and F, B, E suppose segments of circles to be passed whose centres are G and

H respectively. Join GH. Then, since the triangle ABC is, by hypothesis, the *greatest possible* triangle that can be circumscribed about DEF and be equiangular to MNO, AFB must be the *longest possible* line that can be drawn through the point F and terminate in the circular segments FAD and FBE. Wherefore, by the last problem, AFB must be parallel to the line GH, joining the centres of these segments. Whence this

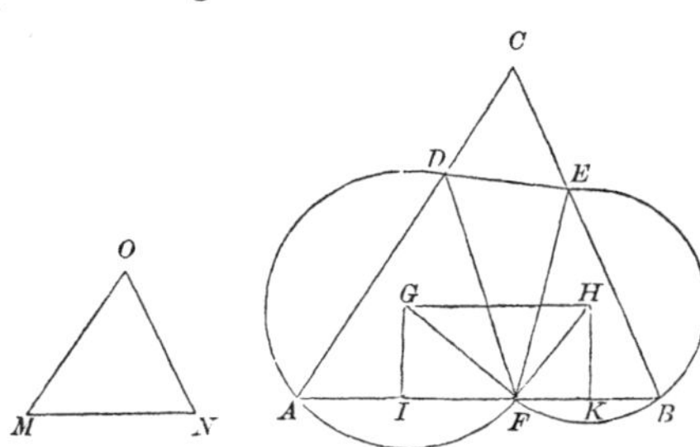

Construction.—On DF and FE (III. Prob. 16*) describe segments to contain the angles M and N respectively, of which the centres are G and H. Join GH, parallel to which draw AFB, and let fall on it the perpendiculars GI and HK Join AD and BE, and produce them to meet in C. Then ABC is the maximum triangle required.

Demonstration.—Since, by last problem, AFB is longer than any other line, as PFQ, it is the *greatest possible side*, and ABC is the *greatest possible triangle*. Q. E. D.

Calculation.—Join GF and HF; then (III. Prob. 16*) the angle DFG is equal to the difference between the angle M and 90°, the angle EFH is equal to the difference between the angle N and 90°, and the angle DFE is given because the triangle DEF is given. Hence we know the angle $GFH = DFG + DFE + EFH$.

1. In triangle GFH, Case 3, find $GH = KI$; then side $AB = 2\,KI = 2\,GH$.

2. In triangle ABC, Case 1, find the sides AC and BC.

Note 1.—See Theorem 13, Scholiums 3 and 4.

Note 2.—This problem and the last are placed here as belonging to the general subject, although they do not strictly come under the proposed mode of analysis.

* III. 33.

PROBLEM VII.—In a right-angled triangle there is given the side of the inscribed square, to determine the sides when the hypothenuse is a *minimum*.

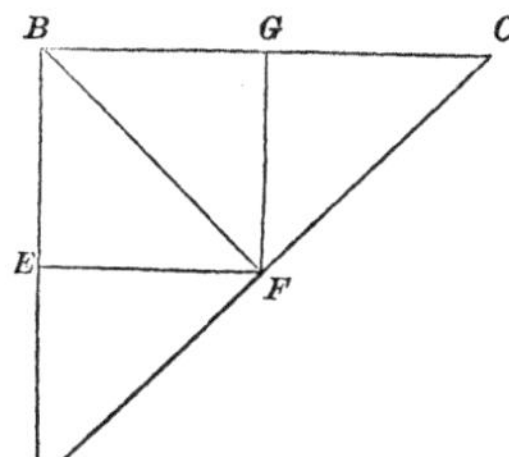

Given $\begin{cases}\text{the square } BEFG, \text{ and the hy-} \\ \text{pothenuse } AFC \text{ to be a } minimum.\end{cases}$

Analysis by the differential calculus.—Let ABC represent the required triangle. Put $AB = x$, $BC = y$, and BG or $BE = a$; then $AE = x - a$, and, by similar triangles, we have $x - a\,(AE) : a\,(EF) :: x\,(AB) : y\,(BC)$; hence $xy - ay = ax$. Also, by the problem, $x^2 + y^2 = AC^2 =$ min. By differentiating these two equations, we have $xdy + ydx - ady = adx$, and $2xdx + 2ydy = 0$, or $dy = -\frac{xdx}{y}$. Putting this for dy in the first differential equation, we have $-\frac{x^2dx}{y} + ydx + \frac{axdx}{y} = adx$. Whence, $-x^2 + y^2 + ax = ay$, or $y^2 - ay = x^2 - ax$; wherefore, by the "Properties of Equations," $x = y$.* That is, the hypothenuse or line AFC will be the shortest *possible* when $AB = BC = 2\,BG = 2\,BE$, and the four constituent triangles are all equal. Whence this

Construction.—Join BF, and draw AFC at right angles to BF; then angle $A =$ angle $C = \frac{1}{2}$ a right angle. Hence $BC = BA$.

Calculation. — $AC = \sqrt{AB^2 + BC^2} = \sqrt{(2\,BE)^2 + (2\,BG)^2} = \sqrt{8\,BE^2} = 2\,BE\sqrt{2}$.

Solution geometrically.—Referring to the figure to Problem IV., "Analysis by Algebra," EL will always equal EI when $BI =$ the hypothenuse AC. Now, as BE is constant, EL or EI will be the shortest when BI is the shortest, and *vice versa.* But EL is a minimum when FL is a minimum, that is, when OC is a minimum, or when OC is perpendicular to GI, and, hence, when $OC = FG = CG$; and, consequently, $AB = BC = 2\,BG = 2\,BE$, as before.

* This equality may also be shown thus: To each side of the equation $y^2 - ay = x^2 - ax$ add $\frac{1}{4}a^2$, and we have $y^2 - ay + \frac{1}{4}a^2 = x^2 - ax + \frac{1}{4}a^2$. Take the square root of each member of this equation, and we have $y - \frac{1}{2}a = x - \frac{1}{2}a$; whence $y = x$.

PROBLEM VIII.—In a right-angled triangle there is given the side of the inscribed square, to determine the triangle when the *sum* of the two legs is a *minimum*.

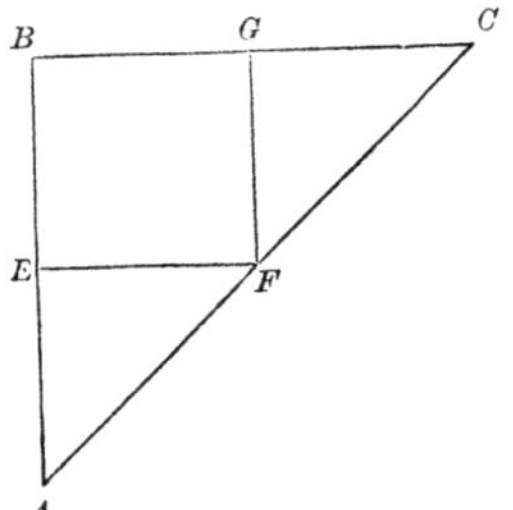

Given $\begin{cases} \text{the square } BEFG, \text{ and the } sum \text{ of} \\ AB + BC \text{ to be the } least\ possible. \end{cases}$

Analysis by the differential calculus.—Put BG or $BE = a$, $AB = x$, and $BC = y$; then $AE = x - a$, and, by similar triangles, we have $x - a\ (AE) : a\ (EF) :: x\ (AB) : y\ (BC)$; hence $xy - ay = ax$; also, by the problem, $x + y = \text{min.}$ By differentiating, $xdy + ydx - ady = adx$, and $dx + dy = 0$, or $dy = -dx$. Putting $-dx$ for its equal dy in the preceding differential equation, we have $-xdx + ydx + adx = adx$; whence $-xdx + ydx = 0$, or $x = y$. Hence $AB + BC$ will be a minimum when $AB = BC = 2\,BG$ or $2\,BE$. Whence the

Construction and *calculation* are precisely as in the last problem.

Solution geometrically.—Referring to the figure to Problem IV., "Analysis by Algebra," $AB + BC$ will evidently be a minimum when $AE + GC = \text{min.}$; that is, when $HL + FH = \text{min.}$, or when $FL = \text{min.}$, which, as shown in the geometrical solution to last problem, is when $AB = BC = 2\,BG$, or when $AB + BC = 4$ times the side of the given square.

PROBLEM IX.—It is required to find a point in the base of a given right-angled triangle, from which if two lines be drawn, one to one extremity of the hypothenuse, and the other parallel to the perpendicular to meet the hypothenuse, and from the point in which this line meets the hypothenuse a line be drawn parallel to the base to meet the one drawn to the extremity of the hypothenuse, the length of this last line shall be a *maximum*.

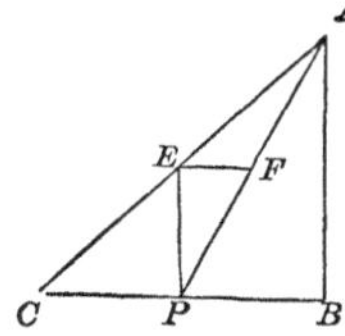

Given { the right-angled triangle ABC, to find the point P such that if PA be joined, and PE be drawn parallel to AB, and EF to BC, the length of EF may be the *greatest possible.*

Analysis by the differential calculus.—Let ABC represent the required triangle, and P the required point in the base, from which a line is to be drawn to A, and PE be drawn parallel to AB, and EF to BC. It is evident that the triangles ABP and PEF are similar. Now, put $AB=a$, $BC=b$, and $CP=x$; then $BP=b-x$; and, by similar triangles, we have $BC : CP :: AB : PE :: BP : EF = \frac{BP \times CP}{BC} = \frac{(b-x)x}{b} = \text{max.}$ That is, $(b-x)x = bx - x^2 = \text{max.}$ By differentiating, $bdx - 2xdx = 0$, or $b = 2x$, whence $x = \frac{1}{2}b = \frac{1}{2}BC$, and $EF = \frac{(b-x)x}{b} = \frac{\left(b-\frac{b}{2}\right)\times\frac{b}{2}}{b} = \frac{b}{4} = \frac{1}{4}BC.$ Whence this

Construction.—Bisect BC in P, which will be the required point. Join PA; draw PE parallel to AB, and EF to BC.

Demonstration geometrically.—We have found above that $EF = \frac{BP \times CP}{BC}$, which is to be a maximum. Now, BC being a *given* line, and hence constant, EF will be a maximum when $BP \times CP$ is a maximum. But the rectangle of the two parts of a given line is greatest when those two parts are equal (II. 5, Euclid*). Hence P must be the middle point of BC. Q. E. D.

* This property may be proved thus: Let P be the middle of the line BC, and O *any other* point in the line; then $CO \times OB = (CP + PO) \times (CP - PO) = CP^2 - PO^2$; that is, $CP^2 = CO \times OB + PO^2$. Hence CP^2 or $CP \times PB$ is greater than the rectangle of *any* other two parts of the line by the square of the difference between the middle and the other point.

C———P——O——B

PROBLEM X.—In a given right-angled triangle to inscribe the greatest possible rectangle.

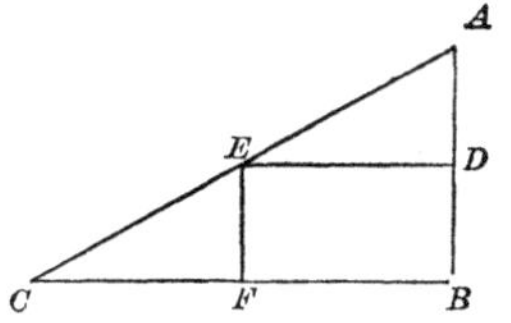

Given $\left\{\begin{array}{l}\text{the right-angled triangle } ABC, \text{ to} \\ \text{inscribe therein the maximum rec-} \\ \text{tangle.}\end{array}\right.$

Analysis by the differential calculus.—Put $AB = a$, $BC = b$, $BF = x$, $BD = y$; then $CF = b - x$. By similar triangles, $b - x$ $(CF) : y\,(FE) :: b\,(BC) : a\,(AB)$; whence $ab - ax = by$. Also, by the problem, $xy = \text{max}$. By differentiating, $-adx = bdy$, or $dy = -\frac{adx}{b}$. And $xdy + ydx = 0$. In this last equation, for dy put its equal $-\frac{adx}{b}$, and we have $-\frac{axdx}{b} + ydx = 0$; whence $by = ax =$ (from above) $ab - ax$. Hence $2ax = ab$, or $x = \frac{1}{2}b$, and $y = \frac{ax}{b} = \frac{1}{2}a$. Wherefore the sides of the maximum rectangle are respectively *halves* of the sides of the given triangle. Whence this

Construction.—Bisect BC in F; draw FE parallel to AB; then $CE = EA$; and draw ED parallel to BC; then $AD = DB$, and the triangles ADE and EFC are equal, and each half of $BDEF$, or $\frac{1}{4} ABC$.

Calculation.—Area of $BDEF = BF \times BD = \frac{1}{2} BC \times \frac{1}{2} AB = \frac{1}{4} BC \times AB =$ half the area of the triangle ABC.

Demonstration geometrically.—By similar triangles, $AB : BC ::$ $AD : DE$ or $BF = \frac{BC}{AB} \times AD$. The area $BDEF = BF \times BD = \frac{BC}{AB} \times AD \times BD$. But $\frac{BC}{AB}$ is constant. Hence $BDEF$ will be a maximum when $AD \times BD$ is a maximum; that is, by foot-note to last problem, when $AD = BD$. Then, also, $BF = FC$, and $AE = EC$. Q. E. D.

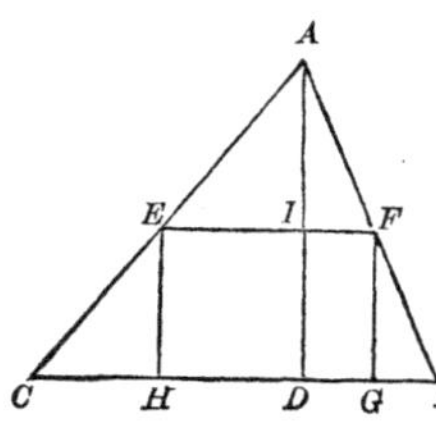

Scholium.—In *any* triangle, the maximum inscribed rectangle is half the triangle. For (in adjacent figure), put $BC = b$, $AD = a$, EH or $DI = x$, and EF or $GH = y$; then $AI = a - x$. By similar triangles, $b\,(BC) : a\,(AD) :: y\,(EF) : a - x\,(AI)$; whence $ab - bx = ay$; also, $xy = \text{max}$. By differentiating, we have $-bdx = ady$, or $dy = -\frac{bdx}{a}$; and $xdy + ydx = 0$. Putting, in this last equation, for dy its equal $-\frac{bdx}{a}$, we have $-\frac{bx}{a}dx + ydx = 0$; whence $ay = bx =$ (from above) $ab - bx$. Wherefore $2bx = ab$, or $x = \frac{1}{2}a$, and $y = \frac{bx}{a} = \frac{1}{2}b$. Wherefore the sides of the maximum inscribed rectangle are respectively halves of the base and perpendicular height of the triangle, and, consequently, its area is half the area of the triangle. Whence this

Construction.—Bisect AD in I. Parallel to BC draw EIF, and parallel to AD draw EH and FG.

Demonstration geometrically.—By similar triangles, $AD : BC :: AI : EF = \frac{BC}{AD} \times AI$. Area of $EFGH = EF \times ID = \frac{BC}{AD} \times AI \times ID$. But $\frac{BC}{AD}$ is constant. Therefore $EFGH$ will be a max. when $AI \times ID$ is a max.; that is, by foot-note to last problem, when $AI = ID$, then $EF = \frac{1}{2}BC$, and area $EFGH = \frac{1}{2}BC \times \frac{1}{2}AD = \frac{1}{2}ABC$. Q. E. D.

PROBLEM XI.—To inscribe the greatest possible rectangle in a given semicircle.

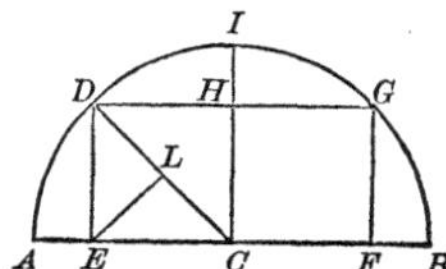

Given $\begin{cases} \text{the semicircle } AIB, \text{ to inscribe there-} \\ \text{in the greatest possible rectangle} \\ EFGD. \end{cases}$

Analysis by the differential calculus.—Suppose $EFGD$ to be the required rectangle. Erect the perpendicular CHI, and join CD. Put $CD = r$, EC or $CF = x$, and ED or $FG = y$; then we have $x^2 + y^2 = r^2$, $2xy = \max$. By differentiating, $2xdx + 2ydy = 0$, and $2xdy + 2ydx = 0$, or $dy = -\frac{ydx}{x}$. By substituting this value of dy, we have $xdx - \frac{y^2dx}{x} = 0$; whence $x^2 = y^2$, or $x = y$; that is, in the maximum rectangle $CF = CE = ED$, the angle $ACD =$ half a right angle $= 45°$, and the arc $AD = \frac{1}{2}$ the quadrant AI. Whence this

Construction.—Bisect the quadrantal arc AI in D; draw DHG parallel to AB, and DE and GF parallel to CI.

Demonstration geometrically.—On CD let fall the perpendicular EL. The area of $EFGD = 2\,EC \times ED = 2\,CD \times EL$. Now, since $2\,CD$ is a constant quantity, $EFGD$ will be a max. when EL is a max.; that is, evidently, when L is the centre of a semicircle described on the radius CD, and $LE = LC = LD$, or when $CE = ED$, and angle $ECD = 45°$. Q. E. D.

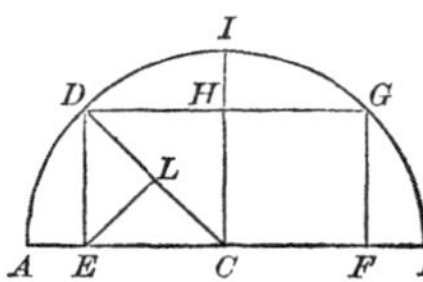

Calculation.—We have $2\,EC^2 = CD^2 =$ the area of the maximum rectangle $EFGD$, $EC = \frac{CD}{\sqrt{2}} = \frac{1}{2}\,CD\sqrt{2} = ED$, and $EF = 2\,EC = CD\sqrt{2}$.

Scholium 1.—The square $DECH$ is the maximum rectangle inscribed in a given quadrant. For the area of $DECH = EC \times ED = CD \times EL$. But $CD \times EL$ is a maximum, as shown in the demonstration, when EL is a maximum; that is, when $LE = LC = LD$, or when $CE = ED$, and angle $ACD = 45°$.

Scholium 2.—If the circle were completed, and another maximum rectangle inscribed in the other semicircle, it would be equal to $EFGD$, and the sides perpendicular to AB would be equal to DE and GF, or they would be DE and GF produced. Hence each of these chords would $= 2\,DE = 2\,GF = DG$, and the whole rectangle would be a square. Hence the maximum rectangle inscribed in a given circle is its inscribed square, the side of which $= DG = EF = CD\sqrt{2}$, as in the calculation.

Problem XII.—Having given the base and vertical angle of a plane triangle, to construct it when the rectangle of the sides including the given angle is a maximum.

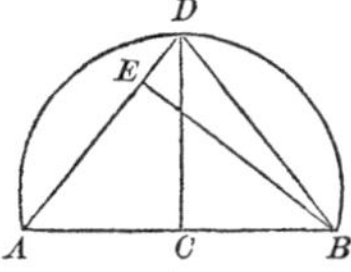

Given $\left\{\begin{array}{l}\text{the base } AB \text{ and the vertical angle } ADB,\\ \text{to construct the triangle such that } AD \times\\ DB \text{ shall be the greatest possible.}\end{array}\right.$

Analysis by the differential calculus.—Let ADB represent the required triangle. On AD let fall the perpendicular BE. Put $AB = a$, $BD = x$, $AD = y$, and angle $ADB = \theta$; then $DE = x\cos\theta$. And (IV. 12 or 13*) $AB^2 = AD^2 + BD^2 - 2\,AD \times DE$; that is,

* II. 12 or 13.

$a^2 = y^2 + x^2 - 2yx \cos \theta$. Also, by the problem, $xy = \text{max.}$ By differentiating, $2ydy + 2xdx - 2x \cos \theta\, dy - 2y \cos \theta\, dx = 0$, and $xdy + ydx = 0$, or $dy = -\frac{ydx}{x}$. Substituting this value for dy in the preceding equation, we have $-\frac{2y^2dx}{x} + 2xdx + \frac{2xy \cos \theta\, dx}{x} - 2y \cos \theta\, dx = 0$. By reduction, $-y^2 + x^2 + xy \cos \theta - xy \cos \theta = 0$; whence $x^2 = y^2$, or $x = y$. That is, in order for $AD \times DB$ to be a maximum, AD and DB must be equal, and the triangle ADB must be isosceles. Whence this

Construction.—On the base AB (III. Prob. 16*) describe a circular segment to contain the given angle. Bisect AB in C; erect the perpendicular CD, and join AD and DB; then $AD = DB$.

Demonstration geometrically.—By Problem I. "In Relation to Areas," the double area of the triangle $ADB = AD \times DB \times \sin ADB$. But (IV. 6†) the double area of $ADB = AB \times CD$. Hence, as AB is constant, $AB \times CD$ will be greatest when CD is greatest. Now, DC is the greatest perpendicular that can be let fall from a point in the circular segment ADB upon AB, and, consequently, ADB is the greatest possible triangle that can be formed on AB with its vertex in the segment ADB. Hence, since $AD \times DB \times \sin ADB = AB \times CD$, and $AB \times CD = \text{max.}$, we have $AD \times DB \times \sin ADB = \text{max.}$ But $\sin ADB$ is a constant quantity; therefore $AD \times DB$ will be a maximum when the area is a maximum; that is, when $AB \times CD$ is a maximum; that is, when CD is a maximum, or when CD is erected from the middle point of AB, and consequently when $AD = DB$. Q. E. D.

Calculation.—In triangle ADB, Case 1, find AD, DB, and CD; then the area $= \frac{1}{2} AB \times CD$.

* III. 33. † VI. 1.

PROBLEM XIII.—On a given line to describe two circles touching each other, such that *twice* the area of one circle added to the area of the other circle shall be the *least possible.*

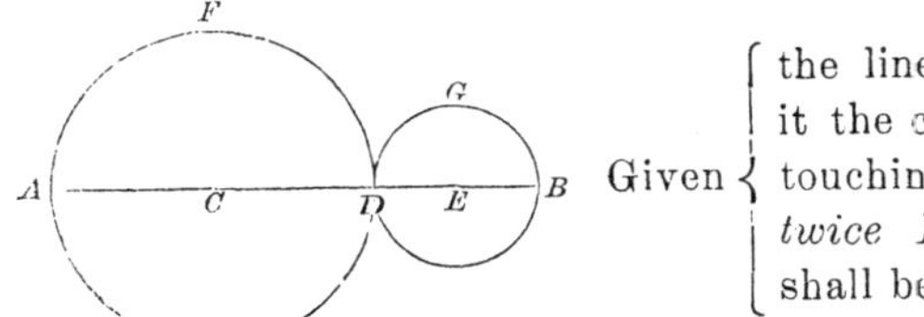

Given { the line ADB, to describe on it the circles AFD and DGB, touching each other, such that *twice* DGB added to AFD shall be a minimum.

Analysis by the differential calculus.—Let AFD and DGB represent the required circles. Put $AB = a$, $AD = x$, $DB = y$, and put $p = .78539+$; then we have $x + y = a$, and $px^2 + 2py^2 =$ min., or, since p is constant, $x^2 + 2y^2 =$ min. By differentiating, $dx + dy = 0$, or $dy = -dx$, and $2xdx + 4ydy = 0$. By putting, in this equation, $-dx$ for its equal dy, we have $2xdx - 4ydx = 0$; whence $x = 2y$, and hence $x + y = 3y = AB$, or $y = \frac{1}{3} AB = DB$; that is, $DB = \frac{1}{2} AD = \frac{1}{3} AB$, and $AD = \frac{2}{3} AB$. Whence this

Construction.—Divide the given line AB into three equal parts in the points C and D. On AD and DB describe the circles AD and DB respectively; then will $AFD + 2\,DGB =$ min.

Demonstration geometrically.—Area $AFD = p \,.\, AD^2$, and area $DGB = p \,.\, DB^2$. Hence $AFB + 2\,DGB = p \,.\, AD^2 + 2p \,.\, DB^2 =$ min. Wherefore, as p is constant, $AD^2 + 2\,DB^2 =$ min. This is equal to $4\,AC^2 + 2\,DB^2 = 2\,AC^2 + 2\,CD^2 + 2\,DB^2 =$ min. Hence $AC^2 + CD^2 + DB^2 =$ min. Now $AD = AC + CD$. Hence $AD^2 = AC^2 + CD^2 + 2\,AC \times CD$. Hence $AC^2 + CD^2 = AD^2 - 2\,AC \times CD$. Wherefore, in order for $AC^2 + CD^2$ to be a minimum, $2\,AC \times CD$ must be a maximum; that is, by foot-note to Problem IX. of this division, when $AC = CD$.

In like manner it may be shown that, in order for $CD^2 + DB^2$ to be a minimum, CD must be equal to DB. Hence, in order that $AC^2 + CD^2 + DB^2 =$ min., or $AD^2 + 2\,DB^2 =$ min., AC, CD, and DB must all be equal, and each $= \frac{1}{3} AB$. Q. E. D.

Corollary.—To divide a given line into any number of parts such that the sum of the squares of the several parts shall be a *minimum*, the parts *must all be equal.*

PROBLEM XIV.—Given, the sum of the base and perpendicular of a right-angled triangle, to construct it such that the square of the hypothenuse added to the square of the base shall be a *minimum*.

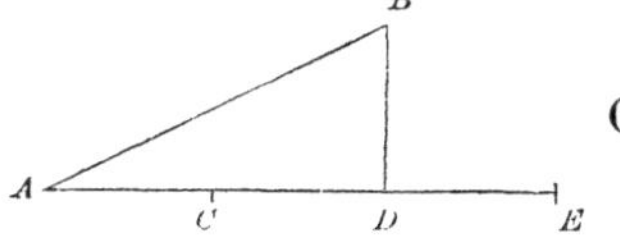

Given $\left\{\right.$ in the right-angled triangle ADB, the sum of $AD + DB = AE$, and $AB^2 + BD^2$ to be the *least possible*.

Analysis.—Let ADB represent the required triangle, having $AD + DB =$ the given line AE, and $AB^2 + BD^2 =$ min. Now, $AB^2 = AD^2 + BD^2$; hence $AB^2 + BD^2 = AD^2 + 2\,BD^2 =$ min., which, by the last problem, will be when $AD = 2\,BD = 2\,DE$, and hence $ED = \frac{1}{3}\,AE$, the given sum.

PROBLEM XV.—Having given the perimeter of a triangle, to construct it such that the sum of the squares of the three sides shall be a *minimum*.

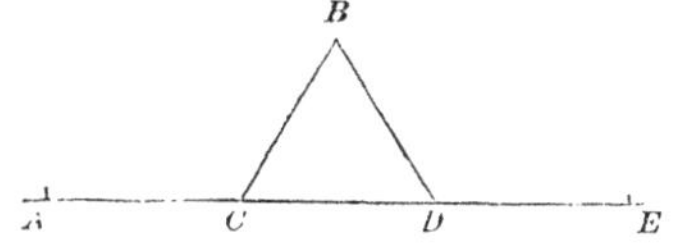

Given $\left\{\right.$ the sum $BC + CD + DB = AE$, and $BC^2 + CD^2 + DB^2$ to be the *least possible*.

Analysis.—Let BCD represent the required triangle. Produce CD both ways, and make $CA = CB$, and $DE = DB$; then $AE =$ the given perimeter, and $BC^2 + CD^2 + DB^2 = AC^2 + CD^2 + DE^2 =$ min., which, by Problem XIII., cor., will be the case when AC, CD, and DE are all equal, and each $= \frac{1}{3}\,AE$. Hence the required triangle BCD is equilateral.

PROBLEM XVI.—To divide a given line such that the sum of n times the area of the circle described on one part, added to the area of the circle described on the other part, shall be minimum.

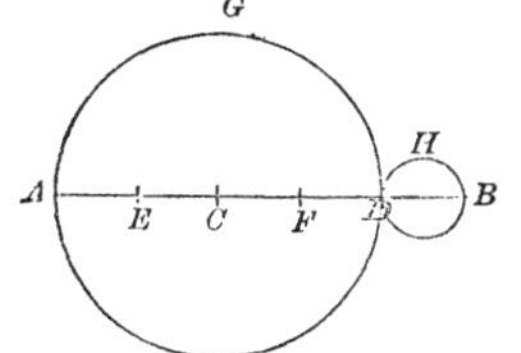

Given $\left\{\right.$ the line AB, to divide it in D so that the sum of n times the area of the circle described on DB, added to the area of the circle described on AD, shall be the *least possible*.

Analysis by the differential calculus.—Put $AB = a$, $AD = x$, and $DB = y$. Then, putting $p = .78539+$, the area of the circle on $DB = py^2$, and the area of the circle on $AD = px^2$. Hence we have $px^2 + npy^2 = \text{min.}$; that is, $x^2 + ny^2 = \text{min}$, and $x + y = a$. By differentiating, $2xdx + 2nydy = 0$, and $dx + dy = 0$, or $dy = -dx$. For dy substituting its value $-dx$ in the preceding equation, we get $2xdx - 2nydx = 0$. Whence $x = ny$, and $x + y = ny + y = (n+1)y = AB$, and $y = \frac{AB}{n+1}$.

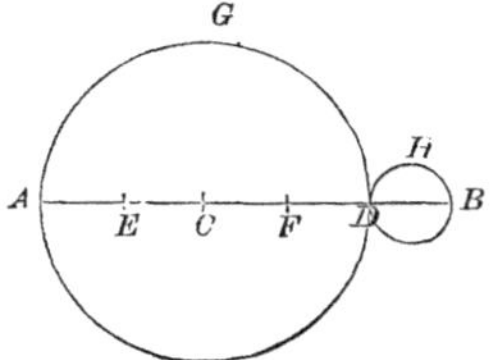

Construction.—Divide AB into $n + 1$ equal parts in the points D, F, C, etc.; then $AD = n$ times DB. On DB and AD describe the circles DHB and AGD respectively, and they will be the circles required, having the area of $AGD + n$ times the area of DHB, a minimum.

Demonstration geometrically.—$AD = nDB$, and $AD^2 = n^2DB^2$. Now, $pAD^2 + npDB^2 = \text{min.}$ Hence $AD^2 + nDB^2 = n^2DB^2 + nDB^2 = n(nDB^2 + DB^2) = \text{min.}$ And, since n is constant, $nDB^2 + DB^2 = \text{min.}$ But $nDB^2 = DF^2 + FC^2 + CE^2 + EA^2$, etc. to n terms. Hence $nDB^2 + DB^2 = DB^2 + DF^2 + FC^2 + CE^2 + \text{etc.}$; that is, is equal to the sum of the squares of the $n + 1$ parts into which the given line AB is divided, which is to be a minimum, and this, as shown in corollary to Problem XIII., will be the case when the $n + 1$ parts are all equal. Q. E. D.

PROBLEM XVII.—From two given points to draw two lines to meet in a straight line given in position, such that the sum of these two lines may be a *minimum*.

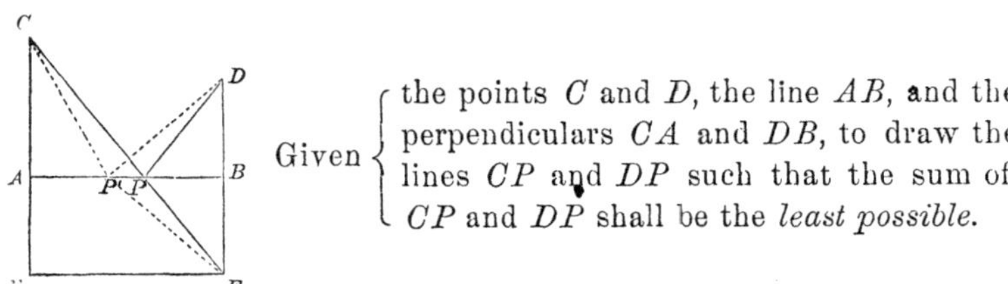

Given $\left\{\begin{array}{l}\text{the points } C \text{ and } D\text{, the line } AB\text{, and the}\\ \text{perpendiculars } CA \text{ and } DB\text{, to draw the}\\ \text{lines } CP \text{ and } DP \text{ such that the sum of}\\ CP \text{ and } DP \text{ shall be the } \textit{least possible.}\end{array}\right.$

Analysis by the differential calculus.—Suppose CP and DP to be the required lines whose sum is a minimum. Put $AC = a$, $BD = b$, $AB = c$, $AP = x$, and $BP = y$; then $CP = \sqrt{a^2 + x^2}$, and $PD = \sqrt{b^2 + y^2}$. Hence we have $\sqrt{a^2 + x^2} + \sqrt{b^2 + y^2} = \text{min.}$, and

$x + y = c$. By differentiating, we have $\frac{xdx}{\sqrt{a^2 + x^2}} + \frac{ydy}{\sqrt{b^2 + y^2}} = 0$, and $dx + dy = 0$, or $dy = -dx$. By substituting $-dx$ for its equal dy, we obtain $\frac{xdx}{\sqrt{a^2 + x^2}} - \frac{ydx}{\sqrt{b^2 + y^2}} = 0$. Whence $\frac{x}{\sqrt{a^2 + x^2}} = \frac{y}{\sqrt{b^2 + y^2}}$, or $\frac{x^2}{a^2 + x^2} = \frac{y^2}{b^2 + y^2}$. By clearing of fractions, $b^2x^2 + x^2y^2 = a^2y^2 + x^2y^2$, which gives $b^2x^2 = a^2y^2$, or $bx = ay$; that is, $a : b :: x : y$, or $AC : BD :: AP : BP$; wherefore (IV. 20*) the triangles APC and BPD are similar, having the angles APC and BPD equal, and, if the triangle BPD be revolved about BP so that D comes to F, CPF will be a straight line.

Construction.—Produce DB so that $BF = DB$; draw CPF, and join PD; then P is evidently the point required. For angle DPB = (I. 5†) angle FPB = angle APC, because they are vertical angles.

Demonstration geometrically.—We have $DP = PF$; hence $CP + DP = CP + PF = CF$. Now, in the line AB take *any other* point, as P', and join P' with the points C, D, and F; then $DP' = P'F$. Hence $CP' + DP' = CP' + P'F$, which are *greater* than the straight line CF, or than its equal $CP + PD$. Q. E. D.

Calculation.—From similar triangles CAP, DBP, we have $AC : BD :: AP : BP$. By composition, $AC + BD : AC :: AP + BP\ (AB) : AP = \frac{AC \times AB}{AC + BD}$. Also, $AC + BD : BD :: AP + BP\ (AB) : BP = \frac{BD \times AB}{AC + BD}$. Then $CP = \sqrt{AC^2 + AP^2}$, and $DP = \sqrt{BD^2 + BP^2}$.

NOTE 1.—Since the angle APC = the angle BPD, if AB were a reflecting surface, a ray of light from C, moving in the direction CP, would be reflected from P, along PD, to D. Hence the reflected ray, in passing from one point to another, goes through the shortest possible distance.

NOTE 2.—If CA be produced to meet FE, drawn parallel to AB, in E, the two last proportions in the calculation may be obtained *directly* from the similar triangles CEF, CAP, and BPD.

* VI. 6. † I. 4.

PROBLEM XVIII.—In a right line are given three points, and from one of the extreme points a perpendicular is erected. It is required to draw from the other two points two lines to meet in this perpendicular, such that the angle between them at the point where they meet shall be a *maximum.*

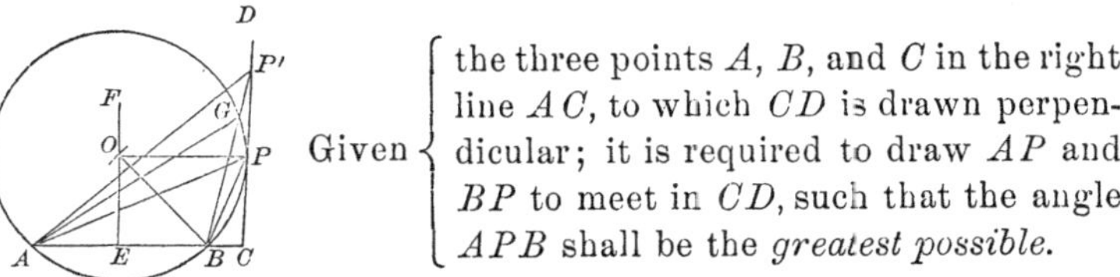

Given $\left\{\right.$ the three points A, B, and C in the right line AC, to which CD is drawn perpendicular; it is required to draw AP and BP to meet in CD, such that the angle APB shall be the *greatest possible.*

Analysis by the differential calculus.—Let P represent the required point, and join AP and BP. Put $AC = a$, $BC = b$, and $CP = x$; then tangent-angle CPB to radius unity $= \frac{b}{x}$, and tangent-angle $CPA = \frac{a}{x}$. Hence, since (Plane Trig., Art. XXV.) tang $(a - b) = \frac{R^2(\text{tang } a - \text{tang } b)}{R^2 + \text{tang } a \times \text{tang } b}$, we have, taking $R = 1$, tang APB

$$= \frac{\frac{a}{x} - \frac{b}{x}}{1 + \frac{ab}{x^2}} = \frac{a - b}{x + \frac{ab}{x}} = \text{max. by the problem.}$$

But, as the numerator $a - b$ of this fraction is constant, for the fraction to be a *maximum* its denominator $x + \frac{ab}{x}$ must be a *minimum.* Differentiating, we have $dx - \frac{abdx}{x^2} = 0$, or $x^2 = ab$; that is, $CP^2 = CA \times CB$; wherefore (IV. 30*) CP must be a *tangent to a circle passing through the points A and B.* Whence this

Construction.—On the middle of AB erect the perpendicular EF, to which apply $BO = EC$, and with centre O and radius OB or EC describe a circle, and it will evidently pass through the points A and B, and touch CD in P. Join AP and BP, and APB will be the maximum angle required. (It is evident (IV. 30†) that $CP^2 = CA \times CB$; that is, that $x^2 = ab$.)

Demonstration geometrically.—In CD take *any other* point, as P', and draw AP' and BGP', the latter cutting the circumference

* III. 37. † III. 36.

of the circle at G. Join AG; then (III. 18, cor. 1*) angle APB $=$ angle AGB. But (I. 25, cor. 6†) the angle AGB is *greater* than $AP'B$. Hence the angle APB is *greater* than $AP'B$, and therefore the angle APB is greater than *any other angle drawn as required*, or it is the *maximum angle*. Q. E. D.

Calculation.—Join OP, which is equal to EC or OB. In triangle OBE, $OE^2 = CP^2 = OB^2 - BE^2 = EC^2 - BE^2 = (EC + EB) \times (EC - EB) = CA \times CB$. In triangle CPB, Case 3, find BP, and angle CPB. In triangle CPA, Case 3, find AP, and angle CPA. Then the *maximum* angle $APB = CPA - CPB$.

PROBLEM XIX.—From two given points in a given right line to draw two lines to meet in another right line given in position, so that the angle included between these lines at the point where they meet shall be a *maximum*.

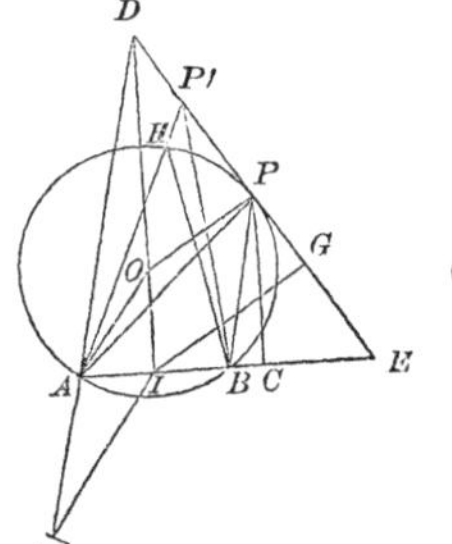

Given $\left\{\right.$ AE, BE, and the angle AED, to draw from the given points A and B the lines AP and BP to meet in ED, such that the angle APB may be the *greatest possible*.

Analysis by the differential calculus.—Let P be the required point at which the angle APB will be a maximum. Put $AE = a$, $BE = b$, $EP = x$, and angle $E = \theta$; then, letting fall the perpendicular PC, $EC = x \cos \theta$, $PC = x \sin \theta$, and $BC = b - x \cos \theta$. Also, tang CPB to radius unity $= \frac{BC}{PC} = \frac{b - x\cos\theta}{x \sin\theta}$, and tang APC $= \frac{AC}{PC} = \frac{a - x\cos\theta}{x\sin\theta}$. Hence, by Trig., Art. XXV., the tangent of

their difference, $APB = \dfrac{\dfrac{a - x\cos\theta}{x\sin\theta} - \dfrac{b - x\cos\theta}{x\sin\theta}}{1 + \dfrac{(a - x\cos\theta) \times (b - x\cos\theta)}{x^2 \sin^2\theta}} =$

* III. 21. † I. 16.

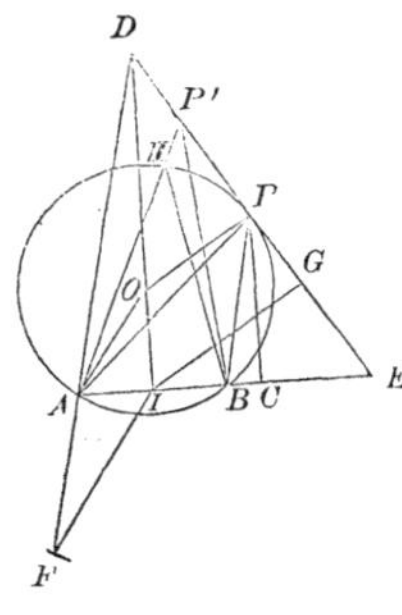

$$\frac{a - x\cos\theta - b + x\cos\theta}{x\sin\theta + \dfrac{(a - x\cos\theta)\times(b - x\cos\theta)}{x\sin\theta}} =$$

$$\frac{a - b}{x\sin\theta + \dfrac{(a - x\cos\theta)\times(b - x\cos\theta)}{x\sin\theta}} = \text{max.}$$

Now, since the numerator $a - b$ is constant, for the *fraction* to be a *maximum* its denominator must be a *minimum*. Hence $x\sin\theta + \dfrac{(a - x\cos\theta)\times(b - x\cos\theta)}{x\sin\theta} = \text{minimum}$, which reduces to $\dfrac{x^2\sin^2\theta + ab - ax\cos\theta - bx\cos\theta + x^2\cos^2\theta}{x\sin\theta} = \dfrac{x^2(\sin^2\theta + \cos^2\theta) + (ab - ax\cos\theta - bx\cos\theta)}{x\sin\theta} =$ (since $\sin^2\theta + \cos^2\theta = 1$) $\dfrac{x + \dfrac{ab}{x} - a\cos\theta - b\cos\theta}{\sin\theta} = \text{min.}$ Hence, omitting constants, $x + \dfrac{ab}{x} = \text{min.}$ By differentiating, we have $dx - \dfrac{abdx}{x^2} = 0$; whence $x^2 = ab$. That is, $EP^2 = AE \times EB$. Wherefore (IV. 30*) ED is a tangent at the point P to a circle passing through the points A and B. Whence this

Construction.—By Problem IX. of "The Circle," through the points A and B describe a circle, touching the line ED in P, thus: From the middle of AB erect the perpendicular ID; let fall the perpendicular IG; join DA, and to it, produced, apply $IF = IG$; draw AO parallel to IF, and OP parallel to IG; then $OP = OA =$ the required radius. For, by similar triangles, we have $DI : IF$ or $IG :: DO : OA$ or OP; whence $OP = OA$ or OB, and a circle described with O as a centre, with the radius OA, will pass through B, and touch ED in P.

Demonstration geometrically.—In ED take *any other* point, as P'. Draw BP' and AHP', the latter cutting the circumference of the circle at H. Join BH; then (III. 18, cor. 1†) angle $APB = AHB$. But (I. 25, cor. 6‡) angle AHB is *greater* than $AP'B$. Hence angle APB is greater than $AP'B$, and APB is the maximum angle required. Q. E. D.

Calculation.—By IV. 30,§ we have $EP = \sqrt{AE \times EB}$. Then, Case 1, find EC and PC, thence BC, AC, and angles BPC, APC,

* III. 37. † III. 21. ‡ I. 16. § III. 36.

when we have angle $APB = APC - BPC =$ maximum angle required.

PROBLEM XX.—Given, the perpendicular height and the vertical angle of a plane triangle, to determine it, such that the base, and consequently the area, shall be a *minimum*.

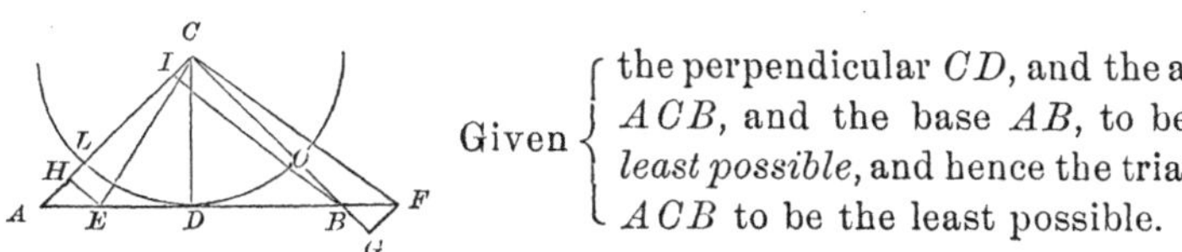

Given $\left\{\begin{array}{l}\text{the perpendicular } CD\text{, and the angle}\\ ACB\text{, and the base } AB\text{, to be the}\\ \textit{least possible}\text{, and hence the triangle}\\ ACB \text{ to be the least possible.}\end{array}\right.$

Analysis by the differential calculus.—Let ABC represent the required triangle. Put the given angle $ACB = \theta$, $CD = a$, $AC = x$, and $BC = y$. On AC, or AC produced, let fall the perpendicular BI; then $CI = CB \cos\theta = y\cos\theta$. Now (IV. 12*), $AB^2 = AC^2 + BC^2 - 2\,AC \times CI = x^2 + y^2 - 2\cos\theta\, xy =$ minimum by the problem. Also, $xy\sin\theta =$ twice the area of $ABC = AB \times CD = a\sqrt{x^2 + y^2 - 2\cos\theta\, xy}$. Hence, since the quantity under the radical sign is a minimum, being equal to AB, its differential is equal to 0; consequently, a times that differential is 0, and therefore the differential of its equal $xy\sin\theta$, or of xy, $= 0$. Putting the differential of $x^2 + y^2 - 2\cos\theta\, xy$ and of $xy = 0$, we have $2xdx + 2ydy - 2\cos\theta\, xdy - 2\cos\theta\, ydx = 0$, and $xdy + ydx = 0$, or $dy = -\frac{ydx}{x}$.

For dy putting its equal $-\frac{ydx}{x}$ in the preceding differential equation, we have $2xdx - \frac{2y^2dx}{x} + \frac{2\cos\theta\, xydx}{x} - 2\cos\theta\, ydx = 0$; whence $x^2 = y^2$, or $x = y$. That is, for AB to be a minimum, AC and CB must be equal. Whence this

Construction.—Draw CD bisecting the given angle ACB, and equal to the given perpendicular. Through D, perpendicular to CD, draw the base ADB; then will $AC = CB$, and ACB will be the required triangle.

Demonstration geometrically.—Take *any other* triangle, as ECF, having angle $ECF = ACB$. On CA and CB, produced if necessary, let fall the perpendiculars EH and FG. Then, since angle $ECF = ACB$, taking ECB from each, we have angle $ECH = FCG$.

* II. 12 and 13.

Hence the triangles ECH and FCG are similar. Also, the triangles EAH and FBG are similar. But, since $CB = CA$, CG is greater than CH; hence FG is greater than EH, and FB greater than EA. Add EB to each, and we have EF greater than AB. Hence AB is the *shortest line possible*. Q. E. D.

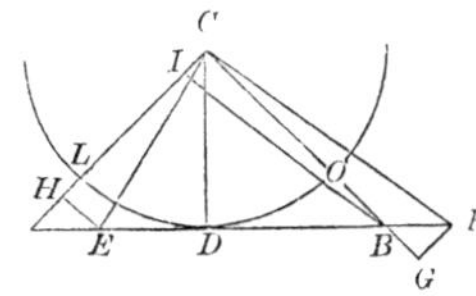

Calculation.—In triangle ACD, Case 1, find AD and AC; then $BC = AC$, and $AB = 2\,AD$.

Corollary 1.—With centre C and radius CD describe the arc LDO. Then, to draw the *minimum* tangent to the arc LDO, which shall have its extremities in the radii CL and CO produced, it must touch the arc LDO at its middle point D, when AD will be equal to DB, and CA to CB.

Corollary 2.—If ACB is a right angle, then AD is the tangent of ACD, and BD is the co-tangent of ACD. Hence, to find an arc the sum of whose tangent and co-tangent shall be a minimum, we see that the tangent and co-tangent must be equal, and hence the arc be half a quadrant, or 45°.

PROBLEM XXI.—The vertical angle of a triangle is given in magnitude, and the vertex of the triangle is in the circumference of a circle given in magnitude and position; also, the base of the triangle is in a straight line given in position, and one extremity of the base is given in position. It is required to determine the triangle when its base is a *minimum*.*

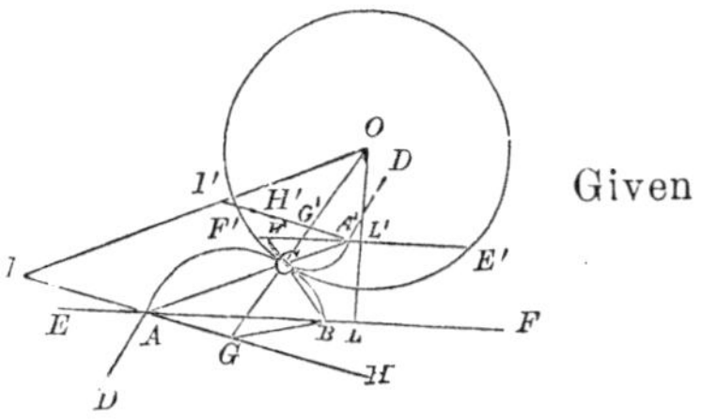

Given { the angle ACB, the vertex C to be in the circumference of the given circle O—$F'E'$, the position of the line $EABF$, and the point A, to construct the triangle ACB, such that the base AB shall be a *minimum*.

Analysis.—Let $F'CE'$ be the given circle, whose centre is O,

* This question was proposed as a "Prize Problem" in the fourth number of the *Mathematical Diary*, edited by Robert Adrain, in 1825, and the above solution was given at the time by the author of this work, and published the next month in No. 5, page 132, and it shared the prize with the late Dr. Charles Farquhar, then of Alexandria, D.C., and J. H. Swale, of Liverpool, England. Dr. Bowditch and Prof. Strong were correspondents of the same mathematical periodical.

EF* the line given in position, A the given point in EF to be one extremity of the base, and FAD equal to the given vertical angle. Draw AH and OL at right angles to AD and EF respectively. Then it is evident (III. Prob. 16†) that the centre of the circle circumscribing the required triangle will lie in AH. In HA, produced, lay AI (in a direction *from* the perpendicular OL) equal to the radius of the given circle. Join IO, and draw AC parallel thereto. Also, draw OC, and produce it till it meets AH in G; then (IV. 15‡), since $AI = OC$, the radius of the given circle, we have $GA = GC$. Hence, with the centre G and radius $= GA$ or GC, describe the segment ACB, and join CB; then will ACB be the triangle required, having its base AB a *minimum*.

Demonstration.—By III. Prob. 16,§ the angle $ACB = BAD =$ the given vertical angle. Now, since the radius of *any other* circular segment whose centre is in AH, to pass through the point A (as must be to make the contained angle equal to FAD) and meet the given circle, must necessarily be greater than GA, it is evident that AB is the *least possible* base. Q. E. D.

Calculation.—Since the position of the line EF, and of the point A, are given, if we join OA, the triangle OAL is given in all its parts. Hence, angle $OAG = OAL + LAG$. In triangle OAG, find angles, and $GA = GC$. In triangle AGB, Case 1, find AB; then (III. 18‖) angle $ABC = \frac{1}{2} AGO$. In triangle ABC, Case 1, find AC and BC.

Corollary.—If the point A coincides with L, AI may be laid either way, and there will then be two similar and equal triangles formed with the *shortest possible* bases.

PROBLEM XXII.—Having given the two parallel sides of a trapezoid, and one diagonal, to construct it such that its area may be a *maximum*. (See figures to scholium under Problem LXIV., "Triangles, Quadrilaterals, and Parallels.")

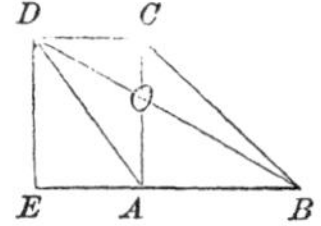

Given $\begin{cases}\text{the parallel sides } AB \text{ and } CD \text{ of the trapezoid } ABCD\text{, and the diagonal } AC\text{, to}\\ \text{construct it such that its area shall be the } \textit{greatest possible.}\end{cases}$

Analysis.—The trapezoid is made up of the two triangles ACB

* If the corresponding marked or "prime" letters are used, the same construction, demonstration, and calculation apply when the given line and base are taken *within* the circle.

† III. 33. ‡ VI. 2. § III. 33. ‖ III. 20.

and ACD, the bases AB and CD of which are given, and hence constant quantities. Therefore, the greater the perpendicular heights of these triangles, the greater will be their areas, and, hence, the area of the trapezoid. But, evidently, their altitudes will be greatest when the diagonal AC is perpendicular to AB, and, consequently, to DC. Whence this

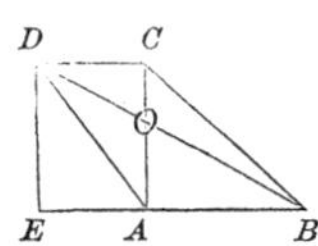

Construction.—In any straight line lay BA = one of the given sides, and AE = the length of the other side. Erect the perpendicular AC = the given diagonal. Draw CD parallel and equal to AE, and join DE, DA, BD, and BC; then $ABCD$ will evidently be the *maximum* trapezoid required.

Calculation.—The area of $ABCD = \frac{1}{2} AC \times (AB + CD)$. The side $CB = \sqrt{AB^2 + AC^2}$, the side $AD = \sqrt{CD^2 + AC^2}$, $BE = BA + AE$, and $BD = \sqrt{BE^2 + ED^2} = \sqrt{BE^2 + AC^2}$; whence the angles may be found at pleasure.

Corollary.—The triangle DEB = triangle $CEB = \frac{1}{2}(AB + AE) \times AC = \frac{1}{2}(AB + CD) \times AC$ = the area of the trapezoid $ABCD$.

NOTE.—See note to Problem VI. of this division, as alike applicable to these two problems.

PROBLEMS

IN RELATION TO

AREAS AND THE DIVISION OF SURFACES.

PROBLEM I.—To find the area of a triangle when two sides and the included angle are given.

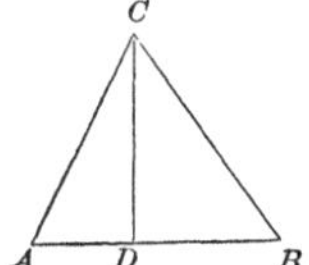

Given $\begin{cases}\text{the sides } AB \text{ and } AC, \text{ and the included} \\ \text{angle } BAC.\end{cases}$

Calculation.—Let fall the perpendicular CD. In triangle ACD, Case 1, find CD; then the area of $ABC = \frac{1}{2}(AB \times CD)$. In practice, the following method is convenient:

Sin D or rad. : sin A : : AC : CD : : (multiplying the last couplet by AB) $AB \times AC : AB \times CD = 2$ area; that is, $2\,ABC = \dfrac{\sin A \times AB \times AC}{\text{rad.}}$, or *rad.* : *sin included angle* : : *product of the sides including that angle* : *double the area of the triangle.*

PROBLEM II.—In a triangle the area, one angle, and a side adjacent to that angle are given, to find the other adjacent side; that is, in the above figure, are given the area of ABC, the angle BAC, and the side AB, to find AC.

Calculation.—By the last problem we have rad. : sin A : : $AB \times AC$: 2 area. Hence rad. $\times$ 2 area $=$ sin $A \times AB \times AC$, where all the terms are known but AC, which is determined by converting this equation into a proportion so as to have AC for the fourth term, thus:—sin $A \times AB$: rad. : : 2 area : AC; that is, as *sin given angle multiplied by the given side* : *rad.* : : 2 *area* : *the other side adjacent to the given angle.*

NOTE 1.—The double area of a triangle being equal to the product of the base by the altitude or perpendicular height, the altitude will be found by reducing the double area to the same denomination as the base and dividing the result by the base. Hence, in the triangle ABC the perpendicular $CD =$ 2 area ABC divided by AB, and the figure for this problem is readily constructed.

NOTE 2.—Since the area of a parallelogram is double the area of a triangle of the same base and altitude, we have *sin given angle* $\times$ *given side* : *rad.* : : *area of parallelogram* : *the other side of the parallelogram adjacent to the given angle.*

NOTE 3.—The perpendicular height of a parallelogram is equal to its area divided by the base, the area and base being reduced to their respective *units of the same name.*

PROBLEM III.—Having given the angles and one side of a triangle, to find the area.

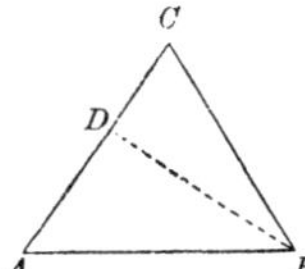

Given $\begin{cases}\text{in the triangle } ABC, \text{ all the angles and the} \\ \text{side } AB, \text{ to find the area.}\end{cases}$

Calculation.—Let fall the perpendicular BD from one extremity of the *given* side on one of the other sides, as AC; then we have

$$\sin C : \sin B : : AB : AC.$$

$$\text{Also, } \sin D \text{ (rad.)} : \sin A : : AB : BD.$$

By multiplying the corresponding terms of these proportions, we have rad. $\times$ sin C : sin A $\times$ sin B : : AB^2 : $AC \times BD =$ the double area of the triangle ABC. That is, *the product of rad.* $\times$ *sin angle opposite to any side* : *the product of the sines of the other two angles* : : *the square of that side* : *double the area of the triangle.*

We have the three proportions,

$$\text{rad.} \times \sin C : \sin A \times \sin B : : AB^2 : 2 \text{ area } ABC. \quad (1.)$$
$$\text{rad.} \times \sin B : \sin A \times \sin C : : AC^2 : 2 \text{ area } ABC. \quad (2.)$$
$$\text{rad.} \times \sin A : \sin B \times \sin C : : BC^2 : 2 \text{ area } ABC. \quad (3.)$$

Corollary.—The fourth term in each of these proportions is evidently the area of a parallelogram constructed with the sides AB and AC and the included angle A.

PROBLEM IV.—Having given the angles and area of a triangle, to find a side. (See above figure.)

Calculation.—Take the proportions in the preceding problem by inversion. To find any side, take the sines of the angles adjacent to that side for the first term. Thus:

To find AB, we have

$$\sin A \times \sin B : \text{rad.} \times \sin C :: 2 \text{ area } ABC : AB^2.$$

To find AC, we have

$$\sin A \times \sin C : \text{rad.} \times \sin B :: 2 \text{ area } ABC : AC^2.$$

To find BC, we have

$$\sin B \times \sin C : \text{rad.} \times \sin A :: 2 \text{ area } ABC : BC^2.$$

NOTE.—The area will always be in the same denomination as the side, and the side as the area. That is, if the *area* is in square chains, perches, yards, or feet, the *side* will be in linear chains, perches, yards, or feet; and, if the *side* is in linear chains, perches, yards, or feet, the *area* will be in square chains, perches, yards, or feet.

PROBLEM V.—Having given the angles and area of a plane triangle, to construct it.

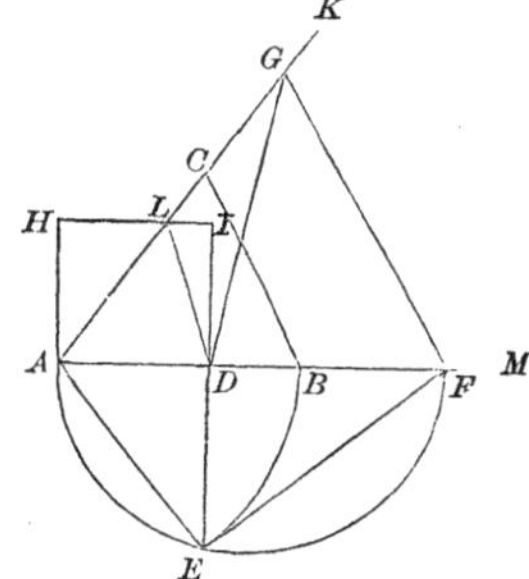

Given { the angles A, B, and C, and the area of the triangle ABC, to construct it.

Analysis and construction.—Draw any line AM, and make the angle MAK = the given angle A. Let the given area, reduced if necessary, be in chains, perches, or some denomination of which there is a *linear unit,* and make AD = the *square root* of the given area. On AD describe the square $ADIH$, which will then contain the given area. Let the side HI, produced if necessary, cut AK in L, and join LD. Then the triangle ALD, being equal to half the square (IV. 2, cor. 1*), contains *half* the given area. Lay LG = AL, and join GD; then, since the bases AL and LG are equal, the triangles ADL and LDG are equal, and $ADG = 2\ ADL$ = the square $ADIH$ = the given area. At G make the angle AGF = the given angle C; then the triangle AGF will be similar to the required triangle. Now, we have to construct a triangle *similar* to one

* VI. 1.

triangle, and *equal* in area to another, both having the same altitude; that is, to construct ABC *similar* to AFG, and *equivalent* to ADG (IV. Prob. 15*). To do which, find a mean proportional between their bases AF and AD by describing a semicircle upon the greater base AF, at D, the end of the less base AD, erecting a perpendicular to meet the semicircle in E, and joining AE, which will be the mean proportional required (IV. 23†). Lay $AB = AE$, and draw BC parallel to GF; then ABC will be *similar* to AFG, and *equivalent* to ADG.

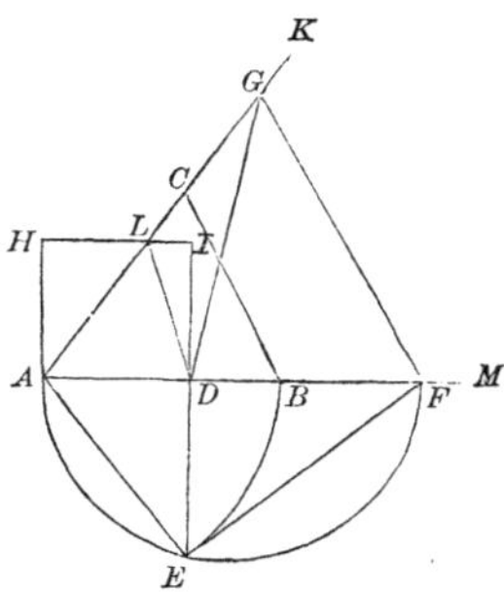

Demonstration.—By IV. 25,‡ we have $AFG : ABC :: AF^2 : AB^2 :: AF^2 : AE^2 :: AF^2 : AF \times AD :: AF : AD :: AFG : ADG$ (IV. 6, cor.§). That is, $AFG : ABC :: AFG : ADG$. Whence $ABC = ADG = 2\ ADL =$ the square $ADIH =$ the given area by construction.

Calculation.—By Prob. IV. A., that is, the last problem, find the sides.

Scholium.—Since (IV. 23, cor.,‖ and the above demonstration) we always have $AF^2 : AE^2 :: AF : AD$, $AF^2 : FE^2 :: AF : FD$, and $AE^2 : EF^2 :: AD : DF$, and similar figures being to each other as the squares of their homologous sides, we have $AF : AD ::$ fig. on AF : sim. fig. on AE or its equal AB, and $AF : FD ::$ fig. on AF : sim. fig. on FE or its equal, and $AD : DF ::$ fig. on AE : sim. fig. on EF or their equals.

Remark.—The preceding five problems are *fundamental problems* in areas, and the principles involved in them are in so frequent requisition that the student will find it to his advantage to be *familiar with them.* In solving the following problems they will be referred to, when used, as Prob. I. A., Prob. II. A., Prob. III. A., etc.; that is, Problem I. in Division on "Areas," Problem II. in Division on "Areas," etc.

* VI. 25. † VI. 8, cor. ‡ VI. 19. § VI. 1. ‖ VI. 8, cor.

PROBLEM VI.—To bisect a *given triangle* by a line drawn from a *given point* in one of the sides.

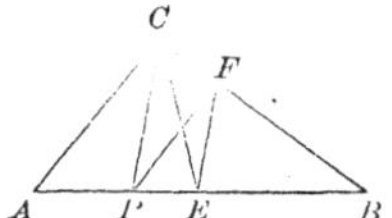

Given $\begin{cases} \text{the triangle } ABC\text{, and the distance } BP\text{,} \\ \text{to draw the line } PF \text{ so that } PFB \text{ shall} \\ \text{be equivalent to } ACFP. \end{cases}$

Analysis.—Suppose PF to be the required line dividing the triangle ABC into two equal parts. Join CP; bisect AB in E, and join CE; then, since $BE = \frac{1}{2}AB$, the triangle $BCE = \frac{1}{2}BCA = BPF$. Take triangle BEF from the first and last of these equals, and we have the triangles EPF and ECF equal; hence (IV. 2, cor. 2*) EF is parallel to CP. Whence this

Construction.—Bisect AB in E; join CP; draw EF parallel to CP, and join PF and CE; then PF will be the division line required.

Demonstration.—By IV. 2, cor. 2,† the triangles EPF and ECF are equal. To each add BEF, and we have $BPF = BEC = \frac{1}{2}BAC$. Q. E. D.

Calculation.—By parallel lines, $BP : BE :: BC : BF$; then $CF = BC - BF$, and $BA : AC :: BP : PF$.

Limits.—If $BP = BE = \frac{1}{2}AB$, the line CP will coincide with CE and be the bisecting line. If BP were less than BE, the point would fall on the line AC, but the *method* of construction and calculation would be the same.

PROBLEM VII.—Having a square given, it is required to construct a rectangle which shall have the same perimeter as the square, and contain half the area.

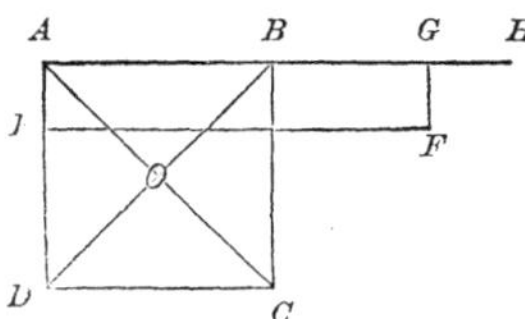

Given $\begin{cases} \text{the square } ABCD\text{, to construct} \\ \text{the rectangle } AEFG \text{ to be equal} \\ \textit{half } ABCD\text{, and have the sides} \\ AG + GF = AB + BC. \end{cases}$

Analysis.—Let $ABCD$ be the given square, of which the diagonals are AC and BD, and $AEFG$ the required rectangle. In AG, produced, take $GH = GF$; then $AH =$ half of the perimeter

* I. 39. † I. 37.

of the rectangle = (by the problem) half the perimeter of the square $= AB + BC$; hence $BH = BC$. Again, the area of $AEFG = AG \times GF = AG \times GH = (AB + BG) \times (AB - BG) = AB^2 - BG^2 =$ (by the problem) $\frac{1}{2} AB^2 = BO^2$ or AO^2. Wherefore $BG^2 = BO^2$, or $BG = BO$; and we have this

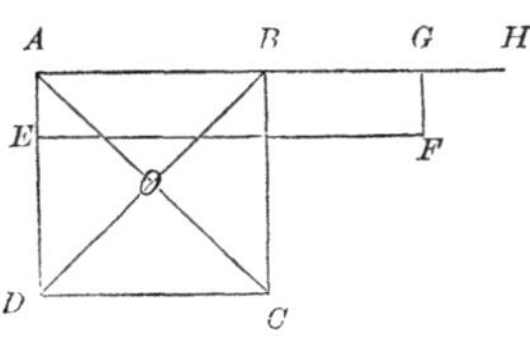

Construction.—Having drawn the diagonals AC and BD to the given square, intersecting in O, produce the side AB, making $BH = AB$; then $AH = \frac{1}{2}$ the perimeter of the required rectangle. On BH lay $BG = BO$, erect the perpendicular $GF = GH$, and draw FE parallel to AH; then $AEFG$ will be the rectangle required.

Demonstration.—We have $AG + GF = AG + GH = AH = AB + BC$. Hence the perimeters are equal. Also, area $AEFG = AG \times GF = AG \times GH = (AB + BG) \times (AB - BG) = AB^2 - BG^2 = AB^2 - BO^2 = AO^2 = \frac{1}{2} AB^2$. Q. E. D.

Problem VIII.—Having given the position of three consecutive zigzag lines, and the lengths of the first two, to draw a line from the starting-point to the third line so as to cut off equal triangles in the two given angles.

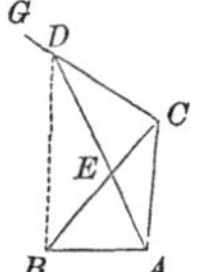

Given $\left\{\begin{array}{l}\text{the angles } ABC \text{ and } BCG\text{, and the lengths of the} \\ \text{sides } AB \text{ and } BC\text{, to draw the line } AED \text{ to make} \\ \text{the triangle } ABE = ECD.\end{array}\right.$

Analysis.—Suppose the line AED so drawn as to make the triangles ABE and ECD equal. Join AC. Add the triangle AEC to each of these equals, and we have the triangles ABC and ADC equal. Whence (IV. 2, cor. 2*) BD is parallel to AC; and we have this

Construction.—Having drawn AB, BC, and CG as given, and joined AC, draw BD parallel to AC, meeting the third side CG in D. Join AD, which will be the required line.

Demonstration.—By IV. 2, cor. 2,* the triangles ABC and ADC are equal. Take AEC from each, and we have the triangles ABE and ECD equal. Q. E. D.

* I. 37.

Calculation.—1. In triangle ABC, Case 3, find AC, and angle ACB.

2. In triangle CBD, Case 1, find CD and BD. Angle $ABD = ABC + CBD$.

3. In triangle ABD, Case 3, find AD, and angle ADB.

4. In triangle BED, Case 1, find DE and EB. Then $AE = AD - DE$, $CE = BC - BE$; and we know every line and every angle, and can readily find the areas of the equal triangles AEB and CED by Prob. I. A. or II. A.

NOTE.—When AB and CG are parallel, CD must equal AB, and $AD = CB$, and the triangles AEB and CED will be equal in all their parts.

PROBLEM IX.—To divide a given triangle into *any number* of equal parts by lines parallel to one of the sides.

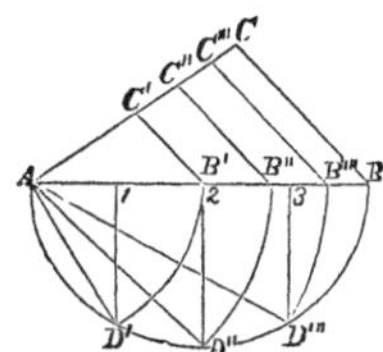

Given $\begin{cases}\text{the triangle } ABC\text{, to divide it into any}\\ \text{number of equal parts, say four.}\end{cases}$

Construction.—In accordance with Problem V. A., scholium, divide AB into the same number of equal parts that the triangle is to be divided into, in the points 1, 2, 3, etc., and at these points erect perpendiculars to meet a semicircle described on AB, in the points D', D'', D''', etc. Join AD', AD'', AD''', etc., and make $AB' = AD'$, $AB'' = AD''$, etc., and parallel to BC draw $B'C'$, $B''C''$, etc., and they will divide ABC as required.

Demonstration.—By scholium to Prob. V. A., $ABC : AB'C'$, $AB''C''$, etc. : : $AB : A1$, $A2$, etc.; hence the parts $AB'C'$, $B'C'C''B''$, $B''C''C'''B'''$, etc. are all equal. Q. E. D.

Calculation. — $AB' = AD' = \sqrt{AB \times A1}$, $AB'' = AD'' = \sqrt{AB \times A2}$, etc. Also, $AB : AB' :: BC : B'C'$, and $AB : AB'' :: BC : B''C''$, etc.

NOTE.—If it is desired to have the parts of the triangle in *any given ratio*, divide AB in *that ratio*, and proceed as above.

Problem X.—To divide a given circle into any number of given concentric rings.

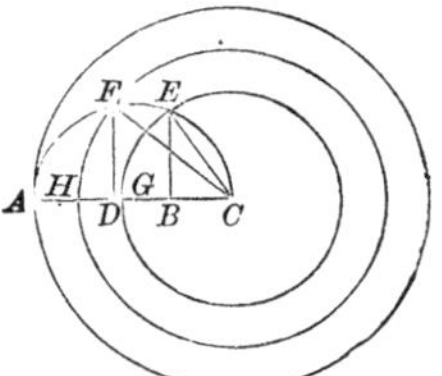

Given { the circle whose radius is CA, to divide it into *any number* of concentric rings whose radii are CG, CH, etc.

Construction.—Divide CA into the same number of equal parts at the points B, D, etc. that the given circle is to be divided into, and at B, D, etc. erect perpendiculars to meet the semicircle on the radius CA, in E, F, etc. Join CE, CF, etc., and with these lines as radii, and centre C, describe circles, and they will divide the given circle as required, as is evident from scholium to Prob. V. A., and the preceding problem.

Note.—If it is desired to have the *rings* in a *given ratio*, divide the *radius* in *that ratio*, and proceed as above.

Problem XI.—In a quadrilateral figure there are given the four angles, and one pair of the opposite sides, to find the area.

Fig. 1.

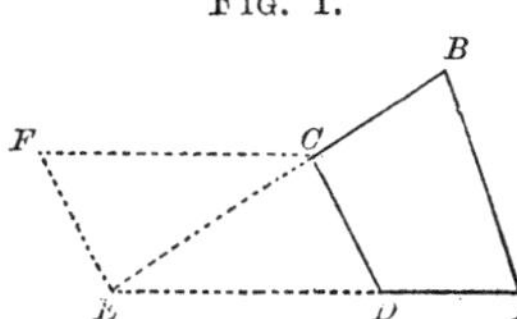

Given, the angles A, B, C, and D, and the sides AB and CD, to find the area.

Fig. 2.

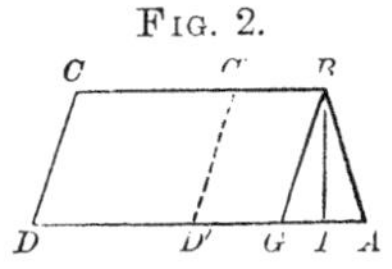

Analysis.—Let $ABCD$ (Fig. 1) represent the quadrilateral figure, of which AB and CD are the *known* sides. Produce the *unknown* sides BC and AD till they meet, as in E. Draw EF parallel to DC, and CF parallel to DE; then $EF = DC$, and the angle AEF = the given angle ADC. Whence this

Construction.—Draw AB = the *longer* given side, and make the angles EAB and ABE = the given angles A and B respectively. At the point E where these lines meet, make the angle AEF = the given angle D, and lay EF equal to the shorter given side. Draw FC parallel to AE, and CD parallel to EF, and $ABCD$ will be the quadrilateral required.

The *demonstration* is evident from the analysis.

Calculation.—By Prob. III. A., in triangle AEB, we have rad. $\times \sin E : \sin A \times \sin B :: AB^2 : 2$ area ABE, and in triangle DEC, rad. $\times \sin E : \sin D \times \sin C :: DC^2 : 2$ area DEC. *Half the difference* of these two results gives the area of $ABCD$, in the same denomination as the sides AB and CD.

If the lengths of the sides AD and BC are desired, in the triangle AEB, Case 1, find AE and BE; and in triangle DEC, Case 1, find DE and CE; then $AD = AE - DE$, and $BC = BE - CE$.

Limits.—If the two given sides AB and CD are parallel, as in Fig. 2, the problem is *unlimited;* for, if we draw *any line*, as $C'D'$, parallel to CD, the quadrilaterals $ABC'D'$ and $ABCD$ will be equiangular, and have the sides AB and $C'D'$ respectively equal to AB and CD, and hence *either of them* will fulfil the conditions of the problem.

PROBLEM XII.—In a quadrilateral are given two angles, and the three including sides, to find the area. (See above figure, where the angles A and B and the sides DA, AB, and BC are given.)

Construction.—Draw $AB =$ to the *second* given side, and make the angles A and $B =$ the given angles. Lay $AD =$ the *first* given side, and $BC =$ the *third*, and join CD, and $ABCD$ will evidently be the figure required.

Calculation.—In triangle ABE, by Prob. III. A. and Case 1, find double area, and sides AE and BE; then $ED = AE - DA$, and $EC = BE - BC$. Also, in triangle ECD, Prob. I. A., find double area ECD; then $ABCD = \frac{1}{2}(2\,ABE - 2\,ECD)$.

Limit.—When BC and AD are parallel, as in Figure 2, let fall the perpendicular BI, and in triangle ABI, Case 1, find BI; then area $ABCD = \frac{1}{2}BI \times (AD + BC)$.

PROBLEM XIII.—Having given the position of three consecutive zigzag lines, and the length of the second one, to draw a line making a given angle with the first side and cutting off equal triangles in the two given angles.

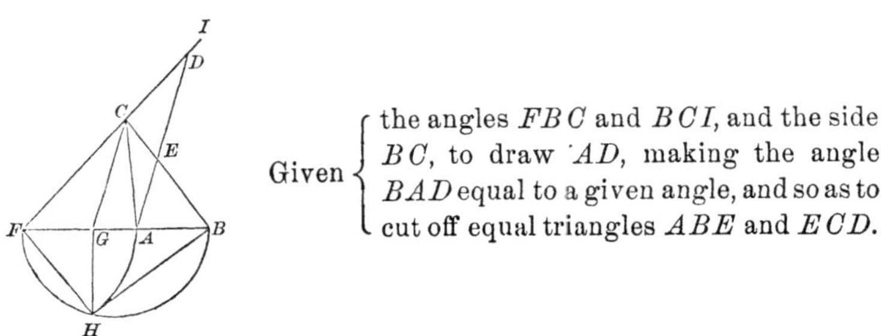

Given $\left\{ \begin{array}{l} \text{the angles } FBC \text{ and } BCI, \text{ and the side} \\ BC, \text{ to draw } AD, \text{ making the angle} \\ BAD \text{ equal to a given angle, and so as to} \\ \text{cut off equal triangles } ABE \text{ and } ECD. \end{array} \right.$

Analysis.—Suppose AD to be the required line, making $BAD =$ the given angle and cutting off the equal triangles ABE and ECD. Produce the first and third sides, BA and DC, to meet in F. To the equal triangles ABE and ECD, add the figure $AECF$, and we have the triangles FCB and FDA equal. Draw CG parallel to DA; then the triangle FCG is similar to FDA. Now, we must construct the triangle FDA similar to FCG, and equal to FCB, they having the same altitude.

Construction.—(See Prob. V. A.) Having drawn the three sides FB, BC, and CI, making BC its given length, and FBC and BCI $=$ the given angles, and produced the sides BF and IC to meet in F, draw CG (Prob. XVIII. "Triangles," etc.), making the angle $BGC =$ the given angle BAD; find a mean proportional FH between the bases FB and FG; make $FA = FH$, and draw AD parallel to CG, which will be the line required.

Demonstration.—The angle $BAD = BGC =$ the given angle. By Prob. V. A., the triangles FDA and FBC are equal. Take the quadrilateral $AECF$ from each, and we have the triangles ABE and ECD equal. Q. E. D.

Calculation.—1. In triangle FBC, Case 1, find FB and FC. In triangle FGC, Case 1, find FG and GC; then $FA = FH = \sqrt{FB \times FG}$, and $AB = FB - FA$.

2. In similar triangles FGC and FAD, we have $FG : FA :: FC : FD$; then $CD = FD - FC$. Also, $FG : FA :: GC : AD$. Now, in triangle ABE, Case 1, find AE and EB, and by Prob. III. A., the area of $ABE =$ area CED; then $CE = BC - EB$, and $DE =$

$AD - AE$, and we know all the sides and angles, and the area of each triangle.

NOTE.—When CI is parallel to FB, bisect BC in E, and through E draw AED (Prob. XVIII. "Triangles," etc.), making the angle $BAD =$ the given angle; then BA will be equal to CD, and the triangles ABE and ECD equal in every respect.

PROBLEM XIV.—To divide a given triangle into two parts having a given ratio to each other, by a line drawn parallel to one of the sides.

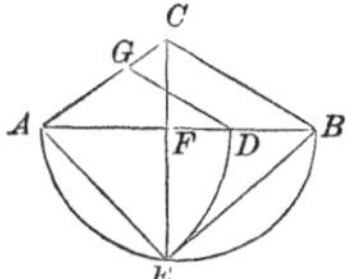

Given $\begin{cases} \text{the sides and angles of the triangle } ABC, \\ \text{to draw } DG \text{ parallel to } BC, \text{ so that } ADG : \\ BDGC :: m : n. \end{cases}$

Analysis.—Divide AB in F so that AF may be to FB in the ratio of $m : n$ (IV. Prob. 1*). Join CF; then ACF and BCF are to each other as m to n, and hence $AGD = ACF$, and $BDGC = BCF$. Wherefore we have to construct a triangle AGD *similar* to ACB and *equal* to ACF, they having the same altitude.

Construction.—(See Prob. V. A.) Find AE a mean proportional between the bases AB and AF, make $AD = AE$, and draw DG parallel to BC, and it will be the division line required.

Demonstration.—By IV. 25,† $ABC : ADG :: AB^2 : AD^2 :: AB^2 : AB \times AF :: AB : AF :: ABC : ACF$. Hence $ADG = ACF$. Take each from ACB, and we have $BDGC = BCF$. Therefore we have $ADG : BDGC : ACF : BCF :: AF : FB :: m : n$ by construction. Q. E. D.

Calculation.—By construction, $m : n :: AF : FB$. By composition, $m + n : m :: (AF + FB) AB : AF = \frac{m}{m+n} \times AB$. Also, $m + n : n :: AB : BF = \frac{n}{m+n} \times AB$. And $AD = AE = \sqrt{AB \times AF} = AB\sqrt{\frac{m}{m+n}}$. Then $BD = AB - AD$, and $AB : AC :: AD : AG$, $AB : BC :: AD : DG$, and $CG = AC - AG$. Now, area $ADG = \frac{m}{m+n} \times$ area ABC, and area $BDGC = \frac{n}{m+n} \times$ area ABC.

Scholium.—If on AB, BD were laid equal to BE, and DG drawn

* VI. 10. † VI. 19.

parallel to AC, the triangle would be divided in the given ratio by a line *parallel to* AC.

If the dividing line is to be parallel to AB, the semicircle must be described on one of the other sides, AC or BC.

When the triangle is to be divided into *equal* parts, AB must be *bisected in* F.

PROBLEM XV.—To divide a given triangle into two parts which shall have a given ratio to each other by a line which shall make a given angle with the base.

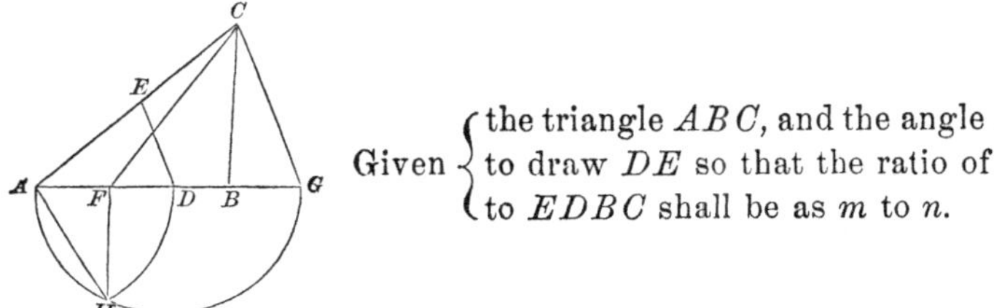

Given $\begin{cases} \text{the triangle } ABC\text{, and the angle } ADE, \\ \text{to draw } DE \text{ so that the ratio of } ADE \\ \text{to } EDBC \text{ shall be as } m \text{ to } n. \end{cases}$

Analysis.—Suppose DE to make the given angle with AB, and to divide the triangle ABC so that $ADE : EDBC :: m : n$. Divide the base AB in F so that $AF : FB :: m : n$, and join CF; then (IV. 6, cor.*) $ACF : FCB :: AF : FB :: m : n$. Hence ACF and FCB are equivalent to the required areas ADE and $EDBC$. Parallel to ED draw CG; then angle $AGC =$ angle ADE, and we have to construct a triangle similar to ACG and equal to ACF, both having the same altitude, by Prob. V. A., thus:

Construction.—Through C (Prob. XVIII., "Triangles," etc.) draw CG, making with AB, produced if necessary, the angle AGC = the given angle ADE. Make $AD = AH =$ a mean proportional between the bases AG and AF, and draw DE parallel to CG, and it will be the division line required.

Demonstration.—By I. 20, cor. 3,† the angle $ADE =$ angle AGC = the given angle by construction. By Prob. V. A., $ACG : AED :: AG^2 : AD^2 :: AG^2 : AG \times AF :: AG : AF :: ACG : ACF$. Hence $AED = ACF$. Taking each from ACB, we have $EDBC = FCB$. Wherefore $AED : EDBC :: ACF : FCB :: AF : FB :: m : n$ by construction. Q. E. D.

Calculation.—We have $AF = \frac{m}{m+n} \times AB$, and $FB = \frac{n}{m+n} \times$

* VI. 1. † I. 29.

AB. Also, in triangle ACG, Case 1, find AG; then $AD = AH = \sqrt{AG \times AF}$, and $DB = AB - AD$. In triangle ADE, Case 1, find AE and ED; then $EC = AC - AE$; area $AED = \frac{m}{m+n} \times$ area ACB, and area $EDCB = \frac{n}{m+n} \times$ area ACB.

Limit.—AH must be less than AB.

PROBLEM XVI.—One side and the two adjacent angles of a quadrilateral figure being given, to lay off a given area by a line which shall make given angles with the two unknown sides.

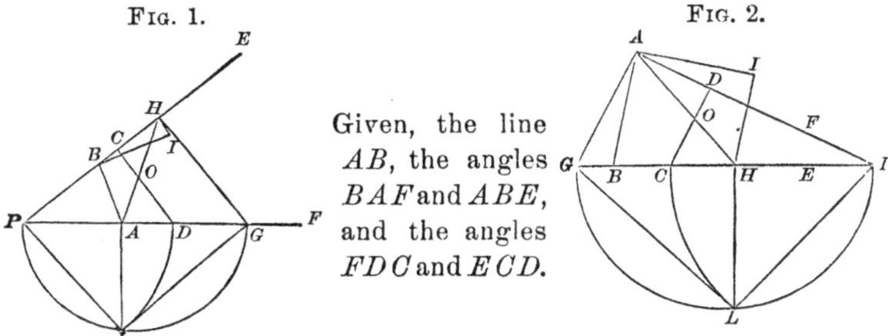
FIG. 1. FIG. 2.

Given, the line AB, the angles BAF and ABE, and the angles FDC and ECD.

Analysis.—Produce the lines EB and FA, either way, to meet in P. Divide twice the given area, reduced to the same denomination as AB, by AB, and on AB erect the perpendicular BI, in Fig. 1, and AI, in Fig. 2, equal to the quotient; draw IH parallel to AB, and join HA; then (IV. 2*) the triangle ABH contains the given area, and is equal to the quadrilateral $ABCD$. Consequently the triangles PAH and PCD are equal. In Fig. 1 draw HG, and in Fig. 2 AG, parallel to CD, meeting the lines EB and FA, produced if necessary. Then we have to construct a triangle PCD equal to PAH, and similar to PHG in Fig. 1 and PAG in Fig. 2, both having the same altitude, by Prob. V. A., thus:

Construction.—Having made the triangle ABH to contain the given area, as in the analysis, and drawn HG (Fig. 1), making the angle $EHG =$ the given angle ECD, take $PD = PL$, a mean proportional between the bases PG and PA, and draw DC parallel to HG. Then $ABCD$ will contain the given area $= ABH$. The construction of Fig. 2 is precisely similar.

Demonstration.—(Prob. V. A.) The triangles PCD and PAH

* VI. 1.

are equal. Take the difference between each and PAB, and we have $ABCD = ABH =$ the given area, and CD is parallel to HG.

FIG. 1.

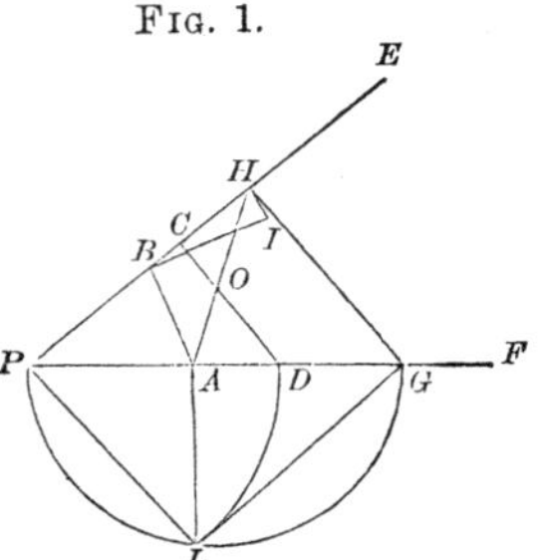

FIG. 2.

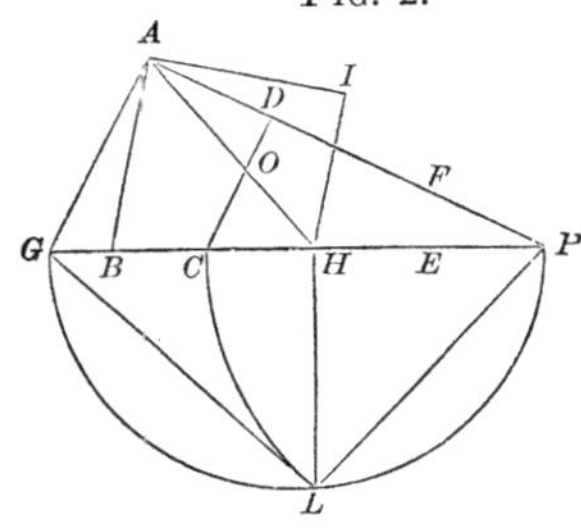

Calculation.—In triangle PBA, Case 1, and Prob. III. A., find PB and PA, and area PBA; whence the area of the triangle PCD is known by adding or subtracting the given area to or from PBA. In triangle PDC, Prob. IV. A., find side DC, and thence, Case 1, the sides PD and PC, from which AD and BC are obtained by subtraction.

Scholium 1.—When CD is to be parallel to AB, HG, in Fig. 1 must be drawn parallel to AB, and AG, Fig. 2, need not be drawn, G coinciding with B.

Scholium 2.—When AF and BE are parallel, bisect AH in O, and through O draw a line CD, making the angle $ECD =$ the given angle, and it will be the division line required. For the triangles AOD and HOC will be equal (I. 5*), and consequently $ABCD = ABH =$ the given area.

PROBLEM XVII.—To divide *any* given quadrilateral figure into two parts which shall have a given ratio to each other, by a line which shall make given angles with the two sides it intersects.

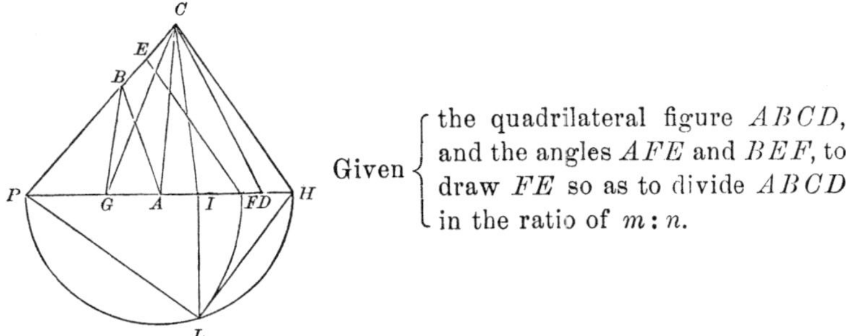

Given $\begin{cases} \text{the quadrilateral figure } ABCD, \\ \text{and the angles } AFE \text{ and } BEF, \text{ to} \\ \text{draw } FE \text{ so as to divide } ABCD \\ \text{in the ratio of } m : n. \end{cases}$

Analysis.—Produce the sides CB and DA which the division line

* I. 4.

FE is to cut, to meet in P. Then reduce the quadrilateral $ABCD$ to an equivalent triangle by joining CA, drawing BG parallel to CA, and joining CG. Then (IV. 2, cor. 2*) triangle $AGC = ABC$. Add ACD to each, and we have $GCD = ABCD$. Divide the base GD in I so that GI may be to ID in the given ratio of m to n, and join CI; then triangle $GCI : ICD :: GI : ID :: m : n$. Hence $ABEF$ must equal GCI, and $FECD$ must equal ICD. Draw CH parallel to the division line EF; then we have to make a triangle PEF equal to PCI and similar to PCH, both having the same altitude, by Prob. V. A., thus:

Construction.—Make the angle $PCH =$ the given angle PEF, and let CH meet AD, produced if necessary, in H. Take $PF = PL =$ a mean proportional between the bases PH and PI, and draw FE parallel to CH, and it will be the division line required.

Demonstration.—We have (Prob. V. A.) $PCH : PEF :: PH^2 : PF^2 :: PH^2 : PH \times PI :: PH : PI :: PCH : PCI$. Hence $PEF = PCI$. Take each from PCD, and we have $FECD = ICD$. But, by the analysis, $ABCD = GCD$. Hence, taking equals from equals, we have $ABEF = GCI$. But $GCI : ICD :: m : n$. Hence their equals $ABEF$ and $FECD$ are to each other as m to n, in the given ratio. Q. E. D.

Calculation.—In triangle PBA, Case 1, and Prob. III. A., find PB and PA, and area PAB. In triangle PCD, find area. Then area $ABCD =$ the difference of these areas; and since $m : n :: ABEF : FECD$, we have, by composition, $m + n : m :: ABCD : ABEF$, and $m + n : n :: ABCD : FECD$. Whence we have $\frac{m}{m + n} \times$ area $ABCD =$ area of $ABEF$, and $\frac{n}{m + n} \times$ area $ABCD =$ area $FECD$. Also, area $PEF = PAB + ABEF$.

Now, in triangle PEF, by Prob. V. A., find EF, and thence, Case 1, find PE and PF; then, by subtraction, find AF, FD, BE, and EC.

Scholium 1.—If the given quadrilateral is a parallelogram, divide either of the sides which the division line is to cut, in the ratio of m to n, and through the point thus obtained draw a line parallel to one of the other sides, and this line will divide the given parallelogram into two parallelograms which have to each other the given ratio of m to n. Now through the middle point of this dividing line draw

* VI. 1.

FE (Prob. XVIII., "Areas," etc.), making the angle $AFE =$ the given angle; then *FE* will be the division line required.

Scholium 2.—If only the lines *BC* and *AD* which the division line is to cut are parallel, divide each of these lines in the ratio of *m* to *n*, the parts corresponding to *m* in each being adjacent to *AB*. Join the points of division, and this line will evidently divide the given trapezoid into two partial trapezoids which shall have to each other the ratio of *m* to *n*.

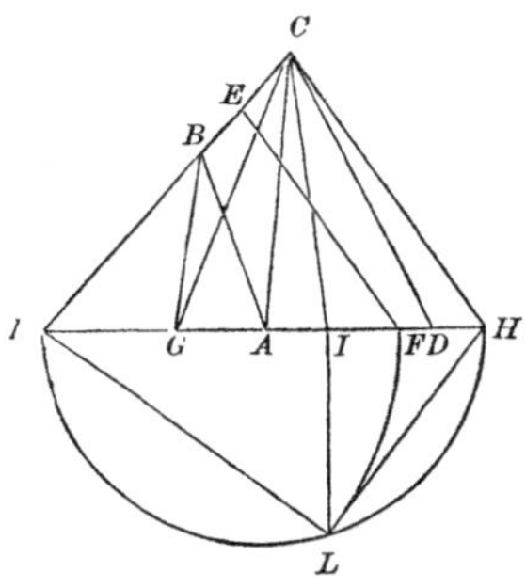

Then, as before, through the middle point of this dividing line draw *FE*, making the angle $AFE =$ the given angle, and *FE* will be the division line required.

Scholium 3.—If *AB* and *CD* are parallel, and *BC* and *AD*, which the division line is to cut, are not parallel, it is the same as this problem.

Scholium 4.—When *EF* is to be parallel to *CD*, *CH* will coincide with *CD*, and the semicircle must be described on *PD*, and *GD* must be divided in *I* so that $GI : ID :: m : n$.

Problem XVIII.—To divide a given trapezium into four equal parts by two lines, one of which shall be parallel to the third side, and the other cut the second and fourth sides.

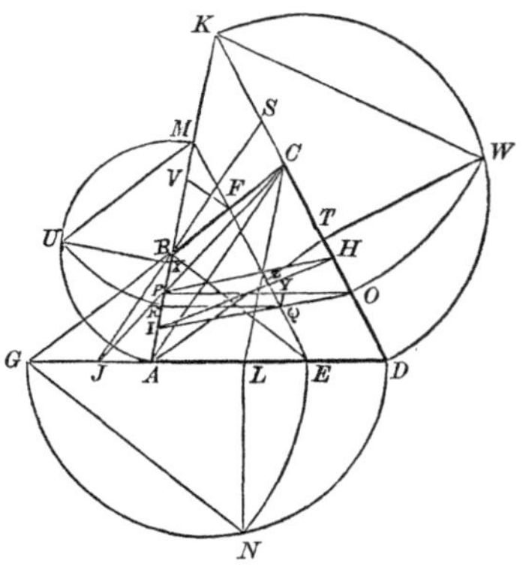

Given { the sides and angles of the trapezium *ABCD*, the sides being numbered in the order of the letters, *AB* being the first, to divide it into four equal parts by the lines *EF* and *IH*, of which *EF* is to be parallel to *CD*.

Analysis.—The same as in the preceding problems.

Construction and *demonstration.*—Produce *AB* and *DC* to meet in *K*, and *DA* and *CB* to meet in *G*. By last problem, scholium 4, divide $ABCD$ into two equal parts by a line EF parallel to CD, thus:—Join CA, draw BJ parallel to it, and join CJ; then $JCD = ABCD$. Bisect JD in L, and join CL; then $JCL = LCD = \frac{1}{2} ABCD$. Now, make a triangle GEF equal to GCL, and similar to GCD (Prob. V. A.), by describing a semicircle on the greater base GD, erecting a perpendicular LN from the end of the less base to meet the semicircle, joining GN, making $GE = GN$, and drawing EF parallel to CD; then EF is one of the required division lines. For (IV. 25*) $GCD : GFE :: GD^2 : GE^2 :: GD^2 : GN^2 :: GD^2 : GD \times GL :: GD : GL :: GCD : GCL$. Hence $GFE = GCL$. Take each from GCD, and we have $EFCD = LCD = \frac{1}{2} ABCD$. Taking these equals from the equals JCD and $ABCD$, we have $ABFE = JCL = \frac{1}{2} ABCD = EFCD$.

In a similar manner, precisely, divide $ABCD$ into two equal parts by the line PO parallel to AD, and divide $ABFE$ into two equal parts by the line RQ parallel to AE. Join OQ, and produce it to meet AB in I, draw PH parallel to IQO, and join IH, which will be the other division line required. For triangle $IHO = IPO$. Add $AIOD$ to each, and we have $AIHD = APOD = \frac{1}{2} ABCD$.

Now, putting Z at the point where the division lines EF and IH intersect each other, we have (IV. 25†) $IHO : IZQ :: IO^2 : IQ^2 :: IPO : IRQ$. But $IHO = IPO$. Hence $IZQ = IRQ$. Add $AIQE$ to each, and we have $AIZE = ARQE = \frac{1}{2} ABFE = \frac{1}{4} ABCD$. Hence $IBFQ$, $FCHZ$, and $EZHD$ are each equal to $\frac{1}{4} ABCD$ also, and the trapezium $ABCD$ is divided into four equal parts, as required, by the lines EF and IH.

Calculation.—Through Q, parallel to AB, draw QY, meeting OP in Y; then $PY = RQ$.

1. In triangle GBA, Prob. III. A. and Case 1, find area GBA, and GB and GA; then $GC = GB + BC$, and $GD = GA + AD$.

2. In triangle GCD, Prob. III. A., find area GCD; then area $ABCD = GCD - GBA$, $ABFE = \frac{1}{2} ABCD$, and $GFE = GBA + ABFE$.

3. In triangle GFE, Prob. IV. A., find FE, and thence, Case 1, GE and GF; then $AE = GE - GA$, $ED = AD - AE$, $BF = GF - GB$, and $FC = BC - BF$.

4. In triangle KBC, Prob. III. A. and Case 1, find area KBC, and sides KB and KC; then $KA = KB + BA$, and $KD = KC + CD$.

5. In triangle KPO ($= KBC + \frac{1}{2} ABCD$) find (Prob. IV. A.) PO, and thence, Case 1, KP and KO; then $AP = KA - KP$.

* VI. 19. † VI. 19.

6. In triangle MBF, Prob. III. A. and Case 1, find area MBF, and sides MB and MF; then $MA = MB + BA$, and $ME = MF + FE$.

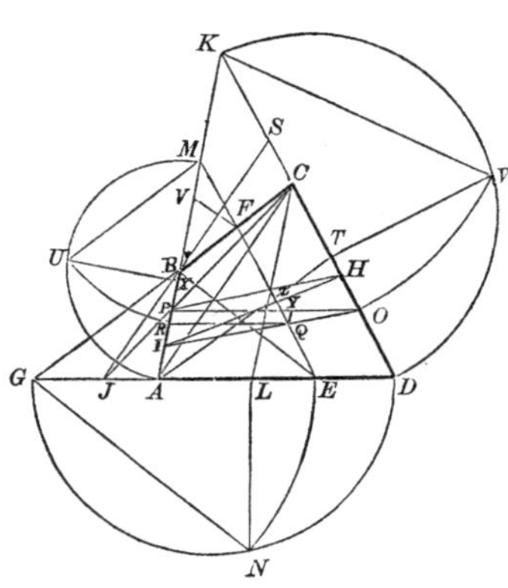

7. In triangle MRQ ($= MBF + \frac{1}{4} ABCD$) find RQ, and thence, Case 1, MR and MQ; then $AR = MA - MR$. Now, $OY = OP - RQ$, and $QY = PR = AP - AR$; then (IV. 18*) $OY : YQ :: OP : PI$. And $AI = AP - PI$. Also, $IB = AB - AI$, and $KI = KB + BI$. By similar triangles KIO, KPH, we have $KI : KP :: KO : KH$. Then $DH = KD - KH$, and $CH = CD - DH$.

8. In triangle KIH, Case 3, find angle KIH, and side IH; then, by parallel lines, we have $KI : MI :: IH : IZ$, and $KI : MI :: KH : MZ$; and thence, by subtraction, we have ZH, FZ, and ZE, and then all the sides, angles, and area of each of the four equal parts are known.

PROBLEM XIX.—Through a given point within a given trapezium to draw a line which will cut off a given area adjacent to the first side.

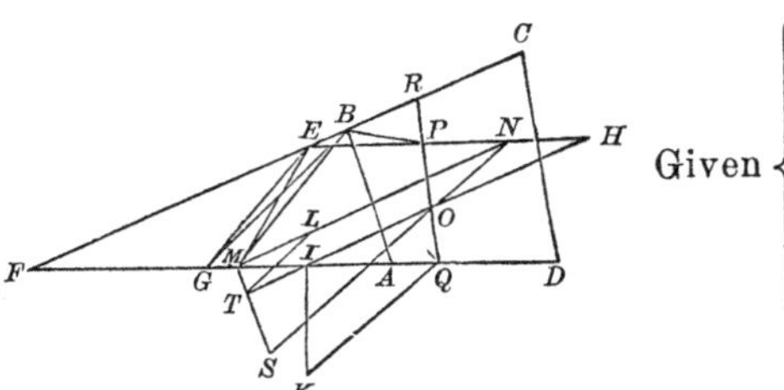

Given { the trapezium $ABCD$, of which AB is the first side, the angle ABP, and the distance BP, to draw RPQ, cutting off the area $ABRQ$ equal to L^2.

L ——

Analysis, construction, and *demonstration.*—Produce the sides DA and CB to meet in F, and through the given point P, parallel to FD, draw an indefinite line EPH. Now, to the line FE apply a parallelogram $FENM$, equivalent to the triangle FBA, thus:—Bisect FA in G, and join BG; then $FBG = ABG = \frac{1}{2} FBA$. Join EG, draw BM parallel to EG, and join EM; then $EMG = EBG$

* VI. 4.

Add FEG to each, and we have triangle $FEM = FBG = \frac{1}{2} FBA$. Draw MN parallel to EF; then parallelogram $FENM$ (I. 28, cor.*) $= 2\ FEM = 2\ FBG =$ triangle FBA.

Next, to the line MN apply the parallelogram $MNHI$ to contain the given area of $ABRQ = L^2$, thus:—Perpendicular to MN draw $MS = L$, and also on MN lay $ML = L$. Join NS, draw LT parallel to NS, and TIH parallel to MN. Then, by similar triangles MNS and MLT, we have $MN : MS\,(L) :: MS\,(L) : MT$. Hence $MN \times MT$ (which is the area of the parallelogram $MNHI$ (IV. 1, cor.†)) $= L^2 =$ the given area. Lastly, we must draw QPR so as to make the triangle $FRQ =$ the parallelogram $FEHI$; then, since $FENM = FBA$, $MNHI$ will be equal to $ABRQ$. To do this, observe that of the three similar triangles IOQ, EPR, and POH, the first two, IOQ and EPR, are portions of the triangle FRQ, but not of the parallelogram $FEHI$, while the third, POH, is part of the parallelogram, but not of the triangle; hence the first two, IOQ and EPR, must be equivalent to the third, POH, and (IV. 27, cor.‡) their homologous sides EP, PH, and IQ will form a right-angled triangle, of which PH is the hypothenuse. Wherefore, perpendicular to FD draw $IK = EP$, from K apply $KQ = PH$, and draw $QOPR$, which will be the line required. For, since the homologous sides IQ, EP, and PH form a right-angled triangle IKQ, by construction, we have (IV. 27, cor.§) $IOQ + EPR = POH$. To each add $FEPOI$, and we have $FRQ = FEHI$. Take the equals $FBA = FENM$ from these, and we have $ABRQ = MNHI = L^2 =$ the given area to be cut off. Q. E. D.

Calculation.—1. In the triangle FBA, Prob. III. A. and Case 1, find area FBA, and the sides FB and FA; then $FG = \frac{1}{2} FA$. Angle $EBP = EBA + ABP$.

2. In triangle EBP, Case 1, find EB and EP; then $FE = FB - EB = MN$.

3. In similar triangles FEG, FBM, we have $FE : FB :: FG : FM$.

4. In parallelogram $MNHI$, we have MN, the area, and angle $M = F$, to find MI (Prob. II. A., Note 2), thus:—$\sin MNI \times MN :$ rad. $::$ area $MNHI : MI$; then $FI = FM + MI = EH$, and $PH = EH - EP$.

5. In triangle IKQ, $IQ = \sqrt{KQ^2 - IK^2} = \sqrt{PH^2 - EP^2}$; then $FQ = FI + IQ$, and $AQ = FQ - FA$, also $QD = AD \backsim AQ$.

6. Area $FQR = FBA + ABRQ$, the given area. In triangle

* I. 34. † VI. 1. ‡ VI. 31. § VI. 31.

FQR, Prob. II. A., find FR, and thence, Case 3, find RQ; then $BR = FR - FB$, and $RC = BC - BR$.

7. In triangle PBR, Case 3, find PR; then $PQ = RQ - PR$.

PROBLEM XX.—Having given the area of any triangle, and the sides and included angle of its inscribed parallelogram, to determine the triangle.

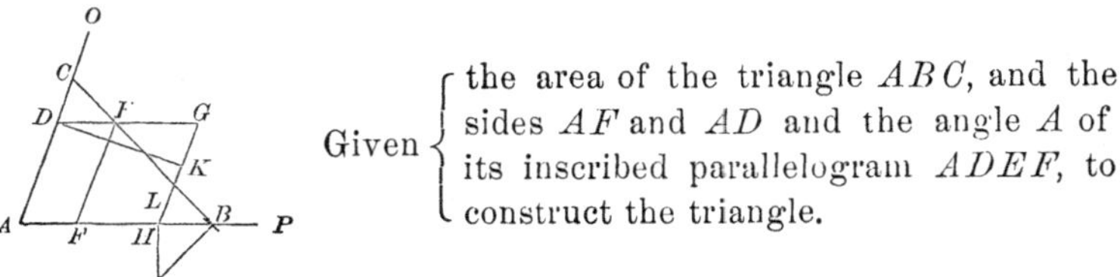

Given { the area of the triangle ABC, and the sides AF and AD and the angle A of its inscribed parallelogram $ADEF$, to construct the triangle.

Analysis.—Produce AF and AD indefinitely to P and O. Divide the given area by AD, draw DK perpendicular to AD and equal to the quotient, and through K, parallel to AD or EF, draw GLH; then the parallelogram $ADGH =$ the triangle ABC. Hence the triangle ELG, which is the part of the parallelogram *without the triangle*, must be equivalent to the sum of the two triangles HLB and DCE, which are the parts of the triangle *without the parallelogram*. Now, these three triangles are manifestly similar; hence (IV. 27, cor.*) the homologous sides HB, DE, and EG will form a right-angled triangle, of which EG is the hypothenuse. Whence this

Construction.—At H erect the perpendicular $HI = AF = DE$; from I apply $IB = EG$ or FH, and through B and E draw the line $BLEC$, and ABC will be the triangle required $= ADGH$.

Demonstration.—Since the homologous sides DE, EG, and HB of the similar triangles CDE, EGL, and LHB are such as to form a right-angled triangle HIB, of which $EG =$ the hypothenuse IB, we have (IV. 27, cor.†) $LHB + CDE = EGL$. To each add the irregular figure $AHLED$, and we have the triangle $ABC =$ the parallelogram $ADGH =$ the given area. Q. E. D.

Calculation.—By Prob. II. A., Note 2, we have $AD \times \sin A$: given area $ADGH$: : rad. : AH; then $HI = AF$, $IB = FH = AH - AF$, and $HB = \sqrt{IB^2 - HI^2}$. Whence $AB = AH + HB$, $FB = FH + HB$, and $FB : AB :: FE : AC$.

Limit.—FH can never be *less* than AF. If FH is *equal* to AF,

* VI. 31. † VI. 31.

AH will be the base of the required triangle, of which one side AB would equal twice AF, and the other AC equal twice AD, and the area of the triangle ABC would be equal to twice $ADEF$.

PROBLEM XXI.—Having given the difference of the radii of two concentric circles, to determine them such that the area of one shall be double that of the other.

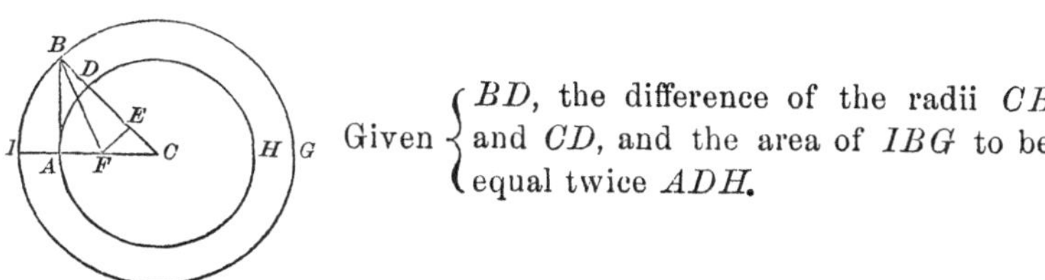

Given $\left\{\begin{array}{l} BD, \text{ the difference of the radii } CB \\ \text{and } CD, \text{ and the area of } IBG \text{ to be} \\ \text{equal twice } ADH. \end{array}\right.$

Analysis.—Let ADH and IBG represent the two circles, of which the area of $IBG = 2\,ADH$, and $BD =$ the given difference of their radii. Now, since the area of $IBG = 2\,ADH$, we have $CB^2 = 2\,CD^2$. Make the angles BCA and CBA each equal to half a right angle; then $CA = AB$, and $CB^2 = CA^2 + AB^2 = 2\,CA^2 = 2\,CD^2$. Hence $CD = CA = AB$. Lay $BE = BA$; then $CE = BD$. Erect EF perpendicular to CB; join BF; then, since $BA = BE$, $AF = FE = EC = BD$, the given difference. Whence this

Construction.—Draw an indefinite line, in which take $CE =$ the given difference of the radii. Erect the perpendicular $EF = EC$, join CF, and produce it, making $FA = FE$. Erect the perpendicular AB to meet the line CE, produced, in B; then $AB = AC$. Join BF, and with the centre C and radii CA and CB, respectively, describe the circles ADH and IBG, and they will be the circles required.

Demonstration.—We have $AB = AC$; hence $CB^2 = AB^2 + AC^2 = 2\,AC^2$. Wherefore the circle $IBG = 2\,ADH$. Again, since $AF = FE$, the triangles BAF and BEF are equal; hence $BE = BA = CA = CD$, and $BD = CB - CD = CB - BE = CE =$ the given difference of the radii by construction.

Calculation.— $CF = \sqrt{CE^2 + EF^2} = \sqrt{2\,CE^2} = CE\sqrt{2}$, and $CA = CF + FA = CF + CE = CD$. And $CB = CD + DB = CF + 2\,FA = CF + 2\,BD$.

Scholium.—BA is a tangent to the circle ADH at the point A.

PROBLEM XXII.—Having given the difference of the sides of two concentric squares, to determine them such that the area of one shall be double that of the other.

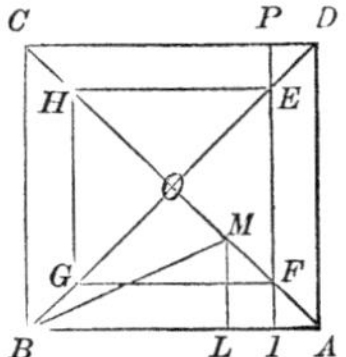

Given $\begin{cases}\text{the difference between the sides } AB \text{ and} \\ GF, \text{ and the area of the square } ABCD \text{ to} \\ \text{be twice } FGHE.\end{cases}$

Analysis.—Suppose $ABCD$ and $FGHE$ to be the concentric squares required. Draw the diagonals FH and GE, and produce them, and they will form the diagonals AC and BD, both pairs of diagonals intersecting in the common centre O.

Now, since $ABCD = 2\,FGHE$, AB^2 must equal $2\,FG^2$. But $AB^2 = AO^2 + BO^2 = 2\,BO^2$; hence $FG = BO$. Make $BL = BO$ or FG, erect the perpendicular LM, and join BM; then, in triangles BOM and BLM, since $BO = BL$, we have $OM = ML = AL =$ the given difference of the sides. Whence this

Construction.—Draw AB and AD at right angles, and AC bisecting the angle BAD. Lay $AL =$ the given difference of the sides, erect the perpendicular LM, lay $MO = ML$, and erect the perpendicular OB; then AB will be the side of the larger square, and BO or AO equal to the side of the smaller square. On AB describe the square $ABCD$, and produce AO and BO, forming the diagonals AC and BD, and $ABCD$ will be the larger square required. Now, take $BL = BO$, bisect AL in I, through I, parallel to AD, draw $IFEP$, draw FG, GH, and HE parallel to AB, BC, and CD respectively; then $FGHE$ will be the smaller required square.

Demonstration.—It is evident that $ABCD$, $FGHE$ are concentric squares. We have to show that $ABCD = 2\,FGHE$, and that $AB - FG = AL = 2\,AI = 2\,IF$.

In triangles BOM and BLM, since $MO = ML$ by construction, we have $BO = BL$. Also $AB = BL + AL = BL + 2\,AI = BL + 2\,IF$. But IP, which is equal to AB, $= EF + IF + EP = EF + 2\,IF$. Hence $BL = EF$, and $AL = AB - BL = AB - EF$; that is, $AB - FG = AL = 2\,AI = 2\,IF$. Also, $AB^2 = BO^2 + AO^2 = 2\,BO^2 = 2\,BL^2 = 2\,FG^2 = 2\,EF^2$; hence area $ABCD = 2\,FGHE$. Q. E. D.

Calculation.—$AM = \sqrt{AL^2 + LM^2} = AL\sqrt{2}$. Also, $AO =$

$AM + MO = AM + AL = EF = GF$. And $AB = BL + AL = AO + AL = GF + AL$.

PROBLEM XXIII.—Having given two parallelograms with a common vertical angle, it is required to draw a line from the remote extremity of a side forming the common angle, cutting the other side of that parallelogram, and both the sides *produced* of the other parallelogram, so as to form two equal trapezoids.

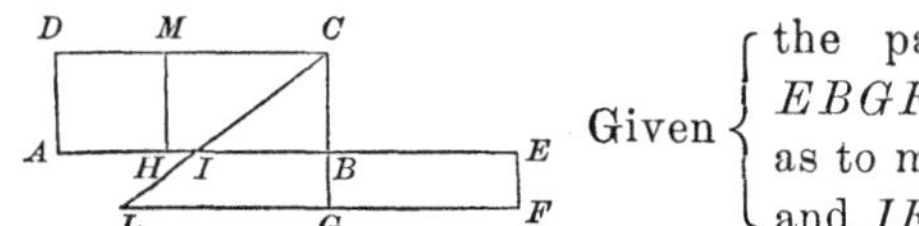

Given $\begin{cases} \text{the parallelograms } ABCD \text{ and} \\ EBGF, \text{ to draw the line } CIL \text{ so} \\ \text{as to make the trapezoids } ADCI \\ \text{and } IEFL \text{ equal.} \end{cases}$

Analysis.—Suppose CIL to be the required line, making $ADCI = IEFL$. Now, if from the larger parallelogram $ABCD$ we cut off $ADMH = EBGF$, we shall have $IBGL = IHMC$. To each add IBC, and we have the triangle $CGL = BHMC$. Whence this

Construction.—Take DM such that $AD : BE :: BG : DM$, and draw MH parallel to AD; then (IV. 24 and 2*) the parallelograms $ADMH$ and $BEFG$ are equal. Now, we have to make the triangle $CGL = BHMC$, to do which, take $CG : 2\,BH :: CB : GL$, and draw CIL; then (IV. 24 and 2†) $CGL = BHMC$, and CIL is the line required.

Demonstration.—By construction, $ADMH = BEFG$, and $CGL = BHMC$. From these last equals take the common part CBI, and we have $IBGL = IHMC$. To these, respectively, add the equals $BEFG$ and $ADMH$, and we have the trapezoids $IEFL$ and $ADCI$ equal. Q. E. D.

Calculation.—By construction, $AD : BE :: BG : DM$; then $BH = CM = CD - DM$. Also, $CG : 2\,BH :: BC : GL$; then $FL = FG + GL$. And, by similar triangles CGL, CBI, we have $CG : GL :: CB : BI$; then $EI = EB + BI$, and $AI = AB - BI$. In triangle CGL, Case 3, find CL; then $CG : CB :: CL : CI$, and $IL = CL - CI$.

* VI. 14. † VI. 14.

PROBLEM XXIV.—Given, the base of a plane triangle, and one angle at the base, to construct the triangle such that its area shall be double the area of its inscribed square.

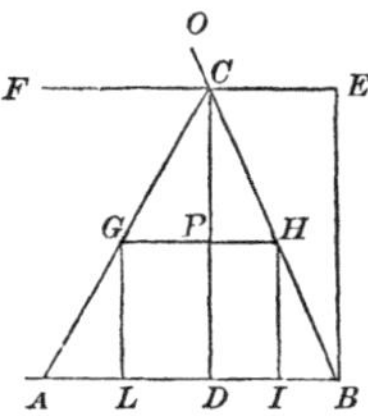

Given $\begin{cases} \text{the base } AB\text{, the angle } ABO\text{, and the} \\ \text{area of the triangle } ABC \text{ to be double} \\ \text{that of the inscribed square } GHIL. \end{cases}$

Analysis.—Suppose the triangle constructed as in the adjacent figure. Let fall the perpendicular CPD; then, as the triangle ABC is equal to twice the square $GHIL$, the four small triangles ALG, GPC, BIH, and HPC must together be equal to the square. The first two of these triangles are similar, and the last two are similar; and when $CP = PD$, the first two are equal, and each half of $GPDL$, and the last two are equal, and each half of $PHID$. Also, $AL = GP = LD$, and $BI = HP = ID$, and hence $AB = 2\,GH = 2\,LI = 2\,PD = CP$. Whence this

Construction.—Make $AB =$ the given base, and angle $ABO =$ the given angle. Erect the perpendicular $BE = AB$, draw ECF parallel to AB, join AC, let fall the perpendicular CD, and bisect it in P. Through P, parallel to AB, draw GPH, and through G and H, parallel to CD, draw GL and HI; then ABC will be the triangle required, and $GHIL$ its inscribed square.

Demonstration.—Since $CD = AB$ and $CP = \frac{1}{2}\,CD$, we have $GH = \frac{1}{2}\,AB = \frac{1}{2}\,CD = GL$; hence $GHIL$ is a square.

Also, area $ABC = AB \times \frac{1}{2}\,CD = 2\,LI \times LI = 2\,LI^2 =$ twice the area $GHIL$.

Calculation.—1. In triangle BDC, Case 1, find BC.

2. In triangle ABC, Case 3, find AC.

$LI = IH = \frac{1}{2}\,AB$. Area $ABC = \frac{1}{2}\,AB \times BE = \frac{1}{2}\,AB^2$; area $GHIL = \frac{1}{2}\,ABC = \frac{1}{4}\,AB^2$.

Scholium.—In any triangle in which a square is inscribed, we have $CD : AB :: CP : GH$. By composition, $AB + CD : AB :: GH + CP$ (or CD) $: GH = \dfrac{AB \times CD}{AB + CD} =$ (when $CD = AB$) $\dfrac{AB^2}{2\,AB} = \frac{1}{2}\,AB = \frac{1}{2}\,CD$.

PROBLEM XXV.—In a plane triangle are given its area, the vertical angle, and the length of the line drawn from the vertical angle to the middle of the base, to construct the triangle.

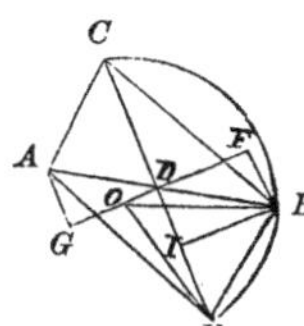

Given $\begin{cases} \text{the area of } ABC\text{, the vertical angle } ACB\text{,} \\ \text{and the line } CD \text{ drawn from } C \text{ to } D\text{, the} \\ \text{middle point of } AB. \end{cases}$

Analysis.—Let ACB represent the required triangle. Complete the parallelogram $ACBE$. Perpendicular to CD draw GDF, meeting AG and BF, drawn parallel to the diagonal CE, in G and F respectively. Now, since $AD = DB$, the triangles ACD and DCB are equal, and each is equal to half the given area of ACB. Hence each of the perpendiculars DG and DF is equal to the given area of ABC divided by CD. Also, the diagonal $CE = 2\,CD$, the given line, and the angle $CBE =$ the supplement of the given angle ACB. Whence this

Construction.—Draw $CE =$ twice the given line, and through D, its middle point, draw the perpendicular GDF, making DG and DF each equal to the quotient of the given area of ACB divided by the given line CD. On CE (III. Prob. 16*) describe the circular segment CBE to contain the *supplement* of the given angle ACB. Draw FB and AG parallel to CE, join AC, CB, BE, and EA; then will ACB be the required triangle.

The *demonstration* is evident from the analysis.

Calculation.—On CE let fall the perpendicular BI, and join E and B with O, the centre of the arc CBE. Then, supposing the circle completed, it is evident (III. 18, and cor. 4†) the angle EOD = the given angle ACB.

1. In triangle ODE, Case 1, find OD and $OE = OB$; then $OF = OD + DF$.

2. In triangle OFB, Case 2, find $FB = DI$; then $CI = CD + DI$, and $EI = ED - DI$.

3. In triangle CBI, $CB = \sqrt{CI^2 + IB^2}$, and in triangle EIB, $BE = \sqrt{EI^2 + IB^2} = AC$.

4. Also, in triangle DIB, $DB = \sqrt{DI^2 + IB^2}$, and $AB = 2\,DB$.

Scholium.—The triangle AEB is similar and equal to ACB in every respect, as is evident from the properties of a parallelogram.

* III. 33. † III. 20 and 22.

PROBLEM XXVI.—Given, three points in a right line, it is required to find a fourth point in that line, such that the *rectangle* of the distances of the fourth point from the first and second points shall be equal to the *square* of its distance from the third point.

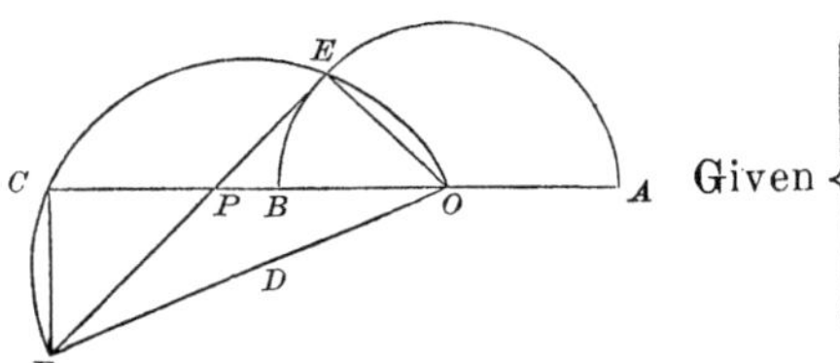

Given $\left\{\begin{array}{l}\text{the three points } A, B, \text{ and } C \text{ in the right line } AC, \text{ to} \\ \text{find a fourth point } P \text{ such that } PB \times PA \text{ shall equal } PC^2.\end{array}\right.$

Analysis.—Suppose the thing done, and that, in the above figure, $PB \times PA = PC^2$. On AB describe a semicircle, to which (Prob. II. of "The Circle") draw the tangent PE; then (IV. 30*) $PE^2 = PB \times PA = PC^2$ by the problem. Hence $PE = PC$; and we have this

Construction.—Erect the perpendicular $CF = OB$, join OF, and on it describe a semicircle $FCEO$, cutting the semicircle on AB in E. Draw OE and FPE; then will P be the point required.

Demonstration.—In similar triangles PCF and PEO, since $FC = EO$, we have $PC = PE$, and $PC^2 = PE^2 =$ (IV. 30†) $PA \times PB$. Q. E. D.

Calculation.—The triangles OEF and OCF are equal in every respect; hence $FE = OC$, and the angle $COF = OFE$.

1. In triangle OCF, Case 2, find angle $COF = POF$; then angle $OPE = OFP + POF = 2\ COF$.

2. In triangle OPE, Case 1, find OP and $PE = PC$; then $PB = BC - CP$, and $PA = PB + BA$.

Scholium.—If we wish CP^2 to be to $PB \times PA$ in a given ratio, say as m^2 to n^2, take $n : m :: OE : CF = \frac{m}{n} \times OE$, then proceed as above. For, $m : n :: CF : OE ::$ (by similar triangles) $CP : PE = \frac{n}{m} \times CP$. Hence $\frac{n^2}{m^2} \times CP^2 = PE^2 = PB \times PA$, and we have $m^2 : n^2 : CP^2 : PB \times PA$.

* III. 36. † III. 36.

PROBLEM XXVII.—To divide a given line into two parts such that the square of one part shall be equal to the rectangle contained by the other part, and a given line.

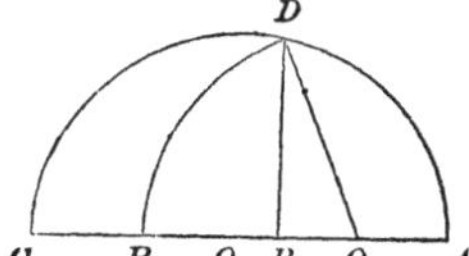

Given $\left\{ \right.$ the lines CB and BA, to find the point P in CB such that $PB^2 = CP \times BA$.

Analysis.—Let the given lines CB and BA be placed contiguous, forming the straight line CA, and suppose P to be the required point. Bisect BA in O. Then, since $PB^2 = CP \times BA$, we have $CP : PB :: PB : BA$. By composition, $CP + PB : PB :: PB + BA : BA$; that is, $CB : PB :: PA : BA$. Hence $CB \times BA = PA \times PB = (PO + OB) \times (PO - OB) = PO^2 - OB^2$. Whence this

Construction.—On CA describe a semicircle, erect the perpendicular BD, bisect BA in O, join OD, and make $OP = OD$; then P will be the point required.

Demonstration.—By analysis and construction, $CB \times BA = BD^2 = DO^2 - OB^2 = PO^2 - OB^2 = (PO + OB) \times (PO - OB) = PA \times PB$. Hence $CB : PB :: PA : BA$. By division, $CB - PB : PB :: PA - BA : BA$; that is, $CP : PB :: PB : BA$. Hence $PB^2 = CP \times BA$. Q. E. D.

Calculation.—Join OD; then $OP = OD = \sqrt{BD^2 + BO^2} = \sqrt{CB \times BA + (\frac{1}{2} AB)^2}$. And $PB = OP - BO$, and $CP = CB - PB$.

PROBLEM XXVIII.—In a plane triangle are given the area, an angle at the base, and a line from the vertex to the middle of the base, to determine the triangle

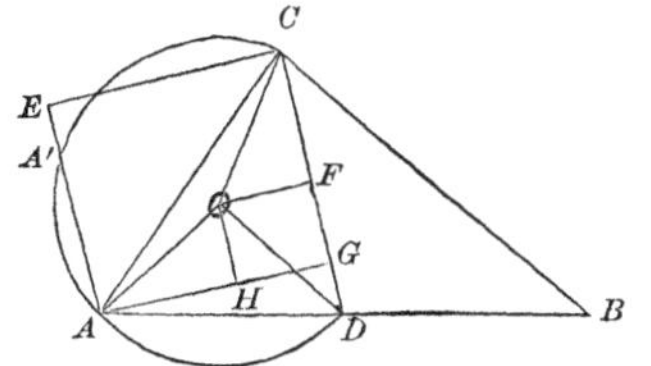

Given $\left\{ \right.$ the area of ABC, the angle BAC, and the line CD from the vertex to the middle of the base, to construct the triangle ABC.

Analysis.—Suppose ABC to be the required triangle. On CD erect the perpendicular CE to meet AE drawn parallel to CD; then CE is known, being equal to the double area of ADC divided by

CD, or equal to the given area of ABC divided by CD. The angle DAC is also known. Whence this

Construction.—Draw CD equal to the given bisecting line, on it (III. Prob. 16*) describe a circular segment, of which the centre is O, to contain the given angle; on CD erect the perpendicular $CE =$ the given area divided by CD; draw $EA'A$ parallel to CD; join CA and AD; produce the latter until $DB = AD$, and join CB; then ACB will be the required triangle.

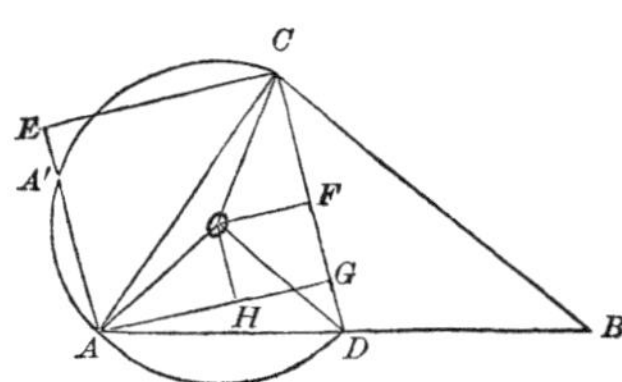

Demonstration.—Since $AD = DB$, area $ACB = 2\,ACD = CD \times CE =$ the given area by construction. Q. E. D.

Calculation.—Join OA, OC, and OD; on CD let fall the perpendiculars OF and AG, and draw OH parallel to CD; then $AG = EC$.

1. In triangle OCF, $CF = \frac{1}{2} CD$, and angle $COF =$ the given angle BAC; find, Case 1, OC, and $OF = HG$; then $AH = AG - HG$. Also, $OH = \sqrt{AO^2 - AH^2} = FG$, and $CG = CF + FG$, and $DG = DF - FG$.

2. $CA = \sqrt{CG^2 + AG^2}$, and $AD = \sqrt{DG^2 + AG^2}$, and $AB = 2\,AD$.

3. In triangle CAB, Case 3, find CB.

Scholium.—If A' had been taken instead of A as the end of the base of the triangle at which the angle is given, and CA' and $A'D$ had been joined, and the latter produced until DB' should equal $A'D$, and CB' been joined, we should have had another triangle $A'CB'$, fulfilling all the conditions of the problem, in which CA' would $= AD$, $A'D = AC$, and $A'B' = 2\,AC$.

* III. 33.

PROBLEM XXIX.—Given, four consecutive sides of a survey in length and position, it is required to draw a line from the remote extremity of the fourth side to cut the second and first sides such that the triangle formed by the first part of this line, the fourth side, and the adjacent segment of the third side, shall be equal to the quadrilateral formed by the remaining part of said line, the second side, and the adjacent segments of the first and third sides.

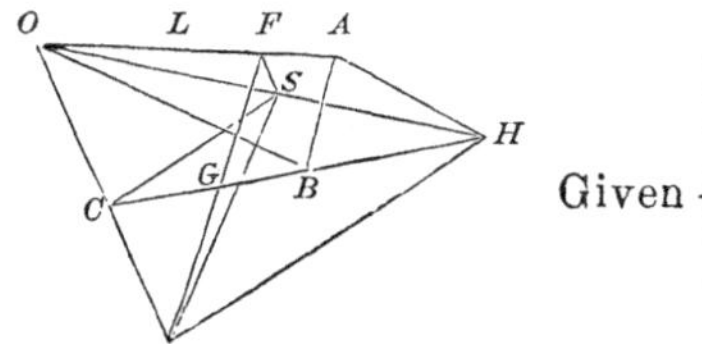

Given { in length and position, the four lines LA, AB, BC, and CD, to draw the line DGF such that the triangle DCG shall be equal to the quadrilateral $ABGF$.

Analysis.—Suppose the figure constructed so that $DCG = ABGF$. Producing DC and AL to meet in O, and adding $OFGC$ to each of these equal quantities, we have the triangle $ODF =$ the quadrilateral $OABC$, a given quantity. Whence this

Construction.—First reduce the quadrilateral $OABC$ to a triangle, thus:—Join OB, draw AH parallel to OB, and join OH; then $OHB = OAB$. Add OCB to each, and we have $OHC = OABC$. Now, it remains to make the triangle $ODF =$ the triangle OHC. To do this, join HD, draw CS parallel to HD, and SF parallel to OD, and draw DGF, the line required.

Demonstration.—Join DS; then (IV. 2, cor. 2*) $ODF = ODS = OCS + SCD = OCS + SCH = OHC = OABC$, as shown in the construction. From the first and last of these equals take the common part $OFGC$, and we have $DCG = ABGF$. Q. E. D.

Calculation.—1. In the quadrilateral $OABC$, we have all the angles, and the sides AB and BC, to find the area, and the sides CO and AO; then $OD = DC + CO$.

2. In the triangle ODF, we have the area $= OABC$, the angle O, and the side OD, to find OF (Prob. II. A.), and thence the side DF, and the angle ODF; then AF, FL, and LO are known.

3. In the triangle CDG, Case 1, find CG, DG, and the area of $CDG = ABGF$; then BG and GF are known by subtraction.

* VI. 1.

PROBLEM XXX.—To construct a lune that shall be equivalent in area to a given isosceles right-angled triangle.

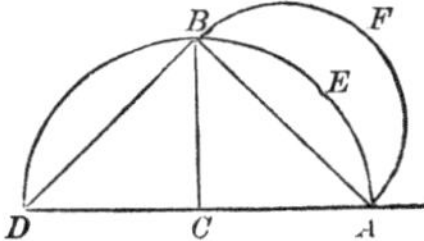

Given $\begin{cases} \text{the isosceles right-angled triangle} \\ ACB, \text{ to construct a lune } AEBF \\ \text{equivalent to the triangle } ACB. \end{cases}$

Analysis.—On the hypothenuse AB of the given isosceles right-angled triangle ACB describe the semicircle AFB, and with centre C describe the quadrantal arc AEB. Now, since the area of the lune $AEBF$ is to be equal to the area of the triangle ACB, by adding the segment $AEBA$ to each, we have the area of the semicircle $ABF =$ the area of the quadrant $CAEB$. Whence this

Construction.—With centre C, and radius CA or CB, describe the quadrantal arc AEB, and on AB describe the semicircle AFB; then $AEBF$ will be the lune required.

Demonstration.—Complete the semicircle ABD, and join BD; then $AB = BD$. Now, since $AB^2 = \frac{1}{2} AD^2$, the semicircle $AFB =$ $\frac{1}{2}$ the semicircle $ABD =$ the quadrant ACB. Take the segment $AEBA$ from each, and we have the lune $AEBF =$ the triangle ACB. Q. E. D.

Scholium.—A lune can be constructed equal in area to any rectilinear figure by reducing the figure to an equivalent triangle, then forming an isosceles right-angled triangle equivalent to this triangle, and, lastly, constructing the lune as in the problem, equivalent to this right-angled triangle.

DEMONSTRATION OF SOME THEOREMS.

THEOREM I.—The lines drawn from each of the three angles of a plane triangle to the middle of the opposite side, intersect in the same point; also, the distance from the middle of either side to this point is one-third of the distance to the opposite angle.

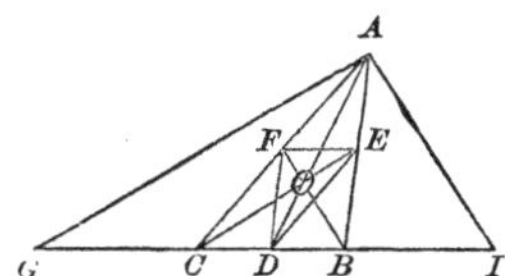

Given $\left\{\begin{array}{l}\text{the triangle } ABC, \text{ and } D, E, F \\ \text{the middle points of the sides; then} \\ AD, BF, \text{ and } CE \text{ intersect at a} \\ \text{common point } O.\end{array}\right.$

Demonstration.—Let ABC be a triangle, D, E, and F the middle points of the sides BC, AB, and AC respectively; then, if AD, BE, and CF be joined, they will intersect in a common point O. Produce BC both ways, making CG and BI each equal to BC, and join AG, AI; then $DG = DI$, and (IV. 16*) AG is parallel to CE, and AI to BF. Also, $DB = \frac{1}{3} DI$, and $DC = \frac{1}{3} DG$.

Now, in the triangle ADI, whose sides are cut by the parallel BF, since $DB = \frac{1}{3} DI$, we have the distance from D to the point in which BF cuts AD equal to $\frac{1}{3} DA$. Also, in the triangle ADG, whose sides are cut by the parallel CE, since $DC = \frac{1}{3} DG$, we have the distance from D to the point in which CE cuts AD equal to $\frac{1}{3} DA$. Hence BF and CE cut AD in the same point, which call O; then all the three bisecting lines intersect in the common point O, and $DO = \frac{1}{3} DA$.

Also, join DF, FE, and DE; then $DF = \frac{1}{2} AB$, $FE = \frac{1}{2} BC$, and $DE = \frac{1}{2} AC$. Now, in similar triangles COB, EOF, since $FE = \frac{1}{2} BC$, we have $FO = \frac{1}{2} OB = \frac{1}{3} FB$, and $EO = \frac{1}{2} OC = \frac{1}{3} EC$. Q. E. D.

NOTE.—See Scholium 3, Theorem VII.

Corollary 1.—Since AD bisects BC, it bisects all lines parallel to BC; hence the centre of gravity of the triangle ABC is in the line

* VI. 2.

AD. Also, since CE bisects AB, it bisects all lines parallel to AB; wherefore the centre of gravity of the triangle ABC is in the line CE. Consequently, as the centre of gravity is in both the lines AD and CE, it must be at their intersection O, and $AO = \frac{2}{3} AD$, $CO = \frac{2}{3} CE$, and $BO = \frac{2}{3} BF$.

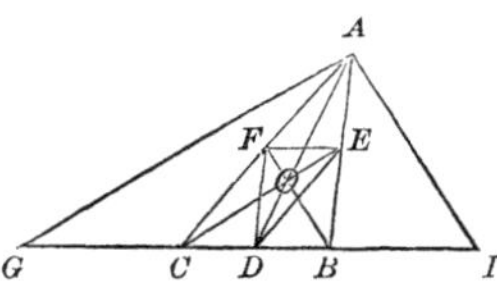

Corollary 2.—If the points F, E, and D be joined, the lines will be respectively parallel to the sides of the triangle ABC, and each side of the triangle DEF will be half the length of the side in the triangle ABC to which it is parallel, and these lines will divide the given triangle into four equal triangles, all similar to each other and to the whole triangle. Moreover, each side of the triangle DEF is the base of two equal rhombuses, which have their upper bases halves of the side to which it is parallel in the triangle ABC, and each rhombus is half of ABC.

THEOREM II.—If a straight line be bisected and produced to any point, the rectangle of the whole line thus produced and the part produced, with the square of half the line bisected, are together equal to the square of the line which is made up of the half and the part produced.*

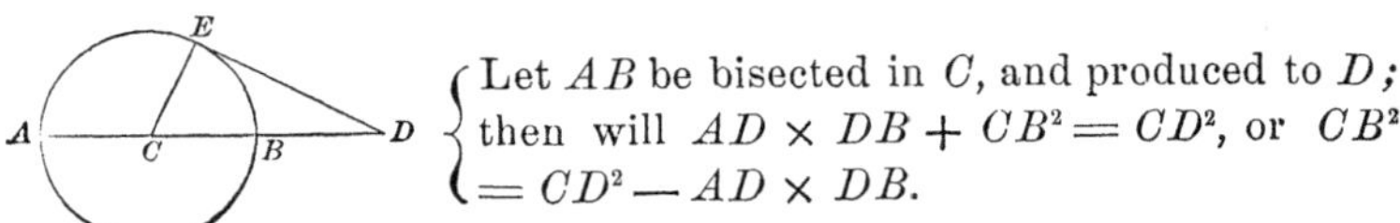

Let AB be bisected in C, and produced to D; then will $AD \times DB + CB^2 = CD^2$, or $CB^2 = CD^2 - AD \times DB$.

Demonstration.—With centre C, and radius CB or CA, describe a circle, to which draw the tangent DE by describing on DC a semicircle DEC, and join CE; then (IV. 30†) $AD \times DB = DE^2$. To each add $CB^2 = CE^2$, and we have $AD \times DB + CB^2 = DE^2 + CE^2 = CD^2$, or $CB^2 = CD^2 - AD \times DB$. Q. E. D.

* This theorem is the Sixth Proposition of the Second Book of Euclid; but, as it is not in Legendre's Geometry, and the property is a very useful one, it is inserted here with a different demonstration from that given in Euclid.

† III. 36.

Theorem III:—If upon the radius of a quadrant a semicircle be described, any radius drawn in the quadrant will intercept equal arcs on the semicircle and on the quadrant

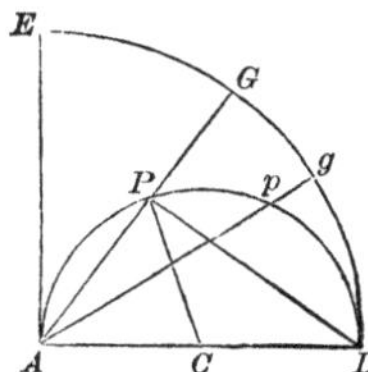

Draw the radii APG and Apg; then the arcs AP and EG are equal; also the arcs Pp and Gg, and pD and gD.

Demonstration.—Since the quadrantal arc EGD and the semicircle APD are equal in length (V. 11*), we have, $90°$: angle EAG : : arc EGD : arc EG. Also, $180°$: angle ACP : : arc APD : arc AP; that is, $90° : \frac{1}{2} ACP \; (= ADP = EAG)$: : arc APD (= arc EGD) arc AP. Hence arc EG = arc AP, and, consequently, arc DG = arc DP, and, in like manner, the arc APp = arc EGg, and arc Dp = arc Dg; then $Pp = Gg$. Q. E. D.

Theorem IV.—If in an isosceles triangle a circle be described, and a tangent be drawn to the circle parallel to the base, then the diameter of the circle will be a mean proportional between the base and the part of the tangent which is intercepted by the sides of the triangle.

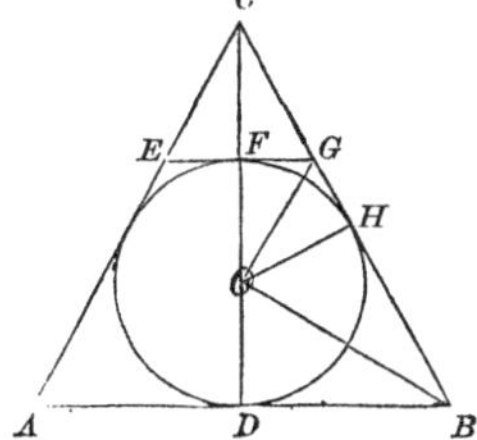

Let the tangent EFG be parallel to AB; then $FD^2 = AB \times EG$.

Demonstration.—Let O be the centre of the circle; join OB and OG, and let fall the perpendiculars CFD and OH on the sides AB and BC respectively; then H will be the point of contact. Now, it is evident that the angles FGH and DOH are equal, because they are respectively the supplements of the equal angles EGC and DBH. Hence their halves OGF and DOB are equal, and the right-angled triangles OGF and DOB are similar, and we have $OF : FG$: : $BD : DO$; whence $2\,OF : 2\,FG$: : $2\,BD : 2\,DO$; that is, $FD : EG$: : $AB : FD$. Hence $FD^2 = AB \times EG$. Q. E. D.

* I. Supp. B.

THEOREM V. ("Theorem of Pappus.")—In any triangle, any parallelograms described upon the two sides are together equivalent to a parallelogram described on the base, and limited by the opposite sides of the parallelograms described on the sides of the triangle, and by parallels to the line which joins the vertex of the triangle and their point of concourse. Prove this, and show how I. 47 of Euclid (or IV. 11, Legendre) can be deduced therefrom.

FIG. 1.

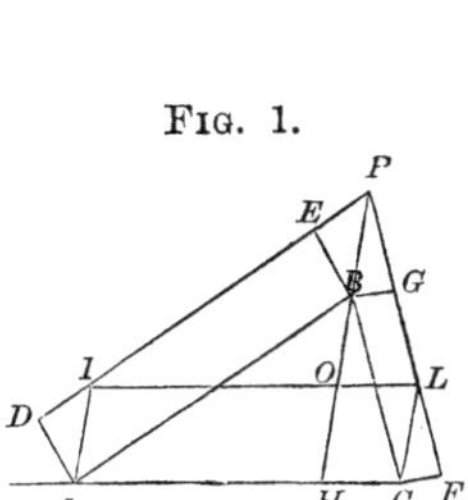

FIG. 2.

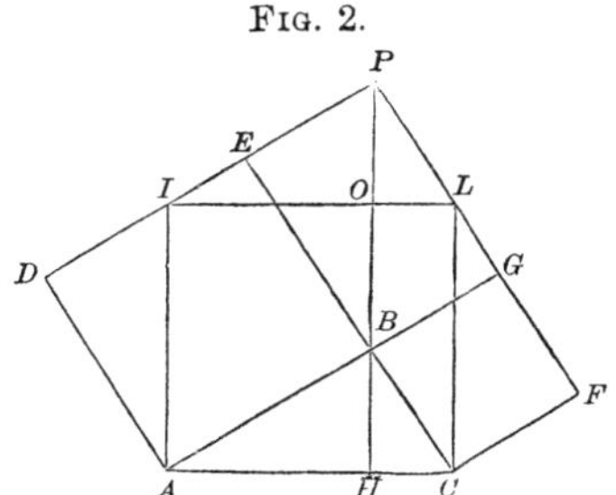

Demonstration.—Let ABC (Fig. 1) be *any* triangle, $ABED$ be *any* parallelogram described at pleasure on AB, and $BCFG$ be *any* parallelogram described at pleasure on BC, and let DE and FG be produced to meet in P, which will be their point of concourse. Draw PBH, and draw AI and CL parallel thereto, meeting DE and FG in the points I and L, and join IL. Then, since the opposite sides of a parallelogram are equal, we have $AI = BP$, and $CL = BP$; hence $AI = CL$, and IL is parallel to AC, and $AILC$ is a parallelogram described on the side AC, and we have to prove that $AILC = ABED + BCFG$. Now (IV. 1*), $ABED = ABPI = AHOI$. Also, $BCFG = BCLP = CLOH$. Therefore, $AILC = AHOI + CLOH = ABED + BCFG$. Q. E. D.

Next, let the triangle ABC (Fig. 2) be right-angled at B. On AB and BC describe the squares $ABED$ and $BCFG$, and let the sides DE and FG be produced and meet in P, which will be the point of concourse. Draw PBH. Then the triangles CBH and PBE are evidently equiangular, and angle $BHC = BEP =$ a right angle, and hence PBH is perpendicular to AC. Draw AI and CL parallel to PBH, meeting DE and FG in the points I and L, and join IL; then $AI = BP = CL$. Whence IL is parallel to AC, and $AILC$ is a rectangle. But the triangles ADI and ABC are evidently similar, and since $AD = AB$, $AI = AC$, and $AILC$ is a square described on the hypothenuse AC, and we have to prove that $AILC$

* VI. 1.

$= ABED + BCFG$. Now (IV. 1*), $ABED = ABPI = AHOI$. Also, $BCFG = BCLP = CLOH$. Hence $AILC = AHOI + CLOH = ABED + BCFG$. Q. E. D.

THEOREM VI. (From *Ladies' Diary*.†)—If through any point within a triangle lines be drawn from the three angles to cut the opposite sides, the product of the three alternate segments, beginning at any angle, and going round *in one direction*, will be equal to the product of the three alternate segments, beginning at the same angle, and going round *in the opposite direction*.

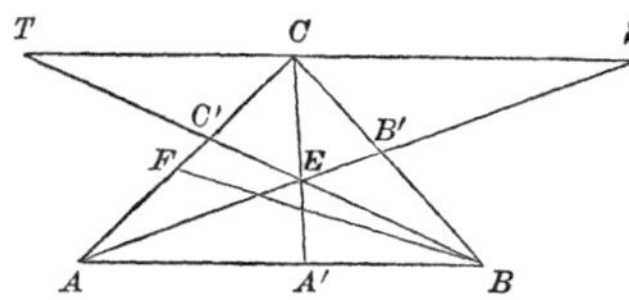

Through the given point E, in the triangle ABC, draw the three lines AEB', BEC', and CEA'. Prove that $AA' \times BB' \times CC' = AC' \times CB' \times BA'$.

Demonstration.—Through C, parallel to AB, draw a line meeting AB' and BC', produced in S and T. Then, by similar triangles TCE and $BA'E$, we have $TC : CE :: BA' : A'E$.

By similar triangles SCE and $AA'E$, we have $CE : CS :: A'E : AA'$.

By multiplying the corresponding terms, we have $TC : CS :: BA' : AA'$. (A).

Again, by similar triangles ABC' and CTC', we have $TC : AB :: CC' : AC'$.

Also, by similar triangles ABB' and SCB', we have $AB : CS :: BB' : CB'$.

By multiplying the corresponding terms, we have $TC : CS :: BB' \times CC' : AC' \times CB'$. (B).

Hence, the first couplet being the same in proportions A and B, we have $BA' : AA' :: BB' \times CC' : AC' \times CB'$.

Wherefore, by equating the products of the extremes and means, we have $AA' \times BB' \times CC' = AC' \times CB' \times BA'$. Q. E. D.

THEOREM VII.—Conversely, when the product of the three alternate segments of the sides of a triangle, taken by going round *in one direction*, is equal to the product of the three alternate segments, beginning at the same place, and going round *in the opposite direc-*

* VI. 1.

† See *Ladies' Mathematical Diary* for 1735–6, Leybourn's Collection, vol. i. p. 246.

tion, then the three lines drawn from the opposite angles and forming these segments *pass through the same point.* That is, if $AA' \times BB' \times CC' = AC' \times CB' \times BA'$, then the three lines AB', BC', and CA' pass through the same point.

Demonstration.—Let AB' and CA' be two of those lines intersecting at E; then, if the third line does not pass through E, when the products of the alternate segments taken in opposite directions are equal, let it have another direction, as BF. Now, in this case we have, by the hypothesis, $AA' \times BB' \times CF = AF \times CB' \times BA'$, and hence $\frac{AF}{CF} = \frac{AA' \times BB'}{CB' \times BA'}$. Join BE, and produce it to cut AC in C'; then, by Theorem VI., $AA' \times BB' \times CC' = AC' \times CB' \times BA'$. Hence $\frac{AC'}{CC'} = \frac{AA' \times BB'}{CB' \times BA'}$. Wherefore $\frac{AF}{CF} = \frac{AC'}{CC'}$, and we have $AF : AC' :: CF : CC'$. But AF is less than AC', being a *part* of it; hence CF must be less than its part CC', which is absurd. Whence the given relation can exist *only* when BF coincides with BEC', and *the three lines pass through the same point.* Q. E. D.

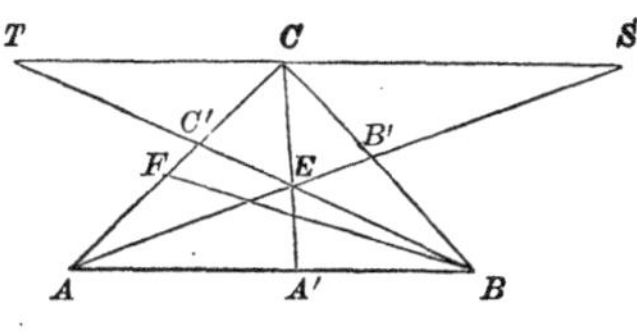

Scholium 1.—When the *three lines* AB', BC', and CA' *bisect the three angles of a triangle respectively, they pass through the same point.*

Demonstration.—Let the three lines AB', BC', and CA' *bisect the angles* A, B, and C respectively; then we have (IV. 17*)

$$AC' : CC' :: AB : BC.$$
$$CB' : BB' :: AC : AB.$$
$$BA' : AA' :: BC : AC.$$

Whence, by multiplying the corresponding terms of these proportions, we have $AC' \times CB' \times BA' : CC' \times BB' \times AA' :: AB \times AC \times BC : AB \times AC \times BC :: 1 : 1$. Hence $AC' \times CB' \times BA' = CC' \times BB' \times AA'$, and, by the theorem, the three lines pass through the same point.

Scholium 2.—When the three lines AB', BC', and CA' are *perpendicular to the opposite sides respectively, they pass through the same point.*

* VI. 3.

Demonstration.—In similar right-angled triangles $AC'B$ and $AA'C$, we have $AC' : AA' :: BC' : CA'$.

In similar right-angled triangles $BA'C$ and $BB'A$, we have $BA' : BB' :: CA' : AB'$.

In similar right-angled triangles $CB'A$ and $CC'B$, we have $CB' : CC' :: AB' : BC'$.

Multiplying these proportions, we have $AC' \times CB' \times BA' : AA' \times BB' \times CC' :: AB' \times CA' \times BC' : AB' \times CA' \times BC' :: 1 : 1$. Hence $AA' \times BB' \times CC' = AC' \times CB' \times BA'$, and *the three lines pass through the same point.* (See Prob. XXVI., "Triangles," etc.)

Scholium 3.—When the three lines AB', BC', and CA' *bisect the opposite sides respectively, they pass through the same point.* (See Prob. XI., "Triangles," etc.)

Demonstration.—We here have $AA' = BA'$, $BB' = CB'$, and $CC' = AC'$; hence, by multiplying these three equations together, we have $AA' \times BB' \times CC' = AC' \times CB' \times BA'$; wherefore, by the theorem, the *three lines pass through the same point.* (See Theorem I.)

THEOREM VIII., being an extension of the property of Theorem VI.—If a straight line be drawn to cut any two sides of a triangle, and the third side, one or all the sides being produced if necessary, thus dividing them into six segments (a *prolonged side* being taken as *one* segment, and its *prolongation* as *another*), then will the product of any three of these segments, *which are not adjacent*, be equal to the product of the other three.

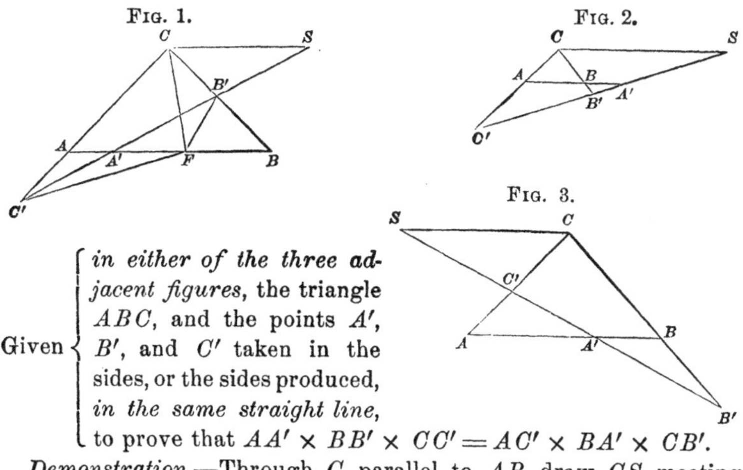

Given { *in either of the three adjacent figures*, the triangle ABC, and the points A', B', and C' taken in the sides, or the sides produced, *in the same straight line*, to prove that $AA' \times BB' \times CC' = AC' \times BA' \times CB'$.

Demonstration.—Through C, parallel to AB, draw CS, meeting the straight line in which the points A', B', and C' are situated, in S.

Then, by similar triangles $C'AA'$ and $C'CS$, we have $AC' : AA' :: CC' : CS$.

Also, by similar triangles $A'B'B$ and $CB'S$, we have $BA' : BB' :: CS : CB'$.

By multiplying the corresponding terms, we have $AC' \times BA' : AA' \times BB' :: CC' : CB'$; whence $AA' \times BB' \times CC' = AC' \times BA' \times CB'$. Q. E. D.

THEOREM IX.—Conversely, if in the three sides of a triangle, or these sides produced, three points be taken such that the product of *any three* of the six segments *not adjacent*, into which the sides with their prolongations are divided, shall be equal to the product of the *other three* segments, then will the three points so taken *be in the same straight line.*

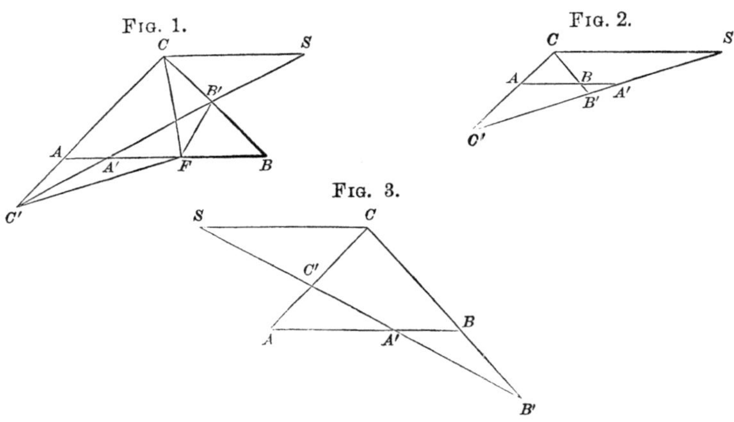

Demonstration.—We have to prove that if $AA' \times BB' \times CC' = AC' \times BA' \times CB'$, then the three points A', B', and C' will be in the same straight line.

Parallel to AB draw CS to meet the line in which the two points B' and C' are situated, in S. Then, if this line does not cut AB in A' (Fig. 1), suppose it to cut AB in some other point, as F. Then, by Theorem VIII., $AF \times BB' \times CC' = BF \times CB' \times AC'$, whence $\frac{AF}{BF} = \frac{CB' \times AC'}{BB' \times CC'}$. But from the given hypothesis $\frac{AA'}{BA'} = \frac{CB' \times AC'}{BB' \times CC'}$. Wherefore $\frac{AF}{BF} = \frac{AA'}{BA'}$, and $AF : AA' :: BF : BA'$.

But AF is greater than AA', wherefore BF is greater than BA', a part greater than the whole, which is absurd. Wherefore the proportion $AF : AA' : BF : BA'$ can be true *only* when F coincides with A'. Hence the three points A', B', and C' are in the same straight line. The same may be proved in like manner by either of the other figures.

NOTE.—It will be observed in each of the figures under Theorems VI. and VIII. that each side has two segments; and bearing in mind that when a side is produced *the whole line thus formed* is *one* segment, and the *produced part* is the other segment, then taking the sides in order, and beginning at any angle (as A), take the segment (AA') terminating there of the first side; then the segment (BB') of the second side, terminating at B; then the segment (CC') of the third side, terminating at C, and place their product equal to the product of the remaining three segments ($BA' \times CB' \times AC'$), and we have $AA' \times BB' \times CC' = BA' \times CB' \times AC'$. By a little attention in this manner, the order in which the segments are to be taken to form the equation in any case, may be readily fixed on the mind of the student.

THEOREM X.*—If a quadrilateral figure be in any manner divided into two quadrilateral figures, and diagonals be drawn to the whole quadrilaterals, and also to the two partial ones, then the points of intersection of these three pairs of diagonals *will be in the same straight line*. Required, the demonstration.

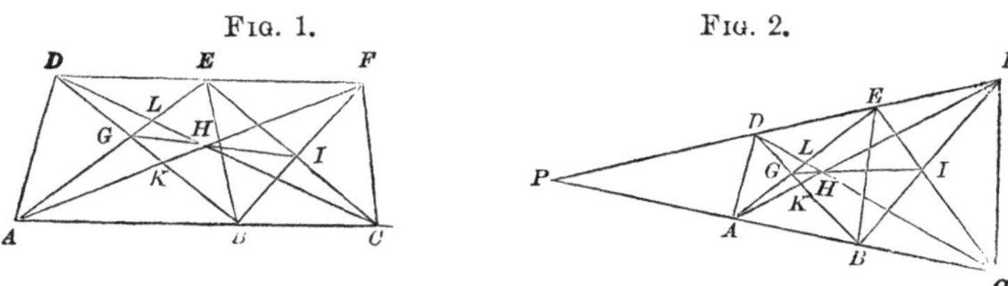

Demonstration.—We have to prove that the points G, H, and I, which are the intersections of the three pairs of diagonals of the quadrilaterals $ABED$, $ACFD$, and $BCFE$, respectively, *are in the same straight line.*

CASE 1.—When the lines DF and AC *are parallel* (Fig. 1).

In similar triangles BIC and FIE, we have $BI : FI :: BC : EF$; hence $BI \times EF = FI \times BC$.

In similar triangles FHD and CHA, we have $FH : AH :: DF : AC$; hence $FH \times AC = AH \times DF$.

* See Silliman's *Journal of Science*, vol. xxiii. p. 224, Second Series. See, also, Gillespie's *Land Surveying*, p. 387.

In similar triangles AGB and EGD, we have $AG:EG::BG:DG$.

By composition, we have $AG:AE::BG:BD$; hence $AG\times BD=AE\times BG$.

Since triangle ABG is cut by the line CLD, we have, Theorem VIII., $AL\times BC\times DG=AC\times BD\times GL$.

Since triangle DEG is cut by the line AKF, we have, Theorem VIII., $GK\times DF\times AE=DK\times EF\times AG$.

Since triangle AGK is cut by the line HLD, we have, Theorem VIII., $AH\times GL\times DK=AL\times DG\times HK$.

Fig. 1.

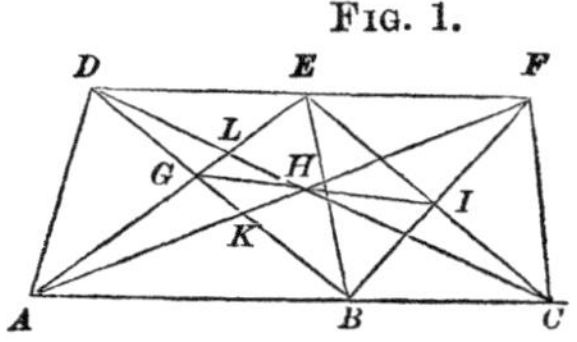

Fig. 2.

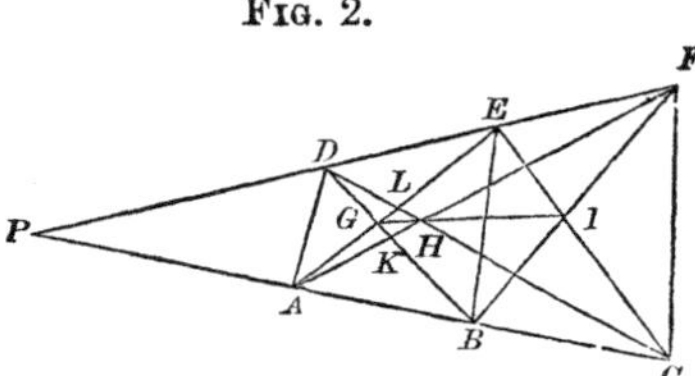

By multiplying the terms on each side of these six equations together, and observing that the twelve terms EF, AC, AG, BD, AL, BC, DG, DF, AE, AH, GL, and DK on one side cancel the same quantities on the other side, we have $BI\times FH\times GK=IF\times BG\times HK$; whence, by Theorem IX., *the points G, H, and I are in the same straight line*, the triangle whose sides are cut being KBF.

Case 2.—When the sides DF and AC are *not* parallel (Fig. 2), let these sides be produced, and they will meet in some point, as P.

Then, since triangle BFP is cut by the line CIE, we have, Theorem VIII., $BI\times EF\times CP=BC\times IF\times EP$.

Since triangle AFP is cut by the line CHD, we have, Theorem VIII., $AC\times FH\times DP=AH\times DF\times CP$.

Since triangle ABG is cut by the line CLD, we have, Theorem VIII., $AL\times BC\times DG=AC\times BD\times GL$.

Since triangle DEG is cut by the line BAP, we have, Theorem VIII., $BD\times EP\times AG=DP\times AE\times BG$.

Since triangle DEG is cut by the line AKF, we have, Theorem VIII., $DF\times AE\times GK=DK\times EF\times AG$.

Since triangle AGK is cut by the line DLH, we have, Theorem VIII., $AH\times GL\times DK=AL\times DG\times HK$.

By multiplying the terms on each side of these six equations respectively together, and observing that the fifteen terms EF, CP,

AC, DP, AL, BC, DG, BD, EP, AG, DF, AE, AH, GL, DK, on one side, cancel the same quantities on the other side, we have $BI \times FH \times GK = IF \times BG \times HK$; whence, by Theorem IX., the three points G, H, and I are in the same straight line, the triangle whose sides are cut being KBF.

NOTE.—The interesting property involved in this theorem was contained in a problem proposed for demonstration about the year 1830, in the last number of the *Mathematical Diary*, a monthly periodical, commenced by Prof. Robert Adrain in 1825, and afterwards conducted by James Ryan, of New York; but, the work being discontinued, no demonstration was published.

THEOREM XI.*—If in each of the three sides of a plane triangle a point be taken at pleasure, and circles be described through each angular point of the triangle, and the points taken in the two sides forming that angle, these three circles will intersect one another at a common point. Required, the demonstration.

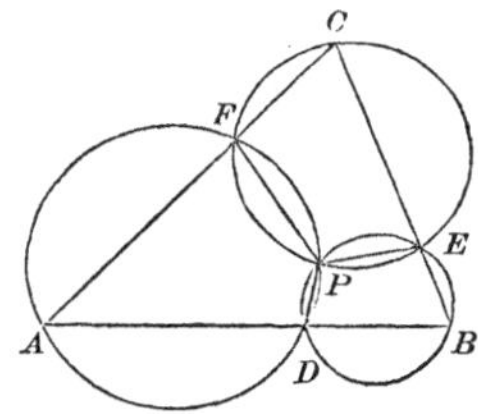

Let the points D, E, F be taken at pleasure in the three sides of the given triangle ABC; then if the circles passing through the points A, D, F and F, C, E intersect each other in the point P, the circle passing through the points E, B, D will pass through P also.

Demonstration.—Join DP, EP, and FP. Then, since $ADPF$ is a quadrilateral inscribed in a circle, its opposite angles A and P are equal to two right angles (III. 18, cor. 4†). For the same reason, the opposite angles C and P of the quadrilateral $FCEP$ are equal to two right angles. Whence these four angles A, C, DPF, and FPE are equal to four right angles. But the three angles DPF, FPE, and DPE are equal to four right angles. Whence the angles $A + C + DPF + FPE = DPF + FPE + DPE$. Wherefore $DPE = A + C$. To each add the angle B, and we have $DPE + B = A + C + B =$ two right angles; consequently the points D, P, E, and B, which designate the quadrilateral $DPEB$, are in the

* In seeking a mode of demonstration for Theorem XIV., following, which was sent to me some years ago by my valued cousin and mathematical correspondent, Benjamin Shoemaker, now of Germantown District, Philadelphia, who was my student in 1826, I discovered the interesting properties in Theorems XI., XII., and XIII. The same properties may, however, have been previously discovered by some one with whose writings I have not had the pleasure to meet.

† III. 22.

circumference of a circle, and hence the circumference of a circle which passes through the three points D, B, and E *must* pass through the point P, the intersection of the circles through the points A, D, F, and F, C, E. Q. E. D.

Corollary.—The angle $DPF =$ the sum of the angles B and C, the angle $FPE =$ the sum of the angles A and B, and the angle $DPE =$ the sum of the angles A and C.

THEOREM XII.—If in each of the three sides of a plane triangle any point be taken at pleasure, and a circle be described through each angular point of the triangle and the two points taken in the sides that form that angle, the triangle formed by joining the centres of these three circles will be equiangular and similar to the first triangle.

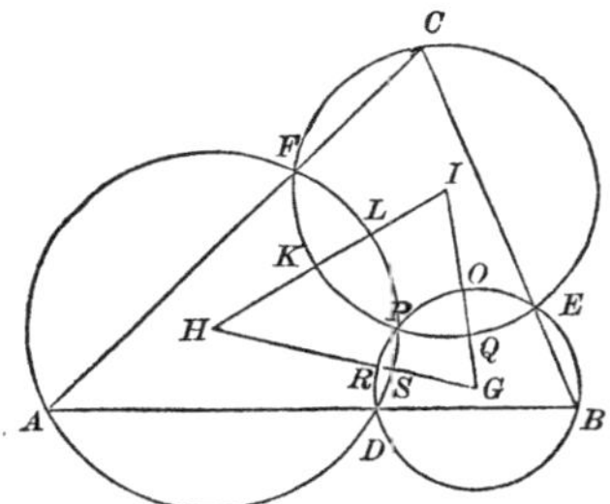

Prove that the triangle HGI formed by joining the centres of the circles $ADPF$, $CFPE$, and $BDPE$ is similar to the triangle ABC. By Theorem XI., the three circles pass through the same point P.

Demonstration.—The arc LP (III. 11*) is half the arc FLP, and the arc PS is half the arc PSD; hence the whole arc LPS, which measures the angle H, is half the whole arc FPD. But (III. 18†) the angle A is measured by half the same arc FPD; hence the angle H of the triangle HIG is equal to the angle A of the triangle ABC.

In like manner the arc RP is half the arc DRP, and the arc PO is half the arc POE; hence the arc RPO, which measures the angle G, is half the arc DPE. But the angle B is measured by half the arc DPE; therefore the angle G of the triangle HIG is equal to the angle B of the triangle ABC.

So, also, arc $KP = \frac{1}{2} FKP$, and arc $PQ = \frac{1}{2} PQE$; hence arc KPQ, which measures the angle I, is equal to half FPE. But $\frac{1}{2} FPE$ measures the angle C. Therefore the angle I of the triangle HIG is equal to the angle C of the triangle ABC. Consequently, the two triangles HIG and ABC are equiangular and similar. Q. E. D.

* III. 3. † III. 20.

THEOREM XIII.—If in each of the three sides of any plane triangle *any* point be taken at pleasure, all triangles whose sides pass through these points and terminate in the circular arcs passing through the angular points of the triangle and the points taken in the sides which form that angle, will be equiangular and similar.

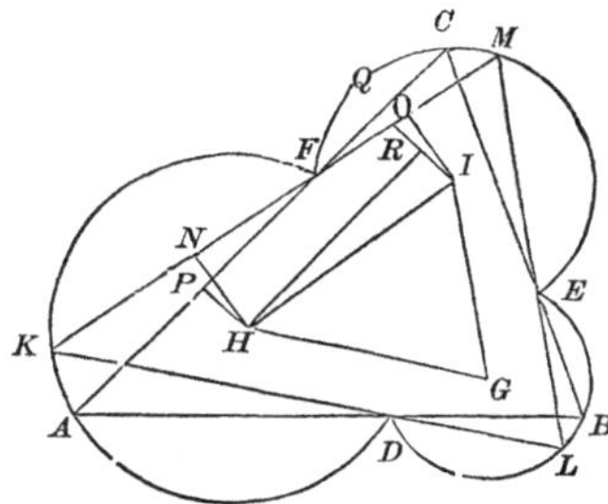

If the points D, E, and F are taken at pleasure in the sides of the triangle ABC, and the arcs FAD, DBE, and ECF described, then *any other* triangle, as KLM, whose sides pass through the points D, E, and F, and are terminated by the circular arcs, will be similar to ABC.

Demonstration.—Through the point D draw *any* line KDL, and through the points F and E draw the lines KF and LE, and let them be produced to meet in a point, as M. Then, since in the triangles ABC and KLM (III. 18, cor. 1*) the angle $K =$ angle A, they being in the same segment, and the angle $L =$ the angle B for the same reason, the angle M *must* be equal to the angle C, and consequently the point M is in the circular arc $FCME$, and the triangle KML is equiangular and similar to ABC. Q. E. D.

Corollary.—Join the centres H, I, G of the circular arcs forming the triangle HIG; then, as the triangle HIG, by Theorem XII., is similar to ABC, all the triangles formed, as KLM, will be similar to HIG and ABC, and hence similar to one another.

Scholium 1.—Of all the sides drawn through one of the points, as F, and terminating in the circular arcs FCE and DAF that intersect at that point, the longest or maximum line will be that (KFM) drawn parallel to the line HI, which joins the centres of those arcs. For let *any other* line, as AFC, pass through F, and on it let fall the perpendiculars HP and IQ. Also, on KM let fall the perpendiculars HN and IO, and draw HR parallel to AC. Then, in the right-angled triangle HRI, HR is less than the hypothenuse HI. Now (III. 6†) $AF = 2\,PF$, and $FC = 2\,FQ$; hence $AC = 2\,PQ = 2\,HR$. Also, $KF = 2\,NF$, and $FM = 2\,FO$; whence $KM = 2\,NO = 2\,HI$. And, since HI is greater than HR, we have KM ($= 2\,HI$) greater than AC ($= 2\,HR$). Q. E. D.

* III. 21. † III. 3.

Scholium 2.—Join KD and ME, and produce them to meet in L. Then, by the theorem, the point L will be in the circular arc DBE, and the triangle KLM will be similar to ABC. But KM is longer than *any other* side, as AC, drawn through F. Hence the area of the triangle KLM is greater than ABC, or than *any other triangle* whose sides pass through the points D, E, F, and terminate in the circular arcs DAF, FCE, and EBD, and it is therefore the *maximum* triangle which can be drawn similar to ABC, and having its sides to pass through the points D, E, and F.

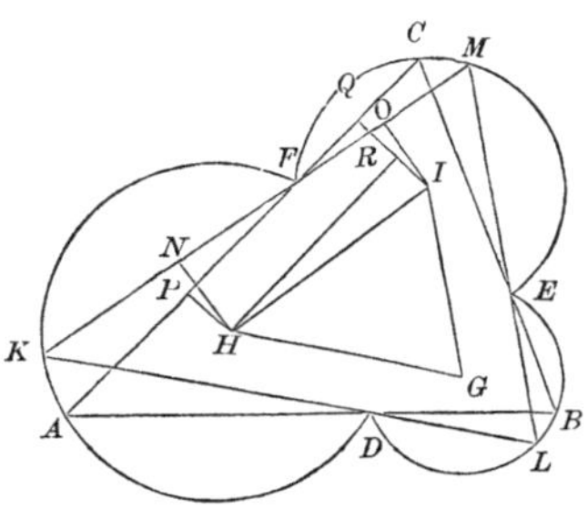

Scholium 3.—If the lines KL, LM, and MK be drawn through the points D, E, and F, respectively parallel to the sides HG, GI, and IH of the triangle GIH, the angular points K, L, and M will be in the circular arcs DAF, FCE, and EBD.

For (IV. 21*) the angles L, M, and K in the triangle LMK are respectively equal to the angles G, I, and H in the triangle GIH. But, by Theorem XII., the angles G, I, and H are respectively equal to the angles B, C, and A of the triangle ABC. Hence the angle L must equal the angle B, the angle $M=$ angle C, and angle $K=$ angle A, and the point K *must* be in the circular arc DAF, M in the arc FCE, and L in the arc EBD.

Scholium 4.—Hence the *maximum triangle* KLM has its sides parallel, respectively, to the sides of the triangle HIG, formed by joining the centres of the circular arcs DAF, FCE, and EBD.

* VI. 5 and 4.

THEOREM XIV. (Proposed by Benjamin Shoemaker.)—If on the three sides of any plane triangle equilateral triangles be described, the triangle formed by joining the centres of these equilateral triangles will be an equilateral triangle.

Demonstration.—On the three sides of the triangle DEF describe the equilateral triangles DBF, FAE, and ECD. Bisect the sides BD and DF in the points O and P, and join the points FO and BP by lines intersecting at H; then will H be the centre of gravity of the triangle DBF, and hence its centre. But FO and BP are perpendiculars from the middle of the lines BD and DF respectively, and hence their intersection H is the centre of a circular arc passing through the points D, B, and F. With the centre H and radius HB or HF describe the arc DBF. In like manner describe the arcs FAE and ECD, whose centres I and G will be the centres of the triangles FAE and ECD respectively. Join the points H, I, and G; it remains to be shown that HIG is an equilateral triangle.

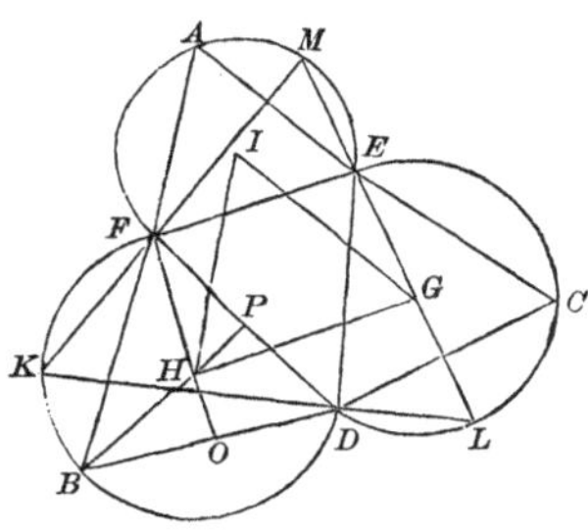

Through the point D draw *any line*, as KDL, and draw KFM and LEM, and the point M, where they meet, will, by Theorem XIII., be in the arc FAE, and the triangle KLM will be equiangular to HIG (cor. to Theorem XIII.). But angle K (III. 18, cor. 1*) = angle B = an angle of an equilateral triangle. For a similar reason the angles M and L, which are equal to the angles A and C, are each an angle of an equilateral triangle. Hence KLM is an equilateral triangle, and, consequently, the triangle HIG, which is similar to KLM (Theorem XII.), is an equilateral triangle also. Q. E. D.

Scholium.—This theorem resolves into Theorem XII., wherein the given triangle KLM, corresponding to ABC in Theorem XII., is equilateral.

* III. 21.

THEOREM XV.—The three perpendiculars, or these perpendiculars produced, let fall from the angular points of a triangle on the opposite sides, produced if necessary, will all intersect in a common point, as O.

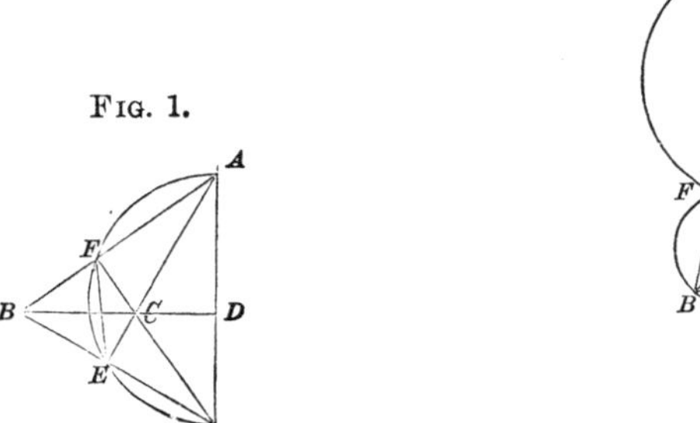

FIG. 1.

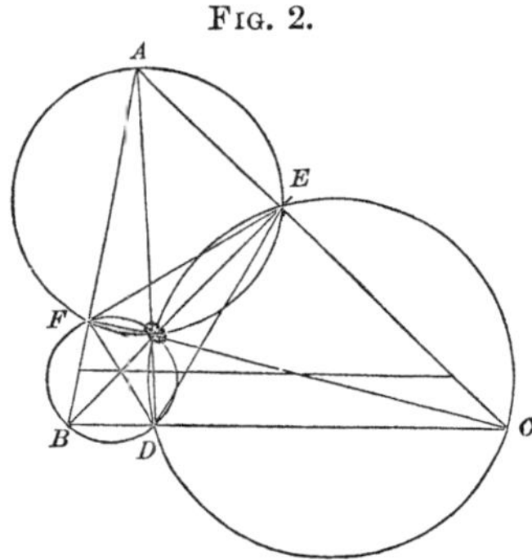

FIG. 2.

Demonstration.—Let ABC (in either figure) represent the triangle. Let fall the perpendiculars BE and CF on the sides AC and AB (produced if necessary), and let these perpendicular lines (produced if necessary) meet in O. Join AO, and let it (produced if necessary) cut BC, or BC produced, in D. Then, since the angles AFO and AEO are right angles, a circle described on AO as a diameter will pass through the points F and E. Let such circle be described, and join the points F and E.

Now, because the angle $FOB = EOC$, and the angle $BFO = CEO$, each being a right angle, the two triangles FOB and EOC are equiangular, and we have (IV. 18*) $OF : OE :: OB : OC$. Hence (IV. 20†) the triangles FOE and BOC are equiangular, and we have angle $CBO = EFO =$ (III. 18, cor. 1‡) EAO. Hence in the two triangles BOD and AOE, we have an angle DBO $(CBO) = EAO$, and angle $BOD = AOE$; wherefore angle $ODB = OEA$. But OEA is a right angle by construction. Hence ODB is a right angle, and AOD is perpendicular to BC, and the three perpendiculars intersect in a common point O. Q. E. D.

NOTE 1.—See Problem XXVI., "Triangles, Quadrilaterals," etc. Also, Scholium 2, Theorem VII.

NOTE 2.—For another and a very neat mode of demonstrating this theorem, and also Theorem I., see Propositions 42 and 43, Book I., of Chauvenet's Treatise on Geometry, an admirable work published in 1871, by J. B. Lippincott & Co., Philadelphia.

* VI. 4. † VI. 6. ‡ III. 21.

Property 1.—Join FD and ED. Since BDO and BFO are right angles, a circle described on BO as a diameter will pass through the points $BDOF$. For the same reason, a circle described on CO as a diameter will pass through the points $CDOE$. Hence, since vertical angles are equal, and angles in the same circular segment are equal, we have angle $BDF = BOF = EOC = EDC$, and angle $BFD = BOD = AOE = AFE$; also, angle $AEF = AOF = COD = CED$.

Therefore, since $BDF = EDC$, a ray of light proceeding from any point in the line ED in the direction of that line, and impinging on the side BC at D, making the angle of reflection equal to the angle of incidence, would be reflected to F; and then, since angle $AFE = BFD$, it would be reflected to E; and again, since $CED = AEF$, it would be reflected along the line ED to the point from which it proceeded. The same is true of every point in each side, and the ray may proceed around in either direction.

Property 2.—Since angle $BDF = EDC$, the complements ADF and ADE are equal; hence DA bisects the angle FDE.

And since angle $AFE = BFD$, their complements CFE and CFD are equal; and hence FC bisects the angle DFE.

In like manner, since angle $CED = AEF$, their complements BEF and BED are equal; hence BE bisects the angle FED.

Hence the three perpendiculars let fall upon the three sides of a triangle bisect the angles of the triangle formed by joining the points where these perpendiculars intersect the sides.*

Remark.—Property 1, above, suggests and solves the following interesting problem:—Determine a point within a given plane triangle from which a ray of light may proceed in the plane of the triangle, and, after being reflected from the several sides successively, making the angle of reflection equal to the angle of incidence, return to the same point again. Also, to find the locus of all such points.

It was shown in Property 1 that the point required is *any* one in either of the sides of the triangle DEF, and hence these three sides are the locus of such point.

* See Chauvenet's Geometry, Appendix I., Proposition 36.

PROBLEM IN TREE-PLANTING.—There are four trees standing, forming an irregular quadrilateral. How may five other trees be planted so that the nine will form ten straight rows with three trees in each row?

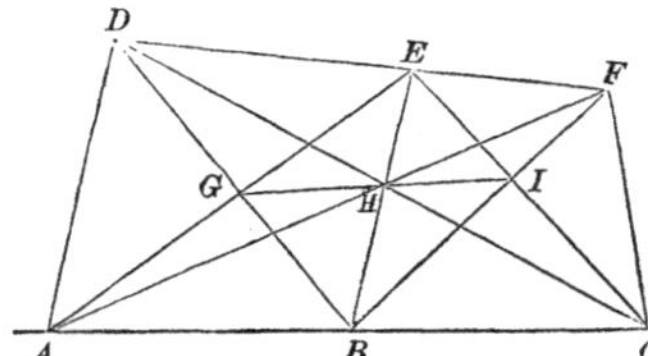

Let A, C, F, D represent the points where the four trees are; it is required to find five other points so that the nine shall form ten straight rows, with three points in each row.

Solution.—It is evident from Theorem X. that if any line, as EB, be drawn through H, the intersection of the diagonals AF and CD, dividing the quadrilateral $ACFD$ into two quadrilaterals, and the diagonals AE and BD, BF and CE, be drawn, intersecting in G and I respectively, G, H, and I will be in a straight line, and hence the points B, E, G, H, I will be the places at which the other five trees must be planted. The rows are ABC, DEF, GHI, EHB, and one row in each of the six diagonals, making ten straight rows, with three trees in each row.

NOTE.—Each of the trees B, H, and E is in four different rows, each of the others in three.

Calculation.—1. In the given quadrilateral $ACFD$, find, Case 3, the diagonal CD, and the angles FDC and FCD. Also, the diagonal AF, and angles DAF and DFA.

2. In triangle DHF, Case 1, find DH and HF; whence CH and HA become known. In like manner find EG and EI, and angles GEH and HEI; for, the point E being assumed in the construction, FE and ED are known.

3. In the triangle EFH, Case 3, find angle $FHE = AHB$, and EH. In triangle AHB, Case 1, find BH and AB; whence BC is known.

4. In triangles GEH and HEI, Case 3, find GH and HI.

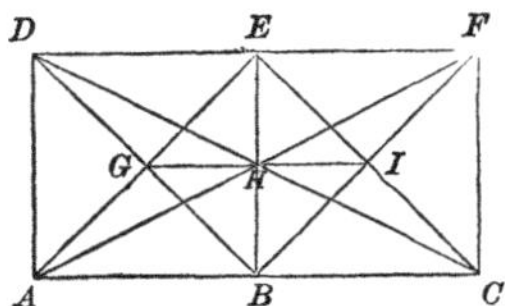

Scholium.—If $ACFD$ is a rectangle, with the length $AC =$ twice the breadth AD, the line EHB will divide it into two equal squares, as in the annexed figure, and the trees would have a symmetrical position. The ten rows are the three parallel lines DEF, GHI, and ABC, the one at right angles to these, and those in the six diagonals. GH and HI each equals $\frac{1}{2}AB$ or BC, and AG, BG, BI, CI, etc., each $=\sqrt{\frac{1}{2}AB^2}$.

PROBLEM IN TREE-PLANTING. (Proposed in *The Agriculturist.*)— Plant a grove of nineteen trees so that they shall constitute nine straight rows, with five trees in each row.

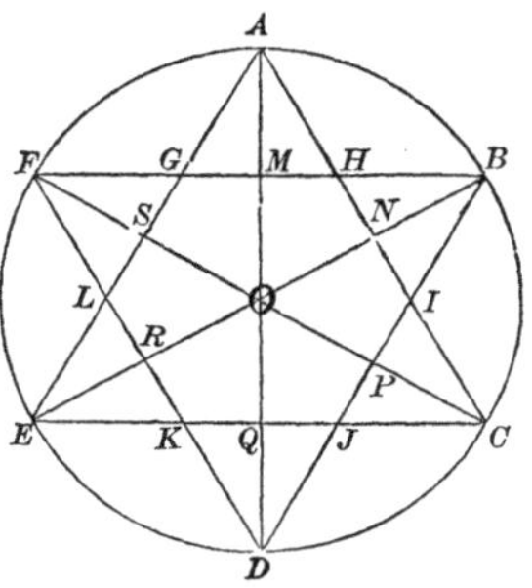

Construction.—On the circumference of the circle whose centre is O lay the radius OA six times, beginning and ending at A, giving the points A, B, C, D, E, F (V. 4*). Draw the six equal chords AC, BD, CE, DF, EA, and FB, forming two equal inscribed equilateral triangles ACE and BDF, and the regular hexagon $GHIJKL$. Draw also the three diameters AOD, BOE, and COF, bisecting the sides of the two inscribed equilateral triangles in the six points M, N, P, Q, R, and S.

Now, if a tree be planted at each point that is lettered, the nineteen trees will be planted as required.

For there are evidently five in each of the sides of the two inscribed equilateral triangles, making six rows, with five trees in each row, and there are five in each of the three diameters, making three rows more; hence there are nine rows, with five trees in each row, as required.

* IV. 15.

DIFFERENTIAL AND INTEGRAL CALCULUS.

AN ATTEMPT AT SIMPLE ILLUSTRATIONS OF SOME PROBLEMS AND PRINCIPLES IN THE DIFFERENTIAL AND INTEGRAL CALCULUS.

Preliminary Remarks.—To find the differential of any power of a variable quantity, *multiply by the index denoting the power, diminish the index by unity, and multiply by the differential of the root.*

The differential of x is written $d.x$, or simply dx; of y, dy; of z, dz; d being merely a *symbol* denoting differential. Hence in problems involving the calculus d is not usually employed by mathematicians to denote a *quantity.*

Examples.—1. The differential of x or x^1, by the rule, is $1 \times x^{1-1} \times dx = x^0 dx = 1 \times dx = dx$.*

2. The differential of $x^3 = 3 \times x^2 \times dx = 3x^2dx$. [The letter d before any quantity implies its differential.]

3. $d.ax^4 = 4 \times ax^3 \times dx = 4ax^3dx$. [A known or constant quantity has no differential; hence $d.ax^4 = a \times d.x^4 = a \times 4x^3dx = 4ax^3dx$.]

4. $d.3ax^5 = 5 \times 3ax^4 \times dx = 15ax^4dx$.

5. $d.(ax^2 + 4x^3 + 5x^4) = 2axdx + 12x^2dx + 20x^3dx$.

6. $d.(by^2 + 3y + a - 4) = 2bydy + 3dy$.

7. $d.\frac{a}{x^3} = d.ax^{-3} = -3 \times ax^{-3-1} \times dx = -3ax^{-4}dx = -\frac{3adx}{x^4}$.

* The division of powers of the same quantity or root is effected by subtracting the index of the divisor from the index of the dividend, and the remainder is the index of the quotient. That is, $\frac{x^5}{x^3} = x^{5-3} = x^2$, $\frac{x^5}{x^5} = x^{5-5} = x^0$. But $\frac{x^5}{x^5} = 1$; hence $x^0 = 1$, $\frac{x^2}{x^4} = x^{2-4} = x^{-2} = \frac{1}{x^2}$. Hence a power may be taken from the numerator to the denominator, or *vice versa*, by changing the sign of the index.

8. $d.\sqrt{a^2 + x^2} = d.(a^2 + x^2)^{\frac{1}{2}} = \frac{1}{2} \times (x^2 + a^2)^{-\frac{1}{2}} \times 2xdx = \frac{xdx}{(x^2 + a^2)^{\frac{1}{2}}} = \frac{xdx}{\sqrt{x^2 + a^2}}$.

9. $d.\frac{a}{\sqrt{a^2 + x^2}} = d.a(a^2 + x^2)^{-\frac{1}{2}} = -\frac{1}{2}a \times (a^2 + x^2)^{-\frac{1}{2}-1} \times 2xdx$ $= -\frac{1}{2}a(a^2 + x^2)^{-\frac{3}{2}} \times 2xdx = -\frac{axdx}{(a^2 + x^2)^{\frac{3}{2}}} = -\frac{axdx}{\sqrt{(a^2 + x^2)^3}}$.

The term employed for this process is to *differentiate, differentiating*, or *differentiation*, according to the connection in which the term is used.

Now, the *reverse* process—that is, from the differential to obtain the quantity from which the differential is supposed to have been or may be derived—is to *integrate* the differential, called, also, *integrating* or *integration.*

The sign or symbol used to indicate this process is $\int$. As the sign $\sqrt{}$ to denote root is derived from the sign or letter R, or $\sqrt{}$, so $\int$, as the sign for *integration*, is derived from a form of the letter S, or $\int$, meaning *the sum of all the increments*, or, in fluxions, *the sum of all possible positions of the flowing quantity.*

To *integrate* a differential, *reverse* the process of differentiating; that is, *increase* the *index* denoting *the power* by *unity*, *divide by the index so increased*, and also *by the differential of the root.*

Examples.—1. Taking the preceding examples, and reversing the process, we have $\int dx = \int x^0 dx = \frac{x^1}{1} = x.$

2. $\int 3x^2dx = \frac{3x^{2+1}dx}{3 \times dx} = x^3.$

3. $\int 4ax^3dx = \frac{4ax^{3+1}dx}{4dx} = ax^4.$

4. $\int 15ax^4dx = \frac{15 \times ax^{4+1}dx}{5 \times dx} = 3ax^5.$

5. $\int(2axdx + 12x^2dx + 20x^3dx) = \int 2axdx + \int 12x^2dx + \int 20x^3dx$ $= ax^2 + 4x^3 + 5x^4.$

6. $\int(2bydy + 3dy) = \int 2bydy + \int 3dy = by^2 + 3y + C.$*

* It will be seen that in the *reverse process* of example 6 we do not obtain the known quantities a and —4, annexed by the sign + or —, which were in the *function* from which this differential was obtained. These, being *constant quantities*, have no differential; and, being connected with the variable terms by the signs plus and minus, they

7. $\int -\frac{3adx}{x^4} = \int -3ax^{-4}dx = -\frac{3ax^{-4+1}dx}{-3dx} = ax^{-3} = \frac{a}{x^3}.$

8. $\int \frac{xdx}{\sqrt{a^2+x^2}} = \int xdx\,(a^2+x^2)^{-\frac{1}{2}} = \frac{(a^2+x^2)^{-\frac{1}{2}+1}xdx}{\frac{1}{2} \times 2xdx} = (a^2+x^2)^{\frac{1}{2}}$
$= \sqrt{a^2+x^2}.$

9. $\int -\frac{axdx}{\sqrt{(a^2+x^2)^3}} = \int -axdx \times (a^2+x^2)^{-\frac{3}{2}} = -\frac{axdx(a^2+x^2)^{-\frac{3}{2}+1}}{-\frac{1}{2} \times 2xdx}$
$= a(a^2+x^2)^{-\frac{1}{2}} = \frac{a}{\sqrt{a^2+x^2}}.$

THE DIFFERENTIAL CO-EFFICIENT.

Definition.—A quantity to differentiate or to integrate is called a *function.*

Example.—If we have a function of the form $u = 5x^3 + 2x^2 + 6x$, we have $du = 15x^2dx + 4xdx + 6dx = (15x^2 + 4x + 6)dx$. Hence $\frac{du}{dx} = 15x^2 + 4x + 6$. Now, this last quantity $(15x^2 + 4x + 6)$ is called the *differential co-efficient* of the *function u*, because it is what the differential, dx, is multiplied by in the value of du. And, as $\frac{du}{dx}$ is what this differential co-efficient is equal to always, the symbol $\frac{du}{dx}$ is adopted as denoting the differential co-efficient. Hence *the differential co-efficient of any function* is the *differential of the function divided by the differential of the variable.*

This quotient expresses, also, the *ratio* of the *differential* of the *function* to the *differential* of the *variable.* The *limit* of this ratio, which is of much use in investigations by the calculus, is determined by supposing the variable, as x in the preceding example, to become 0. Then we have $\frac{du}{dx}$ (that is, the differential of the function u divided by the differential of the variable x) $= 15x^2 + 4x + 6$, which,

necessarily disappear in the differential, and consequently they cannot be restored in the integration, because the differential would be unaffected no matter how many of these constants were in the quantity differentiated connected by the signs + and —, nor of what value they were. The *whole sum of these constants* are, therefore, in integrations, represented by the letter C, implying the sum of these constants, as in example 6. In fact, C is always to be understood as appended to the integral of any function, whether expressed or not. In many problems in the calculus the value of C can be determined by the conditions of the problem.

when $x = 0$, becomes 6, and this is the *limiting ratio* of the differential of the function $5x^3 + 2x^2 + 6x$ divided by the differential of the variable x. But, when x becomes 0, $u = 5x^3 + 2x^2 + 6x$ becomes 0, and du becomes 0, and dx becomes 0, and $\frac{du}{dx} = \frac{0}{0} = 15x^2 + 4x + 6$; and much severe criticism has been bestowed upon a mode of investigation in science, and especially in mathematics, which employs such seemingly intangible nonentities. A learned author has called them the "ghosts of departed quantities," a value remaining *in their relation* when the quantities themselves have vanished.

We must remember, however, that it is not the *quantities themselves*, but *their ratio*, which is the object to be determined, and the *ratio* remains unaltered through whatever changes of proportional increase or diminution *the terms of the ratio* may undergo. To illustrate: The ratio of 3 to 6 is $\frac{3}{6} = \frac{1}{2}$. Now, if each term of this ratio be multiplied by a million, the ratio is still $\frac{1}{2}$. And if each term of the ratio is multiplied by a million for a million of times, or a million million of times, the ratio still remains unaltered, being $\frac{1}{2}$, no more, no less.

Again, if, instead of *multiplying*, the terms be each *divided* by a million, the ratio remains $\frac{1}{2}$. And if each term of the ratio be divided by a million for a million of times, or a million million of times, the ratio continues to be $\frac{1}{2}$, no more, no less. And that which thus maintains its fixed value, eternally, through all conceivable changes of magnitude of its terms possible, large and small, it is reasonable to conclude will retain the same value when the terms each become 0, which is the value of $\frac{0}{0}$.*

* $\frac{0}{0}$ is the *symbol for indefiniteness* in quantities, and it has sometimes been used to perplex young algebraists, as *seeming* to lead to incongruous results. To illustrate:

Put $x = a$, a being *any* number whatever. Then $x^2 = a^2$. Take the first equation from the second, and we have $x^2 - x = a^2 - a$, or, by transposition, $a - x = a^2 - x^2$. Divide by $a - x$, and we have $1 = a + x = 2a$. Now, taking $a = 1, 1\frac{1}{2}, 2, 2\frac{1}{2}, 3, 3\frac{1}{2}$, etc., we have $1 = 2 = 3 = 4 = 5 = 6 = 7$, etc. to infinity.

Again, put $x = a$, as before, then $x^3 = a^3$; subtracting, $x^3 - x = a^3 - a$, or $a - x = a^3 - x^3$. Divide by $a - x$, and we have $1 = a^2 + ax + x^2 = 3a^2$, where a may be *any* value from 1 to infinity. In like manner $a - x = a^4 - x^4$. Divide by $a - x$, and $1 = a^3 + a^2x + ax^2 + x^3 = 4a^3$, where a may be *any* value from 1 to infinity. This process, if continued till the powers of x and a are infinite, will give, in each case, an infinite number of values for the quotient of the first term, $\frac{a - x}{a - x}$.

It will be observed, however, that in all these cases, since $x = a$, $x - a = 0$, $x^2 - a^2$

PROBLEM.—When y' represents a *line* or *area* which moves continually parallel to itself, and its value in any position is known, either definitely or in terms of x, then the value of the surface or solid, generated by such quantity in its movement, will be $\int y'dx$, x being the distance through which the line or surface moves. (See Simpson's *Fluxions*, or Young's *Integral Calculus.*)

Example 1.—To find the area of a parallelogram whose base and height are given.

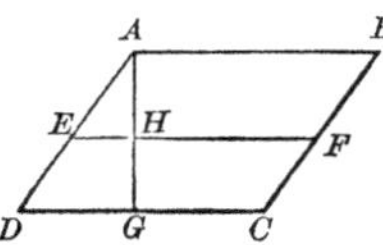

Put the height $AG = a$, the base $CD = b$, $x = AH$, and $y = EF$, which is supposed to begin to move at CD, and to move parallel to itself up the line GA to AB, thus generating the parallelogram $ABCD$. Now, EF in this case is known *definitely*, being equal to $b = y'$. Hence the area $ABCD = \int y'dx = \int bdx = bx =$ (when $x = a$) ab, or the area is equal to the product of the base by the altitude.

NOTE.—If x be taken $= \frac{1}{3}a$, we get $\frac{1}{3}$ the area of $ABCD$; if $x = \frac{1}{4}a$, we get $\frac{1}{4}$ the area, etc.

Example 2.—To find the area of a triangle when the base and height are given.

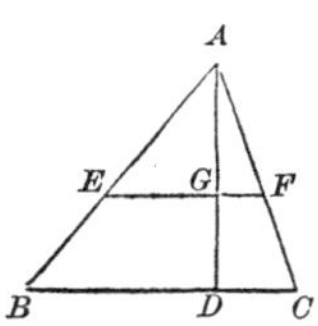

Here the variable line EF, or y', is supposed to move from BC parallel to itself, terminated in every position by the lines AB, AC of the triangle, and thus generating the triangle ABC.

Put $AD = a$, $BC = b$, $AG = x$, $EF = y$; then $a : b :: x : y = \frac{bx}{a} = y'$. Whence the area $ABC = \int y'dx = \int \frac{bxdx}{a} = \frac{bx^2}{2a} =$ (when $x = a$) $\frac{ba^2}{2a} = \frac{ab}{2} = \frac{1}{2}ab$. Hence the

$= 0$, $x^3 - a^3 = 0$, etc., and what we obtain is $\frac{x-a}{x-a}$, or $\frac{0}{0}$, which is not limited to 1, but is a *symbol*, the number of values of which, as shown above, is *infinity multiplied by infinity*, or *infinity's square*. Hence $\frac{0}{0}$ is appropriately termed the *symbol of indefiniteness*. The same principle is familiar in practical arithmetic. Bearing in mind that the product of two factors divided by one factor gives the other factor, taking the product $0 \times 1 = 0$, and dividing by the factor 0, we have $1 = \frac{0}{0}$. Also, $0 \times 9 = 0$; hence $9 = \frac{0}{0}$. And 0×9 millions $= 0$; hence 9 millions $= \frac{0}{0}$, and so on through all numbers, however large or however small.

area of the triangle = half the product of the base by the altitude, or half the parallelogram of the same base and altitude.

Example 3.—To find the area of a parabola.

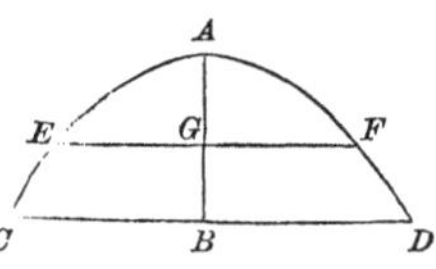

Here EF is the generating line for the *whole* parabola, beginning at CD, moving parallel to itself, and terminating at A; and EG is the generating line for the semi-parabola $CEAB$. Put $AB = a$, $BC = b$, $AG = x$, and $EG = y$; then, by the property of the parabola (see Prob. I., "Parabola," p. 253), we have $a : x :: b^2 : y^2$; whence $y^2 = \frac{b^2x}{a}$, or $y = \frac{b}{\sqrt{a}} \times \sqrt{x} = y'$; and the area of $CEAB = \int y'dx = \int \frac{b}{\sqrt{a}} \times \sqrt{x} \times dx = \int \frac{b}{\sqrt{a}} x^{\frac{1}{2}} dx = \frac{b}{\sqrt{a}} \times \frac{x^{\frac{3}{2}}}{\frac{3}{2}} = \frac{2}{3} \cdot \frac{bx^{\frac{3}{2}}}{\sqrt{a}} =$ (when $x = a$) $\frac{2}{3} \frac{ba^{\frac{3}{2}}}{a^{\frac{1}{2}}} = \frac{2}{3}ba =$ the area ABC. Hence the area of the whole parabola $CAD = \frac{2}{3} AB \times CD = \frac{2}{3}$ the rectangle under AB and CD.

Example 4.—To find the area of a semicircle whose radius is given.

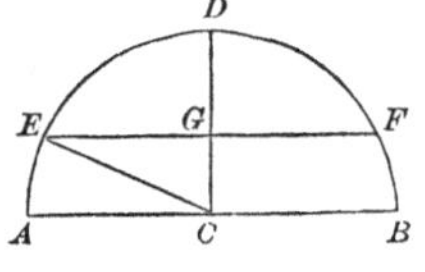

Here the variable line EF, or y, is supposed to move from AB, parallel to itself, limited by the extremity E of the constant line $CE = \frac{1}{2} AB$, thus generating the semicircle ADB. Now, while EF generates the semicircle ADB, EG generates, in like manner, the quadrant ADC. To find the area of the quadrant, therefore, put AC or $CE = r$, and $CG = x$; then $EG = \sqrt{r^2 - x^2} = y = y'$. And the area of the quadrant $ADC = \int y'dx = \int \sqrt{r^2 - x^2} \times dx = \int (r^2 - x^2)^{\frac{1}{2}} dx$ = (by developing $(x^2 - x^2)^{\frac{1}{2}}$ into a series, and multiplying each term by dx) $\int rdx - \int \frac{1}{2} \frac{x^2dx}{r} - \int \frac{1}{2.4} \cdot \frac{x^4dx}{r^3} - \int \frac{1.3}{2.4.6} \cdot \frac{x^6dx}{r^5} - \int \frac{1.3.5}{2.4.6.8} \cdot \frac{x^8dx}{r^7} - \int \frac{1.3.5.7}{2.4.6.8.10} \cdot \frac{x^{10}dx}{r^9}$, etc. $= rx - \frac{x^3}{2.3.r} - \frac{1}{2.4.5} \cdot \frac{x^5}{r^3} - \frac{1.3}{2.4.6.7} \cdot \frac{x^7}{r^5} - \frac{1.3.5}{2.4.6.8.9} \cdot \frac{x^9}{r^7} - \frac{1.3.5.7}{2.4.6.8.10.11} \cdot \frac{x^{11}}{r^9} -$ etc. = (when $x = r$) $r^2 \left[1 - \frac{1}{2.3} - \frac{1}{2.4.5} - \frac{1.3}{2.4.6.7} - \frac{1.3.5}{2.4.6.8.9} - \frac{1.3.5.7}{2.4.6.8.10.11} - \text{etc.}\right] = r^2 \times .78539 =$ the area of the quadrant

AEDC. The area of the semicircle $ADB = 2r^2 \times .78539$, and of the whole circle $= 4r^2 \times .78539 = D^2 \times .78539$. The sum of the series within the brackets (.78539) is the area of a *quadrant* whose *radius* is 1, and hence it is the area of a *circle* whose *diameter* is 1.

NOTE.—The above series converges *very slowly*, requiring some twelve or fifteen terms to obtain the above result; but, with the exception of the part of the denominator of each term which arises from the exponent of x, each term can be obtained from the preceding one with comparatively little labor.

Example 5.—To find the solidity of a cylinder, having the diameter of the base and the altitude given.

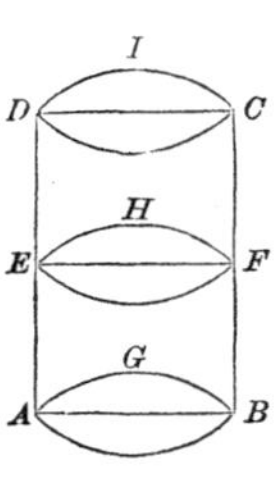

Here the area of the circle EHF is y, or the generating quantity, and it is known *definitely*, being equal to the base AGB.

Put $AD = a$, $AB = b$, $AE = x$, area $EHF = y =$ area $AGB =$ (putting $p = .78539$) $pb^2 = y'$. Then the solidity of the cylinder $= \int y'dx = \int pb^2dx = pb^2x$ $=$ (when $x = a$) $pb^2a =$ the area of the base multiplied by the altitude.

Example 6.—To find the solidity of a cone, having the diameter of the base and the altitude given.

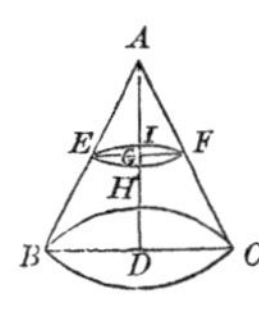

Here the area of the circle EIF is y, the generating quantity. Put $AD = a$, the diameter $BC = b$, $AG = x$; then $a : b :: x : EF = \frac{bx}{a}$, and area $EIF =$ $pEF^2 = p\frac{b^2x^2}{a^2} = y'$. Hence the solidity of the cone $=$ $\int y'dx = \int \frac{pb^2x^2dx}{a^2} = \frac{pb^2x^3}{3a^2} =$ (when $x = a$) $pb^2 \times \frac{1}{3}a =$ the area of the base multiplied by one-third of the altitude, or one-third of the cylinder of the same base and altitude.

Example 7.—To find the solidity of the paraboloid when the base and height are given.

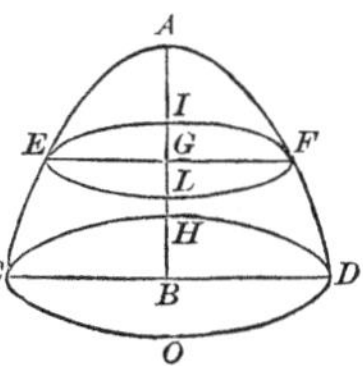

Here the area of the circle $EIFL$ is the generating quantity, or y, beginning at the base $CHDO$, and moving parallel to itself up to A, generating the paraboloid $ACHDO$. Put $AB = a$, the diameter $CD = b$, $AG = x$; then (by Prob. I. of the "Parabola," p. 253) $a : x :: b^2 : EF^2 = \frac{b^2x}{a}$. Now, y = area of circle $EIF = pEF^2 = \frac{pb^2x}{a} = y'$. Hence the solidity of the paraboloid $= \int y'dx = \int \frac{pb^2xdx}{a} = \frac{pb^2}{a}\int xdx = \frac{pb^2}{a} \times \frac{x^2}{2} = \frac{pb^2x^2}{2a} =$ (when $x = a$) $pb^2 \times \frac{1}{2}a$ = the area of the base CHD multiplied by half the altitude AB = half the cylinder of the same base and altitude.

Example 8.—To find the solidity of a hemisphere, and thence of the sphere, having the radius given.

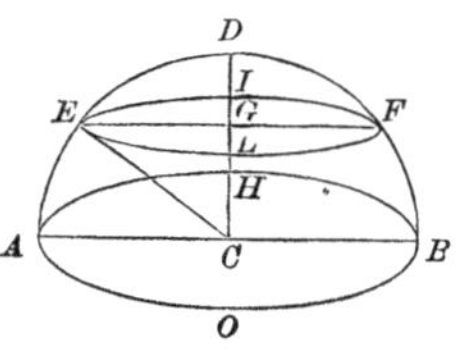

Here the area of the circle $EIFL$ is the generating quantity, or y, beginning at the base $AHBO$, and moving parallel to itself up to D, generating the hemisphere $DAHBO$. Put radius CA or $CE = r$, $CG = x$, $p = 4 \times .78539 = 3.1416$; then $EG^2 = r^2 - x^2$, and area $EIF = pEG^2 = p(r^2 - x^2) = y'$. Hence the solidity of the hemisphere $= \int y'dx = \int p(r^2 - x^2)dx = \int pr^2dx - \int px^2dx = pr^2x - \frac{px^3}{3} =$ (when $x = r$) $p\left(r^3 - \frac{r^3}{3}\right) = p \,.\, \frac{2}{3}r^3$ = (since $D = 2r$, and $D^3 = 8r^3$) $\frac{pD^3}{12} = \frac{p}{6} \times \frac{D^3}{2} = .5236 \times \frac{D^3}{2}$. Hence the solidity of the whole sphere $= D^3 \times .5236$.

Scholium.—Since $p \,.\, \frac{2}{3}r^3$ = the solidity of the hemisphere, the solidity of the whole sphere will be $p \,.\, \frac{4r^3}{3} = p \,.\, 4r^2 \times \frac{r}{3}$ = 4 times the area of a great circle of the sphere multiplied by one-third of the radius. (See Legendre's Geometry, Book VIII., Proposition XIV. and corollaries.)

Example 9.—To find the solidity of a spheroid; that is, of the solid described by an ellipse revolved about either of its axes.

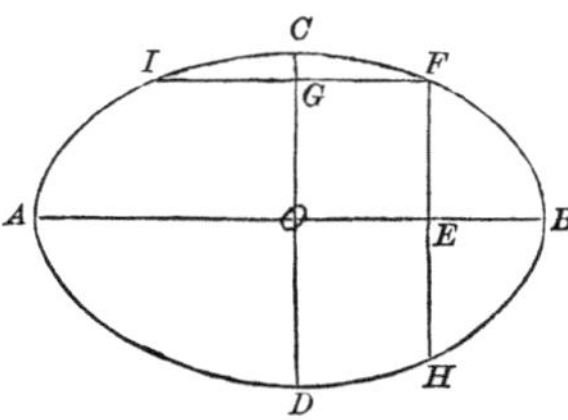

Case 1.—Let the revolution of the ellipse be about the *transverse axis* AB, which gives the *prolate* spheroid; then the ordinate EF will describe a circle, which is the *generating circle*, $= y'$, supposed to move from and parallel to COD to B. The axis *about* which the revolution is made is called the *fixed axe*. The equation of the ellipse (see Prob. I. of "Ellipse," p. 253) is $a^2y^2 + b^2x^2 = a^2b^2$, where $a = OB$, $b = OC$, $x = OE = GF$, $y = EF = OG$. Put $p = 3.1416$, the area of a circle whose radius is 1. Now, the area of the generating circle $FEH = EF^2 \times p = y^2 \times p = \frac{b^2}{a^2}(a^2 - x^2)p = y'$. Hence the solidity $= \int y'dx = \int \frac{b^2}{a^2}(a^2 - x^2)\,pdx = \int b^2pdx - \int \frac{b^2px^2dx}{a^2} = b^2px - \frac{b^2px^3}{3a^2} =$ (when $x = a$) $ab^2p - \frac{ab^2p}{3} = \frac{2}{3}ab^2p =$ the solidity of the semi-spheroid $CODB$. The whole spheroid $ADBC = \frac{4}{3}ab^2p$. Now, putting the major axis $AB = A$, and the minor axis $CD = B$, we have $a = \frac{A}{2}$, and $b = \frac{B}{2}$; substituting these values for a and b in the equation $\frac{4}{3}ab^2p$, it becomes $\frac{4}{3} \times \frac{A}{2} \times \frac{B^2}{4} \times 3.1416 = A \times B^2 \times \frac{3.1416}{6} = A \times B^2 \times .5236 =$ the solidity of the whole prolate spheroid $ADBC$, where the ellipse is revolved about AB.

Case 2.—When the revolution is about the *conjugate* axis CD, then the ordinate GF describes the circle IGF, which is the generating circle, moving from and parallel to the circle AOB to C. Hence, as x is the *moving* line and y the *space moved through*, the solidity in this case will be $\int x'dy$.

The circle $IF = GF^2 \times p = x^2p = \frac{a^2}{b^2}(b^2 - y^2)p = x'$, and $\int x'dy = \int \frac{a^2}{b^2}(b^2 - y^2)pdy = \int a^2pdy - \int \frac{a^2py^2dy}{b^2} = a^2py - \frac{a^2py^3}{3b^2} =$ (when $y = OC = b$) $a^2bp - \frac{a^2bp}{3} = \frac{2a^2bp}{3} =$ the solidity of the semi-spheroid

$ACBO$. The *whole* spheroid $ADBC = \frac{4}{3}a^2bp = \frac{4}{3} \times \frac{A^2}{4} \times \frac{B}{2} \times 3.1416 = A^2 \times B \times \frac{3.1416}{6} = A^2 \times B \times .5236$. In either case the solidity of the *whole spheroid* equals the *square of the revolving axe multiplied by the fixed axe, and the product multiplied by* .5236, which is one-sixth of the area of a circle whose radius is 1.

SOME ELEMENTARY PROBLEMS

IN

ANALYTICAL GEOMETRY.

Introductory Remarks.—In many mathematical investigations the operations are much simplified by designating a point, line, or surface by an *equation*, referring its position to two lines called *axes*, crossing each other at right angles.

This method of representing points, lines, and surfaces by algebraic equations was first introduced by Des Cartes in his Geometry, published in the year 1637; but the principle has been extended and amplified by Lagrange, Delambre, Laplace, and many others, and it has now become one of the most powerful instruments in mathematical research.

It is proposed to give here an idea of the principle, accompanied by a few practical problems as illustrations.

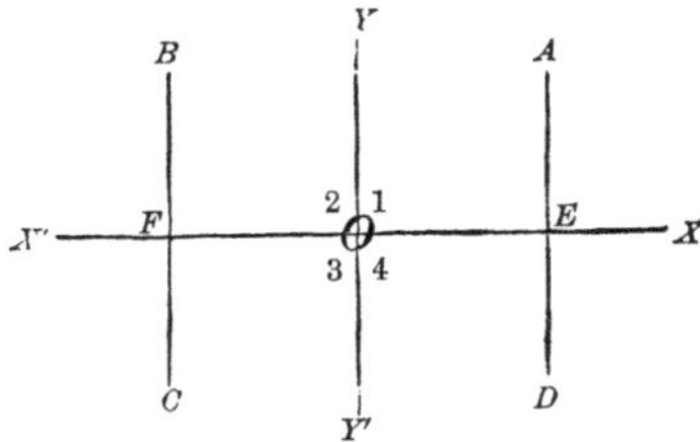

Let XX', YY', in the adjacent figure, represent the lines crossing each other at right angles in the point O, and let these lines be taken as *axes*, to which it is proposed to refer points and lines *on the same plane*. XX' is called the *horizontal* axis, YY' the *vertical* axis, and their point of intersection, O, the *origin of the axes*, or simply the *origin*.

The perpendicular distance, either way, from the vertical axis, measured on the *horizontal* axis, or on a line parallel thereto, is called an *abscissa;* and such distance is generally represented by the *letter* x.

The perpendicular distance, either way, from the horizontal axis, measured on the vertical axis, or on a line parallel thereto, is called an *ordinate*, and such distance is generally represented by the *letter* y.

Thus, designating the four angles at the origin by 1, 2, 3, 4, as in the figure, and drawing AED, BFC parallel to the axis YY', then taking the point A, in angle 1, OE is the abscissa (x), and EA the ordinate (y). In like manner, of the point B, in angle 2, OF is the abscissa (x), and FB the ordinate (y). Of the point C, in angle 3, OF is the abscissa (x), and FC the ordinate (y). Of the point D, in angle 4, OE is the abscissa (x), and ED the ordinate (y).

The *abscissa and ordinate of any point* are called the *co-ordinates of that point*. Thus, OE and EA are the *co-ordinates* of the point A, OF and FB are the *co-ordinates* of the point B, OF and FC are the *co-ordinates* of the point C, and OE and ED are the *co-ordinates* of the point D.

The line XX' is called the *axis of abscissas*, or the *axis of* x; and the line YY' the *axis of ordinates*, or the *axis of* y. The *axes* are sometimes called *diameters*.

Of the two co-ordinates to a point, the abscissa x always represents the *perpendicular distance* of the point from the axis YY', or the axis of y; and the ordinate y represents the *perpendicular distance* of the point from XX', or the axis of x. But these abscissas may measure *either way, right* or *left*, from the vertical axis YY'; and the ordinates may measure *either way, up* or *down*, from the horizontal axis XX'. Hence, in order to designate *which way* they are to be laid off or measured, mathematicians have adopted the notation of marking the abscissas which are to the *right* of the vertical axis YY' with the *plus sign*, $+$, and those to the *left* with the *minus sign*, $-$.

Also, those ordinates which are *above* the horizontal axis XX' are marked *plus*, and those *below*, *minus*. Hence, for any point in angle 1, *x and y will both be plus*. For a point in angle 2, *x will be minus and y plus*. For a point in angle 3, *both x and y will be minus*. In angle 4, *x will be plus and y minus*. For any point on OX or OX', y is 0. For any point on OY or OY', x is 0. At the origin O, $x = 0$ and $y = 0$. Also, if $y = 0$, the *point must be on the axis of* x, and if $x = 0$, the *point must be on the axis of* y.

Hence, if x and y are both given with their signs, the position of the point is determined, and such point is called the point x and y, or the point xy. Wherefore $x = a$, and $y = b$, a and b representing known lines or distances, *is the equation of a point*.

Example 1.—Determine the point whose equation is $x = 9$, $y = 12$.

The required point must be in angle 1, because both co-ordinates are plus. On OX lay $OE = 9$; through E, parallel to YY', draw a line, and on it lay EA, *above* the axis XX', equal to 12, and A will be the point required, whose equation is $x = OE = 9$, $y = EA = 12$.

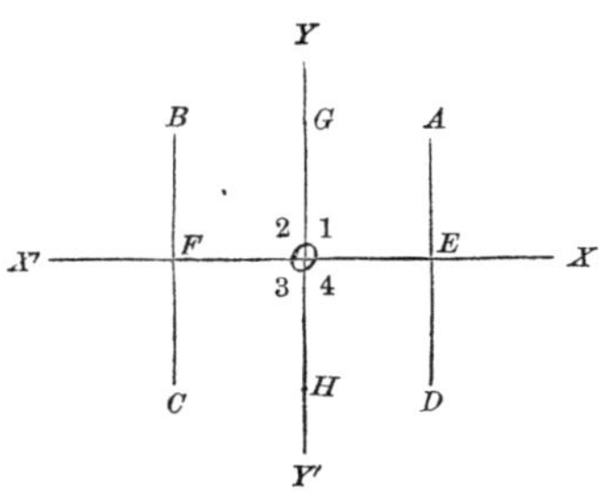

Example 2.—Determine the point whose equation is $x = -9$, $y = 12$.

The required point must be in angle 2, because x is *minus*, and y *plus*. On OX' lay $OF = 9$; through F, parallel to YY', draw a line, and on it lay FB, *above* XX', $= 12$; then B will be the point whose equation is $x = OF = -9$, and $y = FB = +12$.

Example 3.—Determine the point whose equation is $x = -9$, $y = -12$.

This point must be in angle 3, because both co-ordinates are minus. On OX' lay $OF = 9$; through F, parallel to YY', draw a line, on which lay FC, *below* XX', $= 12$; then C will be the point whose equation is $x = OF = -9$, $y = FC = -12$.

Example 4.—Determine the point whose equation is $x = 9$, $y = -12$.

This point must be in angle 4, because x is *plus*, and y *minus*. On OX lay $OE = 9$; through E, parallel to YY', draw a line, and on it, *below* XX', lay $ED = 12$; then D will be the required point whose equation is $x = OE = 9$, $y = ED = -12$.

Example 5.—Determine the point whose equation is $x = \pm 9$, $y = 0$.

This point must be on the axis of x, since $y = 0$. When $x = +9$, $y = 0$, it is E; when $x = -9$, $y = 0$, it is F.

Example 6.—Determine the point whose equation is $x = 0$, $y = \pm 12$.

Since $x = 0$, this point must be on the axis of y. Lay OG, OH each $= 12$: then the equation of the point G is $x = 0$, $y = +12$; and of the point H is $x = 0$, $y = -12$.

Example 7.—Determine the point whose equation is $x = 0, y = 0$. It is the origin O.

Scholium.—The distance of any point (xy) from the origin is always $= \sqrt{x^2 + y^2}$, and, as the square of either a plus or a minus quantity is plus, this distance will always be the same, whatever the *sign* of the co-ordinates, if they are of *the same value*. If D represent the distance, then $D = \sqrt{x^2 + y^2} = \sqrt{OE^2 + EA^2} = \sqrt{OF^2 + OB^2} = \sqrt{OF^2 + OC^2} = \sqrt{OE^2 + OD^2} = OA = OB = OC = OD$. Hence A, B, C, and D are in the circumference of a circle whose radius is OA, OB, etc.

OF THE STRAIGHT LINE.

Every algebraic equation with two unknown quantities of the first degree *represents a straight line on a plane*, of which the unknown quantities are co-ordinates. Hence $y = ax$, and $y = ax + b$, are equations of a straight line, of which x and y are the co-ordinates; and if a and b are known, either integral or fractional, the line can be determined by construction.*

Example 1.—Construct the equation $y = 3x$.

Analysis.—Here, when $x = 0, y = 0$, and *vice versa;* consequently the line must pass through O, the origin of the axes XX', YY'. Also, when x is plus, y is plus; and when x is minus, y is minus.

Construction.—In OX take *any distance* OE. Through E, parallel to YY', draw a line, and on it, *above* XX', lay $EG = 3\,OE$. Through O and G draw the line AOB, and it will be the line required, having its equation $y = 3x$; that is, such that the *ordinate of any point in that line* will be equal to *three times the abscissa of that point.* For, take *any point* in the line AB, as P, and parallel to YY' draw the ordinate PF. Then, in similar triangles OEG, OFP, we have $OE : EG :: OF : FP$. But $EG = 3\,OE$; hence $FP = 3\,OF$; that is, $y = 3x$,

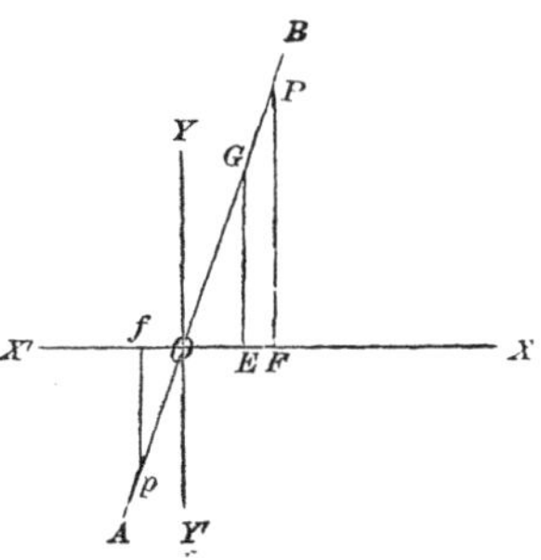

* *Definition.*—The *equation of a line on a plane*, whether the line be straight or curved, is an *algebraic expression* involving two unknown quantities, and such that if *one of the unknown quantities is assumed* of any specific magnitude, the corresponding value of *the other* unknown quantity *is determined from the algebraic expression.*

or the ordinate of the point $P =$ three times its abscissa. If $x = 2$, the equation gives $y = 6$; if $x = 5$, $y = 15$, etc.

In like manner, take any other point, as p, and draw the ordinate pf parallel to the axis YY'. Then, in similar triangles OEG and Ofp, we have $OE : EG :: Of : fp$. But $EG = 3$ times OE; hence $pf = 3\ Of$; that is, $y = 3x$, or the ordinate of the point $p = 3$ times its abscissa. If $x = -1$, $y = -3$; if $x = -4$, $y = -12$, etc.

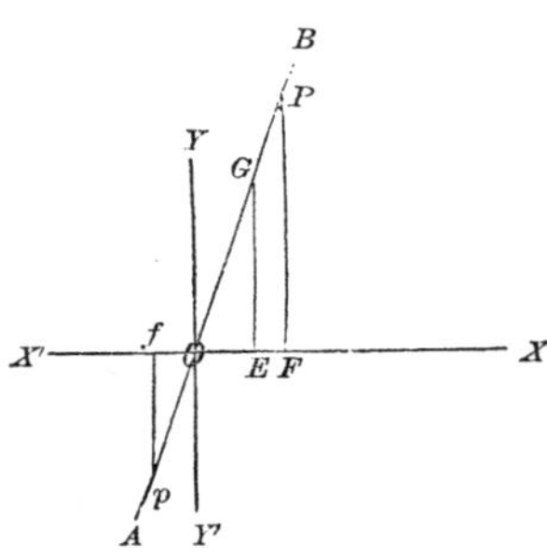

Scholium 1.—It is manifest that for all points in the part OB of the line AB, x will be plus and y plus; and for all points in the part OA, x will be minus and y minus.

Scholium 2.—By trigonometry, in triangle OEG we have $OE : EG ::$ rad. : tang $EOG =$ (when radius is 1) $\frac{EG}{OE} = \frac{3}{1} = 3$. Take the log. of three, and increase its index by 10 for the radius, and we have the logarithmic tangent $= 10.477121$, which gives the angle EOG or $XOB = 71° \ 34'$, nearly. In the equation $y = ax$, a is the tangent of XOB to radius 1.

Example 2.—Construct the equation $2y = -3x$, or $y = -\frac{3}{2}x$.

Analysis.—In this equation, when $x = 0$, $y = 0$, and *vice versa;* hence the required line must pass through the point O, the origin of the axes XX', YY'. We see, also, that when x is *positive*, y will be *negative*, and that when x is *negative*, y will be *positive;* and that $x : y :: 2 : -3$.

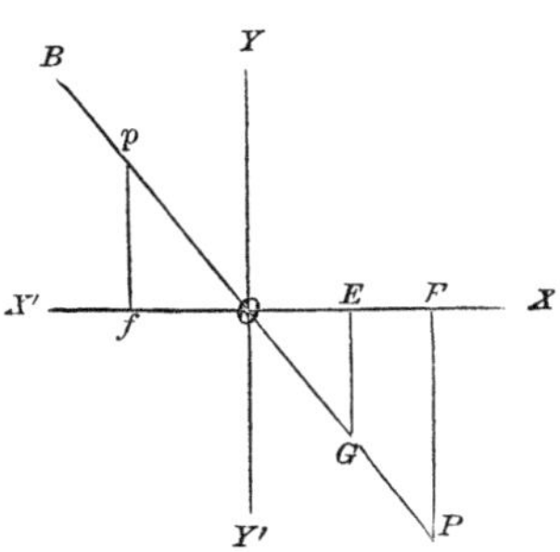

Construction.—In Ox take the abscissa $OE = 2$, and parallel to the axis YY' draw the ordinate EG *below* the axis XX', because y is *negative* when x is *positive*, equal to 3. Through O and G draw the line AOB, which will be the line required, having $y = -\frac{3}{2}x$, or $2y = -3x$.

For, take *any point* in that line, as P, of which the co-ordinates are OF and PF, we have, in similar triangles OEG, OFP, $OF(x) : FP(y) :: OE(2) : EG(-3)$. Hence $2y = -3x$, or $y = -\frac{3}{2}x$, so that if x is known y is known. Also, if we take the point p, whose co-ordinates are Of and fp, we have, in similar triangles OEG, Ofp, $Of\,(x) : fp\,(y) :: OE\,(2) : EG\,(-3)$. Hence $2y = -3x$, or $y = -\frac{2x}{3}$, as before. The same is true for each and every point in the line AOB.

Scholium 1.—It is manifest that for all points in the part OA of the line AB, x will be *positive* and y *negative*, and for all points in the part OB, x will be *negative* and y *positive.*

Scholium 2.—By trigonometry, in triangle OEG we have $OE\,(2) : EG\,(-3) ::$ rad. (1) : tang.-angle EOG or $AOX = -\frac{3}{2}$. To obtain the value of this angle, from the logarithm of 3 subtract the logarithm of 2, and add 10, the logarithm of the tabular radius, to the index of the remainder, and we obtain 10.176091, for the tabular tangent of EOG or AOX, which gives, from the table, 56° 19′, nearly. The tangent being *negative* shows that the line OA falls *below* the axis of x.

NOTE 1.—The co-efficient of x in the value of y $(-\frac{3}{2})$ is properly the tangent of the angle which the required line makes with the axis of x, measuring from OX upwards. When this tangent is negative, it gives the angle XOB, showing that OB falls *to the left* of OY.

NOTE 2.—The student will bear in mind that the point in which a line cuts the *axis of x has $y = 0$*, and where it cuts the *axis of y, it has $x = 0$*. At the origin $x = 0$ and $y = 0$.

Example 3.—Construct the equation $4x + 3y = 24$, or $y = 8 - \frac{4x}{3}$.

Analysis.—The line represented by the equation $4x + 3y = 24$ cannot pass through the origin O, because, when $x = 0$, $y = 8$, and when $y = 0$, $x = 6$, so that, if either y or x becomes 0, the other has a numerical value.

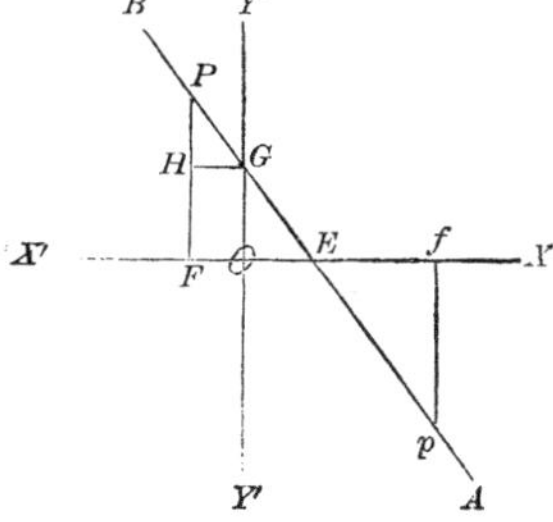

Now, at the point where the line cuts the axis of x, y must $= 0$; hence we have, from the equation, $x = 6$ for the abscissa of that point $= OE$, the distance from the origin at which the required line must cut the axis of x. Hence E is a point in the required line. Also, at the point where the line cuts the axis of y, $x = 0$, and then, from the equation, $y = 8 = OG$, the ordinate of that point, or the distance from the origin at which the required line must cut the axis of y. Hence G is another point in the required line.

Construction.—Lay $OE = 6$, and $OG = 8$, and through the points E and G draw the line $AEGB$, and it will be the one required. For, take *any* point in this line, as P, of which the co-ordinates are OF and FP. Through G draw GH parallel to XX'; then $GH = OF$, and they are both *negative*, because they are to the left of the axis of y. Now, in similar triangles PHG and GOE, we have $-GH$ or $-OF : PH :: OE$ (6) : OG (8). Whence $6\,PH = -8\,OF$, or

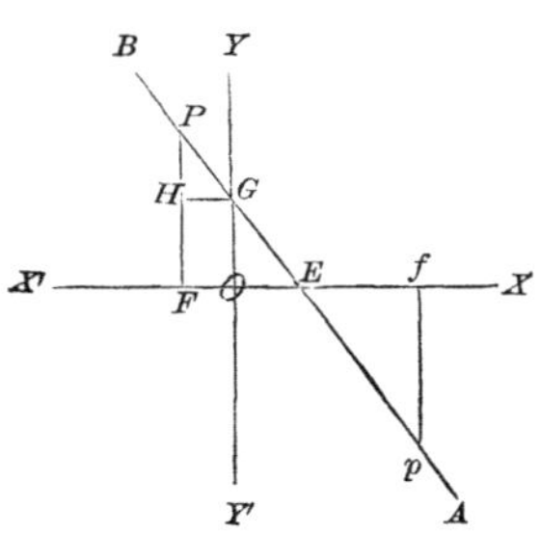

$PH = -\frac{8\,OF}{6} = -\frac{4\,OF}{3} = -\frac{4x}{3}$. And $PF = y = PH + OG = -\frac{4x}{3} + 8$. Wherefore $3y = -4x + 24$, or $3y + 4x = 24$, the equation of the line AB.

In like manner, if we take any other point, as p, of which the co-ordinates are Of and fp, we have $Ef : -fp :: EO\,(6) : OG\,(8)$; whence $Ef = -\frac{6fp}{8} = -\frac{3y}{4}$. Now, $x = Of = OE + Ef = 6 - \frac{3y}{4}$. Hence $4x = 24 - 3y$, or $4x + 3y = 24$, as before.

Scholium 1.—It is manifest that, for any point in the part EA of the line AB, x will be positive and y negative, and for any point in the part GB of the line AB, x will be negative and y positive, while for any point in the part EG, both will be positive.

Scholium 2.—The angle OEG or AEX is found as in Scholium 2 to last problem, its tangent being equal *to the co-efficient of x in* the value of y in the given equation, $= -\frac{4}{3}$; from the logarithm of 4 subtract the logarithm of 3, add 10 to the index of the remainder, and we obtain 10.124939 for the tabular tangent of OEG or AEX, which gives 53° 8′, or 126° 52′, for the angle, and the tangent being *negative* shows that the line EA falls *below* the axis of x, or it is properly the angle XEB.

Example 4.—Construct the equation $3y + 5x = 27$, or $y = 9 - \frac{5x}{3}$.

In this equation, when $x = 0$, $y = 9 = OG$, and when $y = 0$, $x =$

$OE = \frac{27}{5} = 5.4$. Hence make $OE = 5.4$, and $OG = 9$, and through E and G draw the line AB, and it will be the line required, whose equation is $3y + 5x = 27$. For, take *any* point, as P, whose co-ordinates are OF and FP, and draw GH parallel to XX'; then $GH = OF = x$. And by similar triangles GHP and EOG, we have $-GH$ or $-OF\,(-x) : PH :: OE \left(\frac{27}{5}\right) : OG\,(9)$. Hence $\frac{27}{5} \times PH = -9x$, or $PH = -\frac{45x}{27} = -\frac{5x}{3}$. And $y = PF = PH + OG = -\frac{5x}{3} + 9$; that is, $3y = -5x + 27$, or $3y + 5x = 27$, the given equation of the required line.

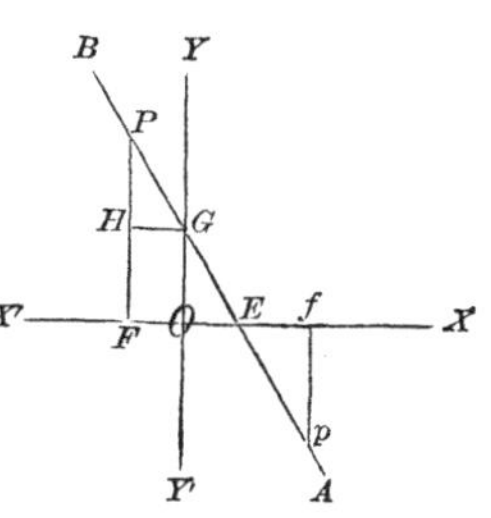

In like manner, if we take the point p, whose co-ordinates are Of and fp, we have, in similar triangles Efp and EOG, $Ef : -fp\,(-y) :: OE \left(\frac{27}{5}\right) : OG\,(9)$. Hence $9\,Ef = -\frac{27}{5}y$, or $Ef = -\frac{3}{5}y$. Now, $x = OE + Ef = \frac{27}{5} - \frac{3}{5}y$; whence $5x = 27 - 3y$, or $3y + 5x = 27$, the same as before.

Example 5.—Construct the equation $3y + 5x = -27$, or $y = -9 - \frac{5x}{3}$.

Here, when $x = 0$, $y = -9 = OG$, and when $y = 0$, $x = -\frac{27}{5} = -5.4 = OE$. Hence lay OE *to the left of* $O = 5.4$, and OG *below* $O = 9$, and through G and E draw the line AB, and its equation will be $3y + 5x = -27$. For, take *any* point, as P, in AB, of which the co-ordinates are OF and FP. Then, in similar triangles GHP and EOG, we have GH or $OF\,(x) : -HP :: EO\left(-\frac{27}{5}\right) : OG\,(-9)$. Hence $+\frac{27}{5} \times HP = -9x$, or $HP = -\frac{45x}{27} = -\frac{5x}{3}$. Now, $y = FP = FH + HP = -9 - \frac{5x}{3}$; or $3y = -27 - 5x$; that is, $3y + 5x = -27$, the given equation of the required line.

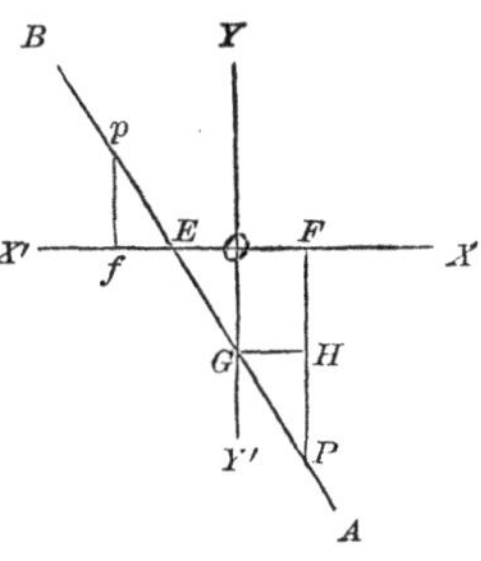

In like manner, take *any other* point, as p, whose co-ordinates are Of and fp. In similar triangles Efp, EOG, we have $-Ef : fp\,(y) :: OE\left(-\frac{27}{5}\right) : OG\,(-9)$. Hence $+9\,Ef = -\frac{27y}{5}$, or $Ef = -\frac{3y}{5}$. And $x = Of = OE + Ef = -\frac{27}{5} - \frac{3y}{5}$, or $5x = -27 - 3y$, or $3y + 5x = -27$, as before.

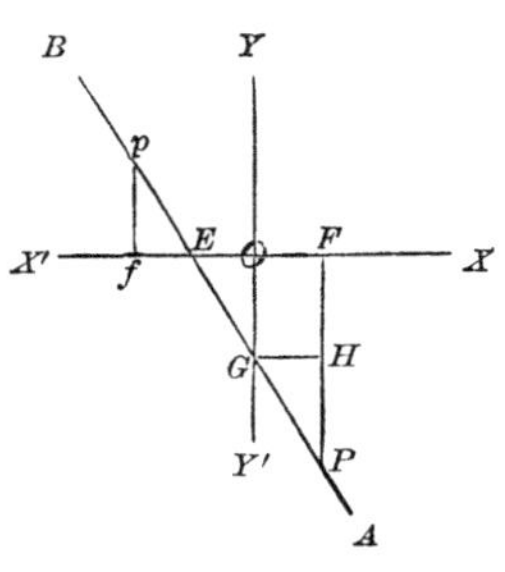

Scholium 1.—In these two examples the triangles EOG are equal. Hence, if the two lines were constructed on the same axes, the angle OGE in *one* would be *equal* and *alternate* to OGE in *the other*, and the two lines would be parallel. Wherefore $ax + by = c$, and $ax + by = -c$, are equations of *two parallel lines* constructed on the same axes.

Scholium 2.—In Example 4, $y = 9 - \frac{5x}{3}$, and in Example 5, $y = -9 - \frac{5x}{3}$. Hence, by Scholium 2 in Examples 2 and 3, the tangent of the angle which the required line makes with the axis of x, in each case, is $-\frac{5}{3}$. To find its amount, from the logarithm of 5 deduct the logarithm of 3, add 10 to the index of the remainder, and we have 10.221849 for the tabular tangent, which gives 59° 2′ for the angle OEG or AEX *below* EX, or, more properly, 120° 58′ for the angle XEB.

Example 6.—Construct the equation $3y - 5x = 27$, or $y = \frac{5x}{3} + 9$.

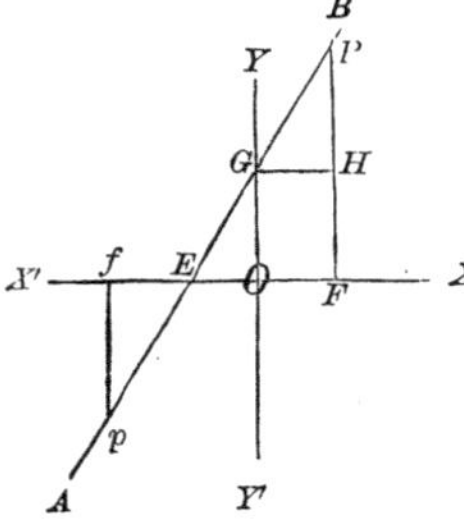

Here, when $y = 0$, $x = -\frac{27}{5} = -5.4 = OE$, and when $x = 0$, $y = \frac{27}{3} = 9 = OG$. Hence lay $OE = 5.4$ *to the left of* O, and $OG = 9$ *above* O, and through E and G draw the line AB, whose equation will be $3y - 5x = 27$. For, taking the points P and p, and proceeding as in Example 4, we have $PH = \frac{5}{3}\,OF = \frac{5}{3}x$, and $y = FP = OG + PH = 9 + \frac{5x}{3}$. Hence $3y = 27 + 5x$, or $3y - 5x$

$=27$, the given equation of the required line. Also, $Ef=\frac{3fp}{5}=\frac{3y}{5}$, and $x=OE+Ef=-\frac{27}{5}+\frac{3y}{5}$. Hence $5x=-27+3y$, or $3y-5x=27$, as before.

Example 7.—Construct the equation $3y-5x=-27$, or $y=\frac{5x}{3}-9$.

In this equation, when $y=0$, $x=\frac{27}{5}=5.4=OE$, and when $x=0$, $y=-9=OG$. Hence lay $OE=5.4$, and $OG=9$, *below the origin* O, and through G and E draw the line AB, and its equation will be $3y-5x=-27$. For, taking the points P and p, and proceeding as in Example 5, we have $EF=\frac{3\,FP}{5}=\frac{3y}{5}$, and $x=OF=OE+EF=\frac{27}{5}+\frac{3y}{5}$, whence $5x=27+3y$, or $3y-5x=-27$, the given equation of the required line. Also, we have $Hp=\frac{-5\times -Of}{3}=\frac{5x}{3}$. And $y=fp=OG+Hp=-9+\frac{5x}{3}$. Hence $3y=-27+5x$, or $3y-5x=-27$, the same as before.

Scholium 1.—In these two figures, also, the triangles EOG are equal. Hence, if the two lines AB had been constructed *on the same axes*, the angle OGE in one would have been *equal* and *alternate* to OGE in the other, and the two lines would have been parallel. If, therefore, the co-efficient of x and y in two equations are *the same*, or *in the same ratio*, the two lines which they represent will be parallel.

Scholium 2.—In the general equation, $ay\pm bx=\pm c$; if $x=0$, $y=\pm\frac{c}{a}=$ the distance from the origin at which the required line cuts the axis of y; and if $y=0$, $x=\pm\frac{c}{b}$, the distance from the origin at which the required line cuts the axis of x, these distances being represented in the preceding figures by the lines OG and OE respectively. Hence, in the construction of a line from its equation of this form, we will always have $OE:OG::\pm\frac{c}{b}:\pm\frac{c}{a}::a:b$; that is, OE is to OG as the co-efficient of y is to the co-efficient of x, *whatever c may be*, the value of c only influencing the *distances* from the origin at which the required line cuts the axes, and *not the angle* it makes with them.

Example 8.—Construct the equations $3y - 5x = -27$, or $y = +\frac{5x}{3} - 9$; and $-3y + 5x = -42$, or $y = \frac{5x}{3} + 14$, on the same axes. The first equation is constructed in the last figure, and gives the line AEB in the figure hereto annexed, where $OE = 5.4$ and $OG = 9$.

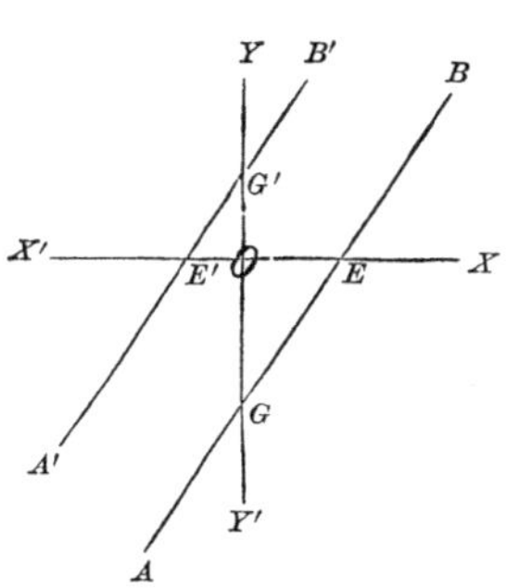

In the second equation $-3y + 5x = -42$, when $y = 0$, $x = -\frac{42}{5} = -8.4 = OE'$; and when $x = 0$, $y = \frac{42}{3} = 14 = OG'$. Now, lay $OE = 5.4$, and $OG = 9$, and draw the line $AGEB$, which represents the equation $3y - 5x = 27$.

Also, lay $OE' = 8.4$ to the *left of* O, and $OG' = 14$, and through E' and G' draw the line $A'B'$, and it will represent the equation $-3y + 5x = -42$, proved as in Example 6.

Now, the triangles OGE and $OG'E'$ are equiangular; for, by construction of AB, $OE : OG :: \frac{27}{5} : 9 :: 27 : 45 :: 3 : 5$; that is, $OE : OG ::$ co-efficient of y : co-efficient of x in the first given equation.

Also, by construction of $A'B'$, we have $OE' : OG' :: \frac{42}{5} : 14 :: 42 : 70 :: 3 : 5$; that is, $OE' : OG' :: 3 : 5 ::$ the co-efficient of y : to the co-efficient of x in the second given equation. Wherefore the angles OGE and $OG'E'$ are equiangular, having the angle $OGE = OG'E'$, and these are *alternate angles*, and hence $A'B'$ is parallel to AB.

Scholium 1.—Although these equations may be constructed on the same axes, they are not *coincident* equations; that is, the values of x and y of the unknown quantities cannot be the same in both equations. In the equation $3y - 5x = -27$, if $y = 6$, $x = 9$. In the equation $-3y + 5x = -42$, if $y = 19$, $x = 3$; and if $x = 9$, $y = 29$, instead of 6 as in the preceding equation, showing that the unknown quantities in each equation have independent integral values.

Scholium 2.—If we have two *independent* equations involving two unknown quantities, and the two unknown quantities *have each a common value* in the two equations, the lines can be constructed, and the common values of the unknown quantities determined, as in the following examples:

Example 9.—Given, the equations $5x - 3y = 27$, and $-2x + 5y = 31$, to find the values of x and y by construction.

Analysis.—Since the co-efficients of x and y in these two equations have not the same ratio, the lines which these equations represent are not parallel, and therefore they must meet and cross each other *at some point*, and of this *common point* of the two lines there will be *common co-ordinates* which will be the values of x and y common to both equations. Now, in the first equation, if $x = 0$, $y = -9$; if $y = 0$, $x = \frac{27}{5} = 5.4$. In the second equation, if $x = 0$, $y = \frac{31}{5} = 6.2$; if $y = 0$, $x = -\frac{31}{2} = -15.5$. Therefore, lay $OE = 5.4$, and OG, *below* O, $= 9$, and through G and E draw the line AB, and it will represent the equation $5x - 3y = 27$.

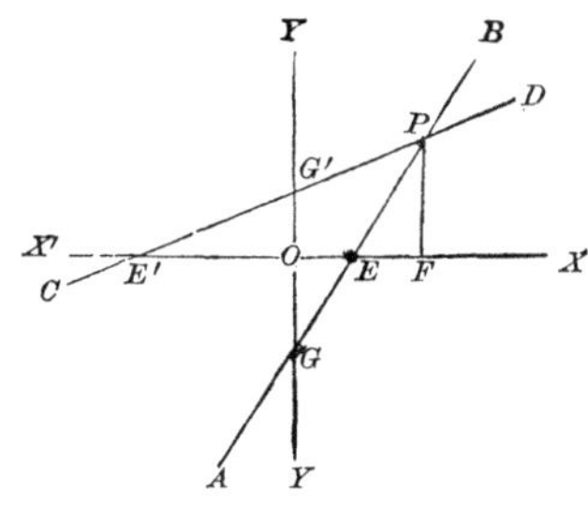

Also, lay OE', $= 15.5$, to *the left of* O, and OG', $= 6.2$, *above* O, and through E' and G' draw the line CD, intersecting the former in the point P. Then CD will be the line represented by $-2x + 5y = 31$.

Now, the co-ordinates OF and FP of the common point P will be the values of x and y respectively, common to the two equations. Measured from the same scale by which OE, OG, etc. were measured, we find $OF = 12 = x$, and $FP = 11 = y$, which are the values found by resolving the equations algebraically. Thus we find algebra and geometry supporting each other, and leading alike to *truth.*

Example 10.—Given, $10x - 4y = 68$, and $-8x + 5y = -40$, to find the values of x and y by construction. Ans. $x = 10$, $y = 8$.

Example 11.—Given, $10x - 4y = 68$, and $-5x + 2y = -26$, to find the values of x and y by construction.

Here the co-efficients of x and y in the two equations have the *same ratio;* that is, $\frac{4}{10} = \frac{2}{5}$. Hence the two lines are parallel (Example 7, Scholium 1), and there can be no common values of x and y, as the lines, being parallel, *can have no common point.*

Example 12.—Given, the equation $x^2 + y^2 = 16^2 = r^2$, to construct the line represented by the equation.

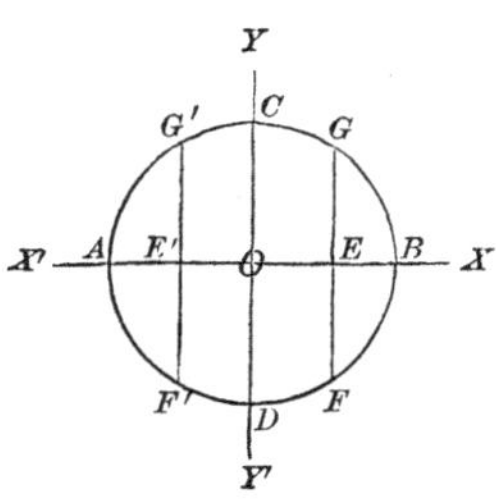

Analysis.—Let XX', YY' be the axes, and O the origin, as in the preceding examples. Now, when $y = 0$, $x = \sqrt{16^2} = +16$ or -16, which shows that the line which the equation represents cuts the axis of x at two points, distant 16 from the origin, on opposite sides of it. On OX and OX', therefore, lay 16 to B and A, and they will be two points in the required line. Also, when $x = 0$, $y = \sqrt{16^2} = +16$ or -16, showing that the line represented by the given equation cuts the axis of y in two points, one above and the other below the origin, and distant 16 from it. Therefore, on OY, OY' lay OC and OD each $= 16$, and C and D will be two other points in the required line.

Again, since the distance of *any point* in a line from the origin (scholium to Example 7 "Of the Point") is the square root of the sum of the squares of the co-ordinates of that point; that is, to $\sqrt{x^2 + y^2}$, and as this distance in the present example is constant, being for every point, as G, F, G' or $F' = \sqrt{x^2+y^2} = \sqrt{256}$ or $\sqrt{r^2}$, it follows that *every point* in the required line is the *same distance from the origin*, and *that* distance is $\sqrt{256} = OB$, OA, OC, $OD = 16$. Hence, if a circle be described with OB or OC as radius, it will be the line represented by the equation $x^2 + y^2 = 16^2 = r^2$.

Scholium 1.—The line represented by the equation $x^2 + y^2 = r^2$ is a circle whose radius is r.

Scholium 2.—Since $y = \pm\sqrt{256 - x^2}$, it is evident that for *every value* of x, as OE or OE', there will be two *equal values* of y, as EG and EF, EG' and EF'.

Also, since $x = \pm\sqrt{256 - y^2}$, for *every value* of y there will be two *equal values* of x. Hence *each axis bisects all the double ordinates drawn parallel to the other axis.*

SECTION ON CURVES:

SHOWING THE

MANNER OF CONSTRUCTING SEVERAL OF THEM,

AND

SOME OF THEIR PROMINENT PROPERTIES.

THE PARABOLA.

PROBLEM I.—To construct a parabola and find its equation.

Draw DR and GL at right angles. Take *any* point F in GL. Bisect FG in A. Parallel to DGR, through F, and any number of points in AL, as 1, 2, 3, 4, 5, etc., *anyhow taken*, draw lines, to which apply the distance $G1$, from F to a and a, on the parallel through the point 1; apply GF, from F to b and b, on the parallel through F; apply $G2$, from F to c and c, on the parallel through 2; apply $G3$, from F to d and d, on the parallel through 3, and so on for each point to H; and apply GH, from F to B and C, on the parallel through H; then the curve passing through the several points A, a, b, c, d, etc., on each side of A, *will be a parabola.* It must be observed that the distance from G to any point on GL must be applied from F to the *parallel through that point.* The curve may be continued in like manner below H indefinitely.

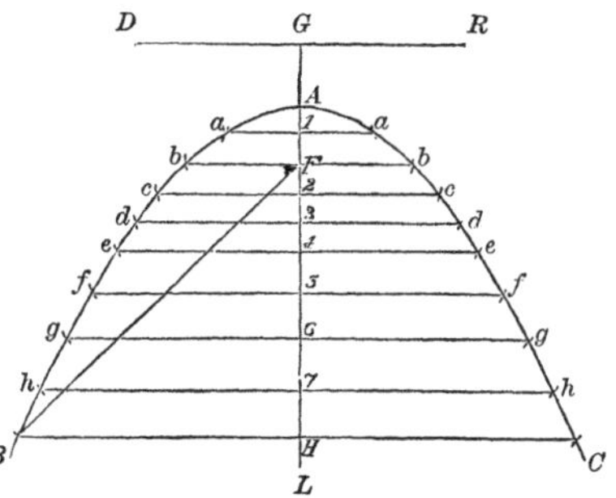

The line DGR is called the *directrix*, AL the *axis*, F the *focus* of the parabola, A the *vertex;* the line bFb, parallel to the directrix, through the focus, is called the *latus rectum* to the axis, or *para-*

meter, and it is equal, as seen by the construction, to $2\,GF = 4\,AF = 4\,AG$.

A line from any point of the curve, perpendicular to the axis, is an *ordinate to the curve* at that point, and the distance from the vertex to the foot of the ordinate is the corresponding *abscissa*.

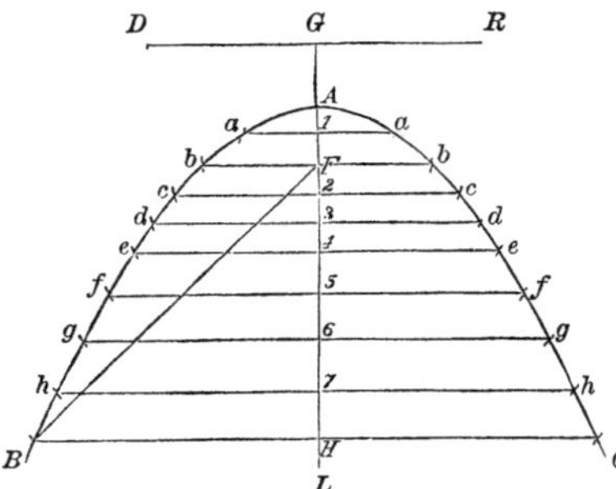

Thus, for the point b, bF is the ordinate, and AF the abscissa. For the point 2, $c2$ is the ordinate, $A2$ the abscissa. For the point B, BH is the ordinate, and AH the abscissa. Also BHC, $h7h$, are *double ordinates* of the points B and h.

Demonstration.—Draw a line from the focus F to the extremity of *any* ordinate, as HB. Then, by the construction, $FB = GH = AH + AF$, and $FH = AH - AF$. Now, $BH^2 = FB^2 - FH^2 = (AH + AF)^2 - (AH - AF)^2 = 4\,AF \times AH = 2\,FG \times AH = bFb \times AH = L \times AH$ (L representing the latus rectum), which is *the property of the parabola*. $L = bFb = 2\,FG = 4\,AF$. Putting $x =$ any abscissa, as AH, $y =$ its ordinate BH, $a =$ latus rectum $= 4\,AF$, we have $y^2 = ax$, which is the *equation of the parabola*, being true for *any* and *every* point of the curve.

Corollary.—The square of *any* ordinate being equal to $4\,AF$ multiplied by its abscissa, and $4\,AF$ *being constant*, the *abscissas are to each other as the squares of the ordinates;* that is, $AH : A4 :: BH^2 : e4^2$; $AH : AF :: BH^2 : bF^2$; $A6 : A3 :: g6^2 : d3^2$, etc.

Problem II.—Given, a parabola, to find its axis, focus, directrix, latus rectum, etc., by construction.

Let BAC be a parabola. Draw any two parallel chords YL and ab, and bisect them in V and c; through the points c and V draw the *diameter* $EcVO$. Perpendicular to EO draw the double ordinate EI, which bisect in S. Through S, parallel to the diameter EVO, draw the *axis* $GASL$. Through E, the extremity of the diameter EO, parallel to the chord YL, draw the *tangent* ET, meeting the axis, produced, in T. Then TS is the *subtangent* to the point E. Draw ER perpendicular to ET; then ER is the *normal* and SR *the subnormal* to the point E. At E make the angle $TEF =$ angle ETL; then F is the *focus* of the parabola. Make $AG = AF$, and perpendicular to GL draw DGR', and it will be the *directrix.*

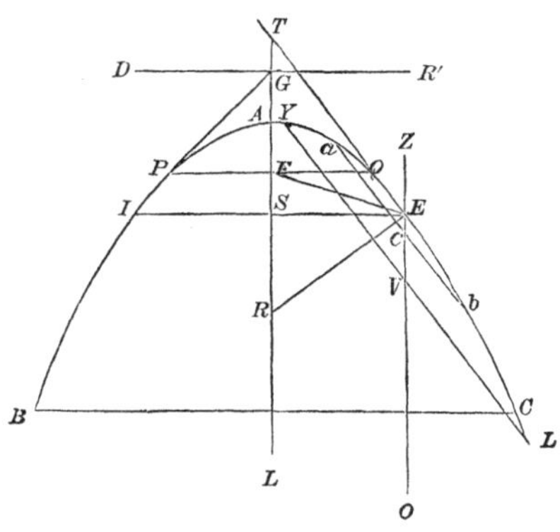

Since angle TEF was made equal to ETF, and TER is a right angle, a semicircle described with centre F and radius FT will pass through the points E and R. Hence FT, FE, and FR are all equal. Produce OE to any point, as Z; then angle $ZET = ETF$ (being alternate angles) $= TEF$; hence the *tangent* ET *bisects the angle* ZEF.

By the construction of the parabola, $GS = FE = FT$. Take GA from the first, and its equal AF from the second, and we have $AS = AT$. Hence the *subtangent to a point* $(ST) =$ *twice the abscissa* (AS). Also, since $FG = 2\,AF$, the tangent at P or Q, the extremity of the latus rectum, *will meet the axis at* G, *its intersection with the directrix.* To draw a tangent to *any point*, as E, make the subtangent $ST =$ twice the abscissa AS, and join TE, the tangent.

Also, since $GS\ (= FE) = FR$, take FS from each, and we have $SR = FG = FP = FQ$; hence the *subnormal* $(SR) =$ *half the latus rectum* (PFQ).

Note.—The area of the parabola, and the solidity of the paraboloid, were found in the article on "The Calculus."

PROBLEM III.—To construct the parabola from its equation $y^2 = ax$.*

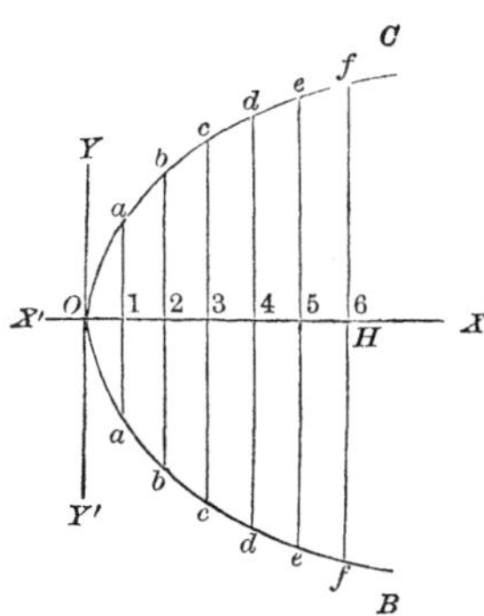

Analysis.—Let O be the origin of the axes XX', YY'. It is evident from the equation $y^2 = ax$ that if $y = 0$, $x = 0$; and if $x = 0$, $y = 0$. Hence the line represented by the equation passes through the origin O. It is evident, moreover, that x cannot be negative, or minus, for, if it were, ax would be minus, and we would have $y = \sqrt{-ax}$, which is impossible. Hence the curve must lie *wholly* on the right of the axis YY'.

Also, since $y = \pm\sqrt{ax}$, every value of x will give two values of y, numerically equal, but of different signs: hence the axis of x bisects all lines drawn parallel to the axis of y and terminated at both ends by the curve; and it bisects, also, the area of any portion of the parabola cut off by a line parallel to the axis of y.

Example 1.—In the equation $y^2 = ax$, take $x = 5$, and it becomes $y^2 = 5x$, or $y = \pm\sqrt{5x}$. By the table of square roots, annexed to this volume, if $x = 1$, $y = \sqrt{5} = \pm 2.24$; if $x = 2$, $y = \sqrt{10} = \pm 3.16$; if $x = 3$, $y = \sqrt{15} = \pm 3.87$; if $x = 4$, $y = \pm 4.47$; if $x = 5$, $y = \sqrt{25} = \pm 5$; if $x = 6$, $y = \pm 5.48$, etc. Now, lay $O1 = 1$, $O2 = 2$, $O3 = 3$, $O4 = 4$, etc., and through the points 1, 2, 3, 4, etc. draw lines parallel to the axis YY'; and, since *all the values of y have both plus and minus signs, they must be laid both ways from the axis of x;* lay $1a = 2.24$, $2b = 3.16$, $3c = 3.87$, $4d = 4.47$, etc., laying each *up* and *down*, and the curve passing through O, a, b, c, d, etc., both ways from O, will be the parabola required.

Example 2.—In the equation of the parabola, take $x = -3$, and it becomes $y^2 = -3x$, or $y = \sqrt{-3x}$.

Here the values of x must be minus in order to have $3x$ plus and y^2 plus. Hence the curve lies wholly to the *left* of the axis YY'.

* See remark to Example 2, Analytical Geometry.

If $x = -1$, $y = \sqrt{3} = \pm 1.73$.
" $x = -2$, $y = \sqrt{6} = \pm 2.45$.
" $x = -3$, $y = \pm 3.00$.
" $x = -4$, $y = \pm 3.46$.
" $x = -5$, $y = \pm 3.87$.
" $x = -6$, $y = \pm 4.24$.
" $x = -7$, $y = \pm 4.58$.
" $x = -8$, $y = \pm 4.90$.
" $x = -9$, $y = \pm 5.20$.

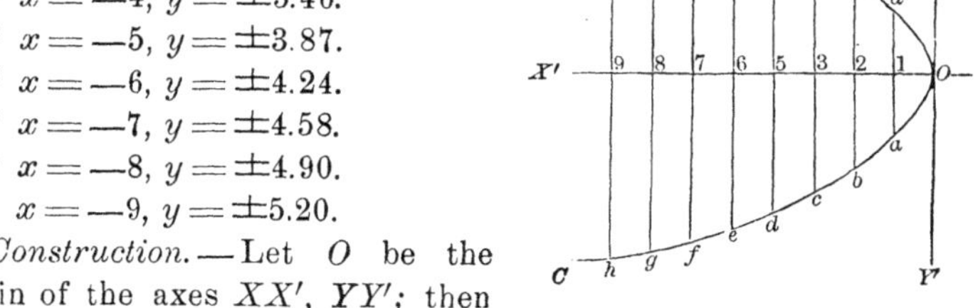

Construction.—Let O be the origin of the axes XX', YY'; then proceed exactly as in Example 1, laying the values of x, 1, 2, 3, etc. *to the left of* YY', because all its values are *necessarily* negative, and hence the curve lies wholly to the left of YY'.

NOTE.—By taking $x = .5$, 1.5, 2.5, 3.5, etc. in these examples, or other intervening quantities, intermediate points at pleasure may be obtained.

THE ELLIPSE.

PROBLEM I.—To construct the ellipse, having the axes given, and determine its equation.

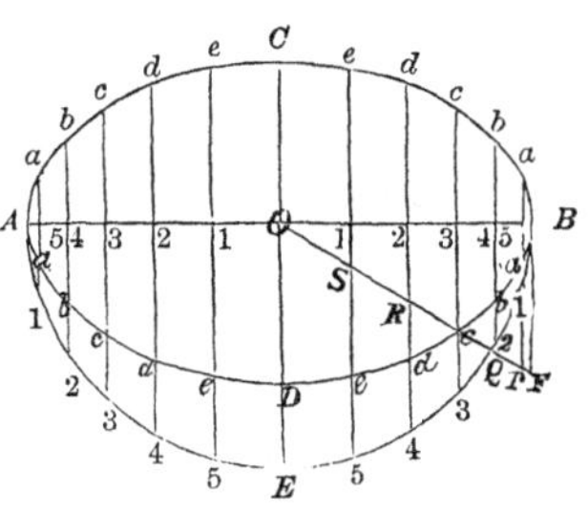

Construction.—Draw the axes AB and CD, bisecting each other at right angles at the point O, which will be the centre of the ellipse. Draw BF parallel and equal to OC or OD, and join OF. With the centre O and radius OA or OB, half the greater axis AB, describe the semicircle AEB, E being the prolongation of CD. Divide the semicircle AEB into six equal parts by laying the radius from A to 4, and from B to 4, and from E, *both ways*, to 2 and 2; then bisect each of these equal parts, and each quadrant will be divided into six equal parts in the points 1, 2, 3, etc., and the whole semicircle into twelve. Through each of these points draw lines parallel to CE, cutting the axis AB in the points 1, 2, 3, etc., and the line OF in the points S, R, c, Q, and P, the point Q coinciding with 2 on the semicircle. Now, take the distance $1S$ in the dividers, and lay it from *each* 5 which is on the axis, *up* and *down*, to a and a, giving four points in the ellipse.

Take the distance $2R$, in like manner, and lay it from *each* 4 on the axis, *up* and *down*, to b and b, giving four more points in the curve. In like manner, take the distance $3c$ (noting that c on OF is a point in the curve), and lay it from *each* 3 on the axis, *up* and *down*, to c and c. Lay the distance $4Q$ from *each* 2 on the axis, *up* and *down*, to d and d; and lay $5P$ from *each* 1 on the axis, *up* and *down*, to e and e; then the curve passing through these twenty-four points A, B, C, D, and each of the points a, b, c, d, e, taken in regular succession from A through C, B, and D, will be an ellipse. It will be noticed that $1S$ is just as far, *measured on the semicircle*, from OE as $5a$ is from B or A; that is, $E5 = B1 = A1$; also, $2R$ as far from OE as $4b$ is from B or A, so that the arcs $E5 = B1$ or $A1$, $E4 = B2$ or $A2$, $E2 = B4$ or $A4$, etc. Also, since $B4$ and $E2$ are equal, each being composed of four equal parts of the semicircle, we have the line $O4$ equal to the line $2d4$.

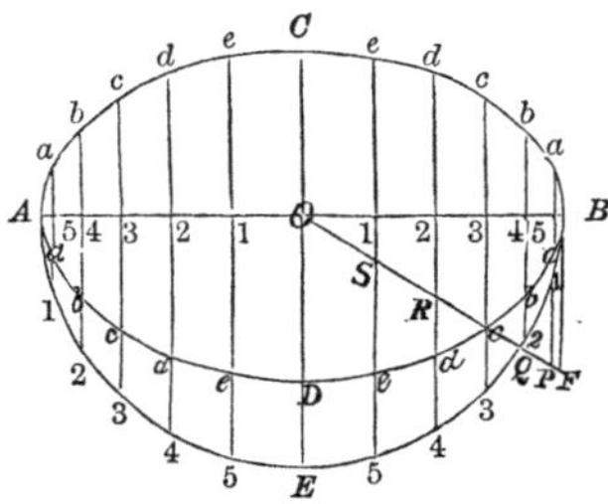

Demonstration.—Taking *any* point in the curve, as d, in the quadrant BD, $2d$, by construction, $= 4Q$. In similar triangles OBF and $O4Q$, we have $O4$ (or $2d4$) : $4Q$ (or $2Rd$) :: OB : BF. Hence $(2d4)^2 : (2Rd)^2 :: OB^2 : BF^2$ or OD^2. But (IV. 23, cor.*) $(2d4)^2 = A2 \times 2B$; hence $A2 \times 2B : (2Rd)^2 :: OB^2 : OD^2$; that is, as the rectangle of the two abscissas of the point d ($A2 \times 2B$) is to the square of the ordinate ($2Rd$), so is the square of the semi-axis major (OB) to the square of the semi-axis minor (OD), which is the *property of the ellipse.*

If $OB = a$, $OD = b$, $O2 = x$, and $2Rd = y =$ the ordinate, then $A2 = a + x$, and $2B = a - x$, the two abscissas, and the above proportion becomes $(a + x) \times (a - x) : y^2 :: a^2 : b^2$, or $a^2 - x^2 : y^2 :: a^2 : b^2$; whence $y^2 = \frac{b^2}{a^2}(a^2 - x^2)$, or $y = \frac{b}{a}\sqrt{a^2 - x^2}$; or $a^2y^2 + b^2x^2 = a^2b^2$, either of which is *the equation of the ellipse.*

PROBLEM II.—Having an ellipse given, it is required to find its axes, foci, focal tangents, and to draw a tangent to *any point* in the ellipse.

Construction.—In the given ellipse, draw *any two* parallel chords, as EG and HI; bisect these in J and K, and through these points

* VI. 13.

draw the line $LJKM$, which will be a *diameter to the ellipse.* Bisect LM in O, which will be the *centre of the ellipse.* With the centre O and radius OL describe an arc, cutting the ellipse in L and N; bisect the arc LN in P, and through P and O draw $APOB$, which will be the *major axis* or *principal diameter* of the ellipse.

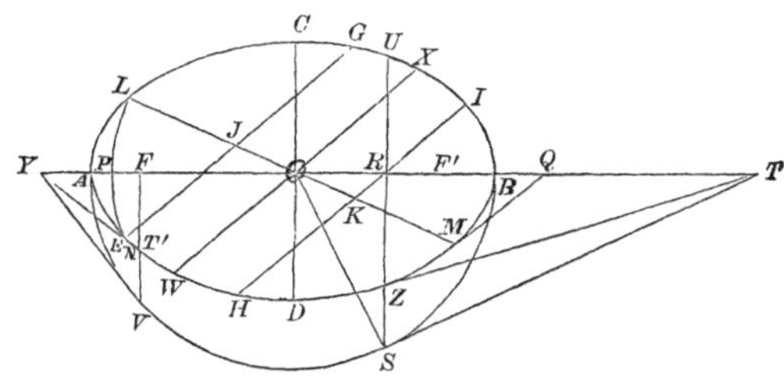

Through O draw COD perpendicular to AB, and it will be the *minor axis.* With C as a centre and radius OA or OB describe arcs, cutting the axis AB in the points F and F', which will be the *two foci* of the ellipse. Through the centre O, parallel to either of the chords EG and HI, draw WOX, and it will be the *conjugate diameter to LOM,* and LM and WX are called a *pair of conjugate diameters*, of which each one *bisects all chords* in the ellipse drawn parallel to the other, such chords being called *double ordinates* to the *diameter* that *bisects them*, and the two parts into which the diameter is divided are the *corresponding abscissas.* The lines through the extremities of either diameter, parallel to its conjugate or double ordinate, are *tangents to the ellipse* at those points. Thus, MQ, drawn through M, parallel to the diameter WOX or the chord HI, is a *tangent to the ellipse* at the point M.

On AB, with O as the centre, describe a semicircle. From the focus F, perpendicular to AB, draw the line $FT'V$, draw VY a tangent to the semicircle, and join $T'Y$; then $T'Y$ will be a tangent to the ellipse at T', and it is called the *focal tangent.*

To draw a tangent to the ellipse at *any given point*, as Z, parallel to COD draw the *double ordinate URZ*, and produce it to meet the semicircle at S. At S draw a tangent ST to the semicircle, meeting the axis AB, produced, in T, and join TZ, which will be *the tangent to the ellipse* at Z. The *tangents* to the *semicircle* and *ellipse*, from points in the same ordinate, *meet at the same point in the axis.*

Since, as shown in last problem, $AR \times RB : RZ^2 :: AO^2 : OD^2$, and $AR \times RB = RS^2$, we have $RS^2 : RZ^2 :: AO^2 : OD^2$, or $RS : RZ :: AO : OD$; that is, the ordinate of the circle at any point of the axis is to the ordinate of the ellipse at the same point, as AO to OD, or as the major axis to the minor axis, in a constant ratio. Hence these ordinates to the circle and ellipse will always be proportionate to each other, and we have $SR : ZR :: VF : T'F$, etc.

PROBLEM III.—To find the area of an ellipse by the calculus. (See problem in the article "Calculus.")

Put $AO = a$, $OC = b$, $OR = x$, $RU = y$. Now, to find the area of the quadrant $BOCUB$ of the ellipse, RU is the moving line, supposed to commence at OC, and moving parallel to itself through the distance OB, to generate the surface of the quadrant $BOCB$. Now, by Problem I., the equation of the ellipse is $y^2 = \frac{b^2}{a^2}(a^2 - x^2)$, or $y = \frac{b}{a}\sqrt{a^2 - x^2}$. Hence the area of the quadrant $BOCUB = \int y dx = \int \frac{b}{a}\sqrt{a^2 - x^2} \times dx = \frac{b}{a}\int\sqrt{a^2 - x^2} \times dx$. But, by Example 4 of the problem in the article "Calculus," $\int\sqrt{a^2 - x^2} \times dx$ is the area of a *quadrant of a circle* whose radius is a. Hence $\frac{b}{a}\int\sqrt{a^2 - x^2} \times dx = \frac{b}{a} \times$ the area of the quadrant of a circle whose radius is a. That is, we have $a : b ::$ area of quadrant of circle : area quadrant of ellipse :: area of circle : area of ellipse, the diameter of the circle being the major axis of the ellipse, $= 2a$. Then the area of the circle $= 4a^2 \times .78539$, and $a : b :: 4a^2 \times .78539$: area of ellipse $= 4ab \times .78539 = 2a \times 2b \times .78539 = AB \times CD \times .78539$. That is, the *area of an ellipse is equal to the product of the axes of the ellipse multiplied by the area of a circle whose diameter is unity*, or it is *equal to the area of a circle whose diameter is a mean proportional between the two axes*, $= \sqrt{4ab}$.

PROBLEM IV.—To construct the ellipse from its equation $y^2 = \frac{b^2}{a^2}\sqrt{a^2 - x^2}$, or $a^2y^2 + b^2x^2 = a^2b^2$.*

Analysis.—It is evident from the equation $a^2y^2 + b^2x^2 = a^2b^2$ that if $y = 0$, $x = +a$ or $-a$, which shows that the curve represented by the equation cuts the axis of x in two points, distant a from the origin, one to the *right*, the other to the *left* of O.

Also, if $x = 0$, $y = +b$ or $-b$, showing that the line cuts the axis of y in two points, distant b from the origin, one *above* and the other *below*.

* See remark to Example 11, Analytical Geometry.

Construction.—Let XX', YY' be the axes, intersecting at the origin O; in OX', lay $OA = a$; and in OX, lay $OB = a$; also in OY and OY', lay OC and OD each equal to b, and A, C, B, and D will be four points in the curve, and AB and CD the two axes. Since x and y enter into the equation of the curve in the second power only, each of them may be either plus or minus, because the *square* in either case would be plus. Also, since $y = \pm\frac{b}{a}\sqrt{a^2 - x^2} = +\frac{b}{a}\sqrt{a^2 - x^2}$, or $-\frac{b}{a}\sqrt{a^2 - x^2}$, it is evident that for every value of x (which may be taken *any* quantity, *plus* or *minus*, numerically less than a) there will be two values of y, *numerically equal*, but *of contrary signs.*

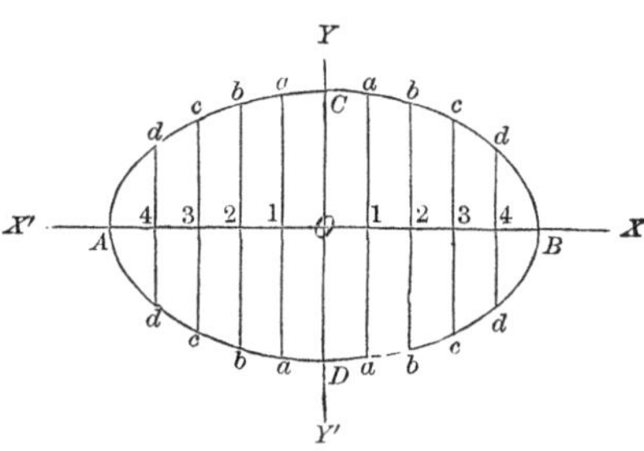

And since $x = \pm\frac{a}{b}\sqrt{b^2 - y^2} = +\frac{a}{b}\sqrt{b^2 - y^2}$, or $-\frac{a}{b}\sqrt{b^2 - y^2}$, every value of y (which may be taken *any* quantity, *plus* or *minus*, numerically less than b) will give two values of x, *numerically equal*, but *of contrary signs.* Hence each axis bisects every line drawn parallel to the other axis and having its extremities in the curve whose equation is $a^2y^2 + b^2x^2 = a^2b^2$.

Now, taking any numerical values for a and b, as $a = 5$, $b = 3$, with these values we have $y = \pm\frac{3}{5}\sqrt{25 - x^2}$. As x occurs in the equation only in the form of x^2, and as the square of $+x$ and of $-x$ is the same, $+x^2$, x may have either sign.

Taking x, therefore, successively ± 1, ± 2, ± 3, etc., we have (using the table of square roots),

If $x = \pm 1$, $y = \pm\frac{3}{5}\sqrt{24} = \pm 2.94$.
" $x = \pm 2$, $y = \pm\frac{3}{5}\sqrt{21} = \pm 2.75$.
" $x = \pm 3$, $y = \pm\frac{3}{5}\sqrt{16} = \pm 2.40$.
" $x = \pm 4$, $y = \pm\frac{3}{5}\sqrt{9} = \pm 1.80$.
" $x = \pm 5$, $y = 0$.

Now, on the axis AB, from O, lay the values taken for x, 1, 2, 3, and 4, *both ways*, to the points 1, 2, 3, and 4, and through these several points, parallel to CD, draw lines. Then lay the value of y, when $x = \pm 1$, from 1 on each side of O, *up* and *down*, to a and a, which will give four points in the ellipse. In like manner lay the values of y, when $x = \pm 2$, ± 3, ± 4, from the points 2, 3, and 4 respectively, on each side of O, *up* and *down*, to b and b, c and c, d and d; then, through these several points A, C, B, D, and a, b, c, d, beginning at A, and taking the points in regular succession, draw the curve, which will be the ellipse required.

NOTE.—By taking smaller values for x, as $\frac{1}{4}$, $\frac{1}{2}$, $\frac{3}{4}$, 1, $1\frac{1}{4}$, $1\frac{1}{2}$, $1\frac{3}{4}$, 2, $2\frac{1}{4}$, etc., the number of points may be increased at pleasure, and the points in the curve being then nearer together, they can be connected with greater correspondence with the curve. This *remark applies to all curves constructed by points from their equations.*

THE HYPERBOLA.

PROBLEM I.—To construct a hyperbola, having the axes given, and determine its equation when referred to its asymptotes. Given, the axes AB and CD.

Construction.—Draw AB equal to the *first axis*, which bisect in O, and through B, perpendicular to AB, draw a line, and on it lay BC and BD, each equal to half the *second axis*. Draw OC and OD, and produce them indefinitely, as to R and Q. Through B draw Ba parallel to OQ, and Ba' parallel to OR; then Oa, aC, aB, Oa', $a'D$, and $a'B$ will all be equal. Lay either of these equal distances on the line OCR, from C to b, c, d, etc., and on ODQ, from D to b', c', d', etc. Through the points C, b, c, d, etc., parallel to OQ, draw lines, and through the points D, b', c', d', etc., parallel to OR, draw lines. Then through the points A and a draw a line, to meet the parallel through C, in E. Through A and C draw a line to meet the parallel, through b, in F; and, in like manner, from A, *through each point* b, c, d, etc., and a', D, b', c', d', etc. *draw a line to meet the parallel at the next point*, in G, H, K, etc. and in E', F', G', H', K', etc.; and the curve passing through these points, B, E, F, G, H, etc., B', E', F', G', H', etc., will be a hyperbola, of which AB and CD are the *axes*, O the *centre*, and OR and OQ the *asymptotes*.

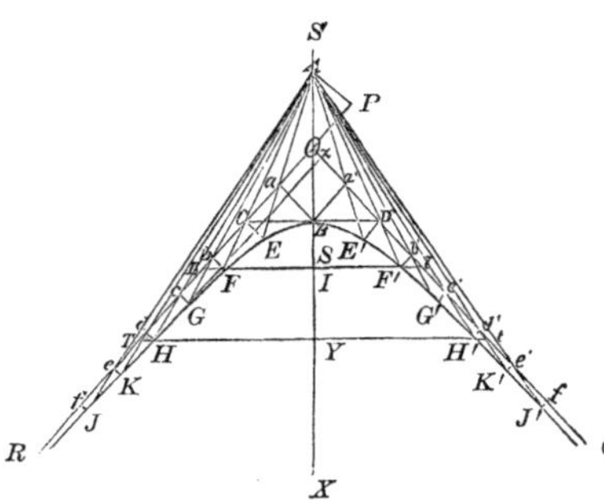

Demonstration.—Through A, parallel to OD, draw a line to meet CO, produced, in P. Then $OP = Oa = \frac{1}{2} OC$, and $AP = OP = aB = Oa'$. Also, $Oa^2 = \frac{1}{4} OC^2 = \frac{1}{4}(OB^2 + BC^2)$. Now, taking *any point*, as F, in the curve, and comparing the similar triangles FCb and ACP, we have $bF : AP :: bC : CP$. But, since $OP = Oa = bC$, we have $bO = CP$. Hence $bF : Oa' :: Oa : bO$; wherefore $bF \times bO = Oa' \times Oa = Oa^2 = \frac{1}{4}(OB^2 + BC^2)$, *which is the equation of the hyperbola referred to its asymptotes.* The same can, in like manner, be proved of *any other* point, as G, H, etc., G', H', etc. We, hence, always have $Oa^2 = Oa \times aB = OC \times CE = Ob \times bF = Oc \times cG = Od \times dH$, etc. $= OD \times DE' = Ob' \times b'F'$, etc. Q. E. D.

Corollary.—Since $Oa \times aB = OC \times CE = Ob \times bF = Oc \times cG$, and Ob is *greater* than OC, bF must be *less* than CE. For a like reason, since the factors on OG, as Oc, Od, etc., become *continually greater*, the other factors, cG, dH, eK, etc., must continually become less and less, so that the *curve approaches nearer and nearer* to the *asymptote.* Yet we see, by the mode of construction, the *curve can never meet the asymptote*, because the line drawn from A to a point in the asymptote *must cross the asymptote*, and be *continued to meet the parallel through the next point, before it can form a point in the curve.**

Scholium 1.—If the asymptotes OR and OQ be regarded as the axes of co-ordinates, and their intersection O the origin, and x represent the abscissa of any point in the curve, and y its corresponding ordinate, then of the point H, $x = Od$, $y = dH$; of the point G, $x = Oc$, $y = cG$; of the point E, $x = OC$, $y = CE$, etc.

Now, putting $OB = a$, and $BC = b$, we have $OC^2 = a^2 + b^2$, Oa^2 or $aB^2 = \frac{1}{4} OC^2 = \frac{1}{4}(a^2 + b^2)$, and $xy = Od \times dH = Oa \times aB = Oa^2 = \frac{1}{4}(a^2 + b^2)$. Hence $xy = \frac{1}{4}(a^2 + b^2)$ *is the equation of the hyperbola referred to its asymptotes as axes of co-ordinates.*

Scholium 2.—In triangle OBC, Case 2, find angle BOC. Twice angle $BOC = COD =$ angle $ROQ =$ angle made by the axes of co-ordinates OR and OQ. When the axes of the hyperbola are equal, we have $OB = BC$, the angles COB and BOD are each half a right angle, and ROQ a right angle. In this case the hyperbola

* It frequently perplexes the young student when he is told that one line can approach nearer and nearer to another, forever, without the possibility of ever arriving at it; but the *principle may be illustrated* by the familiar example of reducing $\frac{1}{3}$ to a decimal $= .33333$, etc., every additional figure bringing the value of the decimal nearer to $\frac{1}{3}$; but if the line of figures were continued around the earth, it could never become *equal* to $\frac{1}{3}$, although it is *getting nearer to that value by every figure that is added.*

is called *equilateral,* and the axes and all the co-ordinates are *rectangular.*

Scholium 3.—$OC = \sqrt{OB^2 + BC^2}$, $Oa = \frac{1}{2} OC = aB$, $OC = 2\,Oa$, $Ob = 3\,Oa$, $Oc = 4\,Oa$, $Od = 5\,Oa$, $Oe = 6\,Oa$, etc., all of which are hence known. Now, since $Oa^2 = OC \times CE$, we have $CE = \frac{Oa^2}{CE} = \frac{Oa^2}{2Oa} = \frac{1}{2}\,Oa$. Also, $Ob \times bF = Oa^2$; hence $bF = \frac{Oa^2}{3Oa} = \frac{1}{3}\,Oa$. In like manner, $cG = \frac{1}{4}\,Oa$, $dH = \frac{1}{5}\,Oa$, $eK = \frac{1}{6}\,Oa$. At the *millionth* point, the ordinate would be *one-millionth* of Oa; at the *billionth* point, *one-billionth* of Oa; and so on, the ordinate always having a value represented by figures, so that the *curve can never meet the asymptote, mathematically regarded.*

Problem 2.—Having given the equation of the hyperbola referred to its asymptotes, to pass to or determine its equation when referred to its axes or principal diameters.

Given, $Od \times dH = aB^2 = Oa^2 = (\frac{1}{4}\,OC^2) = \frac{1}{4}(OB^2 + BC^2)$, to prove that $AY \times YB : YH^2 :: OB^2 : BC^2 :: AB^2 : CD^2$.

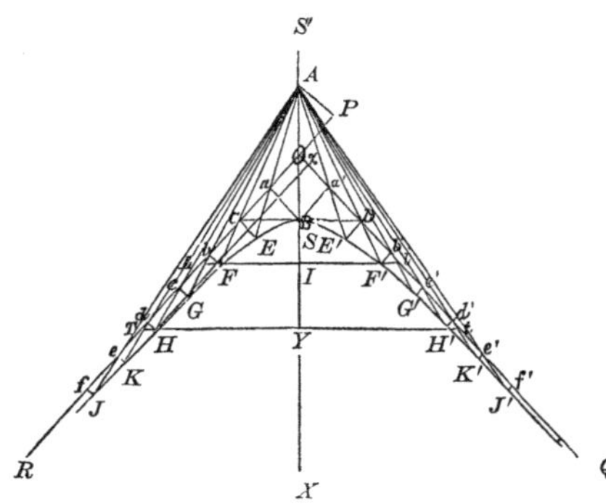

Let the hyperbola be constructed as in Problem I.; then AB is the *first axis,* CD the *second,* and O the *centre* of the hyperbola. Make OS and OS' each equal to OC; then S and S' will be the *foci.* Through any points in the hyperbola, as H and F, draw $THYH't$ and $LFIF'$ parallel to CD, and consequently perpendicular to the axis AX, t and l being the intersections of OQ, with TY and LI produced. Then HYH', FIF' are *double ordinates* to the points H and F respectively, HY and FI the ordinates, AY and BY the abscissas of the point H, and AI and BI the abscissas of the point F. By equal triangles TdH, $td'H'$, and ATY, AtY, we have $TH = tH'$, and $TY = tY$; hence $HY = H'Y$. By similar triangles THd, CBa, and ZHt, CBa, we have $TH : dH :: CB : Ba$, and $Ht : ZH\ (Od) :: CB : Ca\ (Ba)$. By multiplying the corresponding terms of these proportions, we obtain $TH \times Ht : Od \times dH :: CB^2 : Ba^2$ or Oa^2.

But $Od \times dH = Oa^2$; hence $CB^2 = TH \times Ht = (TY - YH) \times (TY + YH) = TY^2 - YH^2$.

Also, Theorem II., $OB^2 = OY^2 - AY \times YB$. Again, by similar triangles OBC, OYT, we have $OY^2 : TY^2 :: [OB^2 : BC^2] :: OY^2 - AY \times YB : TY^2 - YH^2$. By division, $OY^2 : AY \times YB :: TY^2 : YH^2$; or by inversion, $AY \times YB : YH^2 :: [OY^2 : TY^2] :: OB^2 : BC^2 :: AB^2 : CD^2$. Q. E. D.

Scholium 1.—Since $Oa \times aB = OC \times CE = Ob \times bF = Oc \times cG = Od \times dH$, etc., the parallelograms $OdHZ$, and all those under Oc, cG; Ob, bF; OC, CE, etc., are equal to one another, and each equal to the parallelogram $OaBa'$, which is a *property* of the *asymptotes* of the *hyperbola*.

Scholium 2.—It was shown that $TH \times Ht = BC^2$. In the same way it may be shown that $LF \times Fl = BC^2$, and so of *any other point* in the hyperbola. Hence these rectangles are all equal, and we have $TH : LF :: Fl : Ht$. But Ht is *greater* than Fl; hence TH is *less* than LF, and consequently the fact that the *curve continually approaches* the *asymptote* is shown by another method.

Scholium 3.—Putting $OB = a$, $BC = b$, and taking *any* point in the hyperbola, as H, whose ordinate is YH, and abscissas AY and YB, and putting the ordinate $YH = y$, and $OY = x$, then the abscissa $AY = x + a$, and $YB = x - a$, and $AY \times YB = (x + a) \times (x - a) = x^2 - a^2$; then the equation of the hyperbola, $AY \times YB : YH^2 :: OB^2 : BC^2$, becomes $x^2 - a^2 : y^2 :: a^2 : b^2$, whence $y^2 = \frac{b^2}{a^2}(x^2 - a^2)$, or $a^2y^2 - b^2x^2 = -a^2b^2$, which is the *analytical* or *algebraic equation of the hyperbola* referred to its axes.

PROBLEM III.—To construct the hyperbola from its equation referred to its axes, which, by last problem, is $a^2y^2 - b^2x^2 = -a^2b^2$, or $y^2 = \frac{b^2}{a^2}(x^2 - a^2)$.*

Analysis.—It is evident from the equation that if $y = 0$, $x = +a$ or $-a$, showing that the line represented by the equation intersects the axis of x at two points, distant a from the origin, one to the *right*, the other to the *left*.

Also, since when $x = 0$, $y = \sqrt{-b^2}$, an imaginary quantity, the line *cannot intersect the axis of* y.

Since x and y enter into the equation in the *second power* only, each of them may be either plus or minus, because the *square* in

* See remark to Example 11, Analytical Geometry.

either case would be plus. It is evident from the equation $y^2 = \frac{b^2}{a^2}(x^2 - a^2)$ that x may be taken *any* quantity not numerically less than a, plus or minus; and that for every value of x there will be two values of y, *numerically equal*, but of *contrary signs*. Also, for every value of y there will be two values of x, *numerically equal*, but of contrary signs. Hence each axis bisects every line drawn parallel to the other axis, having its extremities in the curve of the hyperbola, or the opposite hyperbolas.

In the expression $y = \sqrt{\frac{b^2}{a^2}(x^2 - a^2)}$, x cannot be less than a, otherwise we would have the square root of a minus quantity, which is impossible.

Example 1.—*Construction.*—Taking $a = \pm 5$, $b = \pm 3$, the equation becomes $25y^2 - 9x^2 = -225$, or $y = \pm\frac{3}{5}\sqrt{x^2 - 25}$.

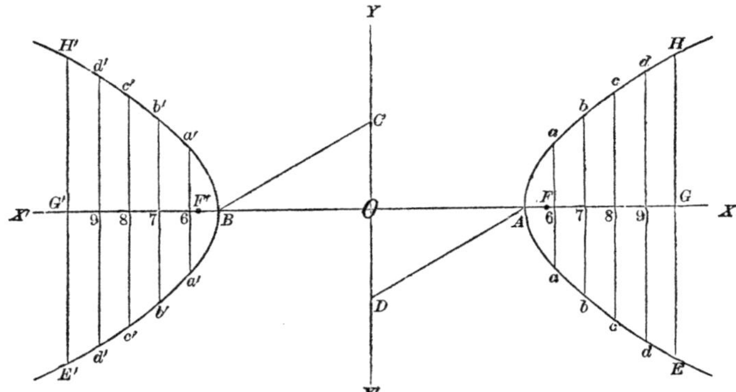

Draw the axes XX', YY', intersecting in O, and take OA, OB, each $= 5$, on opposite sides of O; then A and B will be the points in which the curve represented by the equation $25y^2 - 9y^2 = -225$ cuts the axis of x. Lay also $OC = +3$, and $OD = -3$. Now, as x cannot be less than 5, take $x = \pm 6$, $x = \pm 7$, $x = \pm 8$, $x = \pm 9$, etc., and lay these values of x *both ways*, from O, on the axis XX', to 6, 7, 8, 9, etc., and through these points, respectively, draw lines parallel to YY'. If $x = 6$, $y = \pm\frac{3}{5}\sqrt{11} = \pm 1.99$, which lay on the parallels through 6, up and down, to a, a and a', a', and these will be points in the curve.

If $x = 7$, $y = \pm\frac{3}{5}\sqrt{24} = \pm 2.94$, which lay on the parallels through 7, up and down, to b, b and b', b', and these will be other points in the curve.

If $x = 8$, $y = \pm\frac{3}{5}\sqrt{39} = \pm 3.71$; if $x = 9$, $y = \pm 4.49$; if $x = 10$, $y = \pm 5.20$, etc., which lay on the corresponding parallels through 8, 9, etc. to c, c, d, d, etc., and to c', c', d', d', etc., and then each of the curves passing through the points A, a, b, c, d, etc. both ways from A, and through B, a', b', c', d', etc. both ways from B, will be a hyperbola, and the two together are called *opposite hyperbolas.*

O is the *centre,* AB the *first axis,* CD the *second axis,* and A and B are called the *vertices* of the hyperbolas.

As equal ordinates, or values of y, correspond to equal abscissas, or values of x, the *opposite hyperbolas are similar figures.*

Join BC, and lay the distance BC, from O, on the axis of x, both ways, to F and F'; then these points are the *foci of the hyperbola.* Parallel to CD draw any double ordinate HGE, $H'G'E'$; then, of the point H, GH is the ordinate, and BG and AG the two corresponding abscissas; and of the point H', $G'H'$ is the ordinate, and AG', BG' the two corresponding abscissas.

Scholium.—Join AD, and it will be parallel to BC, because the alternate angles OBC and OAD are equal.

Example 2.—Determine the axis, and construct the hyperbola, whose equation is $y^2 - 3x^2 = -5$.

Multiply the given equation by $\frac{-5}{1 \times -3} = \frac{5}{3}$,* and it becomes $\frac{5}{3}y^2 - 5x^2 = \frac{5}{3} \times -5 = -\frac{25}{3}$. Then $a^2 = \frac{5}{3}$, $b^2 = -5$, and $a^2 \times b^2 = -\frac{25}{3}$, and the equation is of the form $a^2y^2 - b^2x^2 = -a^2b^2$, which is the equation of a hyperbola. We have for the semi-axes, $a = \sqrt{\frac{5}{3}} = \frac{1}{3}\sqrt{15} = \pm 1.29$, $b = \sqrt{5} = \pm 2.24$. We have $y^2 - 3x^2 = -5$; hence $y = \pm\sqrt{3x^2 - 5}$. Now, the *least value* that $3x^2$ can have is 5. Then $x^2 = \frac{5}{3} = \frac{15}{9}$, and $x = \sqrt{\frac{15}{9}} = \frac{1}{3}\sqrt{15} = \pm 1.29$, which lay on the axis of x from O to A and B, and these will be the vertices of the opposite hyperbolas. On the axis of y lay also OC and OD each $= \sqrt{5} = \pm 2.24$.

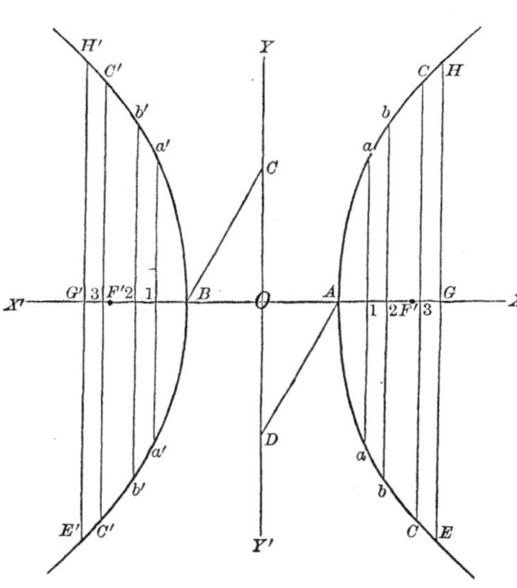

If $x = 1\frac{3}{4}$, $y = \pm 2.05$; if $x = 2$, $y = \pm 2.65$; if $x = 2\frac{1}{2}$, $y = \pm 3.71$. It is evident that x cannot be numerically less than $\sqrt{\frac{5}{3}} = \frac{1}{3}\sqrt{15} = 1.29$.

Lay $O1$, $O2$, $O3$, etc., $=$ the values of x, on XX', to the *right* of O, and also to the *left* of O, to 1, 2, 3, etc., and through the points 1, 2, 3, parallel to CD, draw lines, on which parallels lay the corresponding values of y to a, b, c, etc., and to a', b', c', etc., and these give points in the required curve, any number of which may be obtained by assuming different values for x. Each of the curves

* It will be seen that the multiplier $\frac{-5}{1 \times -3}$ is the absolute quantity -5 divided by the product of the co-efficients of y^2 and x^2, with their respective signs.

passing through the points A, a, b, c, etc., both ways from A, and through B, a', b', c', etc., both ways from B, will be a hyperbola, of which the equation is $y^2 - 3x^2 = -5$. The two together are called *opposite hyperbolas.*

Join AD and BC, and they will be parallel. Lay BC on the axis of x, from O, both ways, to F and F', which points will be the foci, AB is the first axis, and CD the second. If the double ordinates HGE and $H'G'E'$ be drawn, HG is the ordinate and AG and BG the abscissas of the point H; and $H'G'$ the ordinate and AG' and BG' the abscissas of the point H'.

Example 3.—Determine the axes, and construct the hyperbola, whose equation is $2y^2 - 4x^2 = 4$. Multiply the given equation by $\frac{4}{2 \times -4} = -\frac{1}{2}$,* and we get $-y^2 + 2x^2 = -2$, where $a^2 = -1$, $b^2 = 2$, $a^2b^2 = -2$, and the given equation is reduced to the form $a^2y^2 - b^2x^2 = -a^2b^2$, which is the equation of a hyperbola. The first axis $= 2\sqrt{-1}$, which shows that the curve cannot meet the axis of x. If $y = 0$, $x = \sqrt{-1}$; if $x = 0$, $y = \pm\sqrt{2} = \pm 1.41$. Lay OA, OB each equal to 1; then $AB =$ the *first* axis. Lay OC and CD each equal to $\sqrt{2} = 1.41$; then $CD =$ the *second* axis.

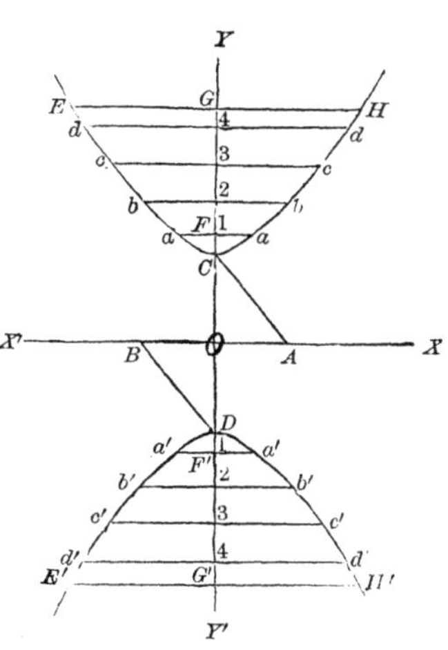

Now, $y^2 = 2x^2 + 2$; hence $y = \pm\sqrt{2x^2 + 2}$. The least value of y is when $x = 0$; then $y = \pm\sqrt{2} = 1.41 = OC$ or OD; $x^2 = \frac{y^2 - 2}{2}$, or $x = \pm\sqrt{\frac{y^2 - 2}{2}} = \frac{1}{2}\sqrt{2y^2 - 4}$.

If $y = 1\frac{1}{2}$, $x = \pm .35$; if $y = 2$, $x = \pm 1.00$; if $y = 2\frac{1}{2}$, $x = \pm 1.46$; if $y = 3$, $x = \pm 1.87$.

Lay the values of y from O, up and down, to 1, 2, 3, 4, etc., and through these points draw lines parallel to the axis XX', and on these parallels, from the points 1, 2, 3, 4, etc., lay the corresponding values of x to a, a; b, b; c, c; d, d, etc., and to a', a'; b', b'; c', c'; d', d', etc.; then each of the curves passing through the points C, a, b, c,

* See foot-note to last example.

d, etc., both ways from C, and through D, a', b', c', d', etc., both ways from D, will be a hyperbola, whose equation is $2y^2 - 4x^2 = 4$.

Join AC and DB, and they will be parallel. Lay AC on the axis of y, both ways from O, to F and F', which points will be the foci. C and D are the vertices of the opposite hyperbolas.

If parallel to XX' the double ordinates HGE and $H'G'E'$ be drawn, HG will be the ordinate and CG and DG the abscissas of the point H; and $H'G'$ the ordinate and CG' and DG' the abscissas of the point H'.

THE CONCHOID OF NICOMEDES.*

To construct the conchoid by points.

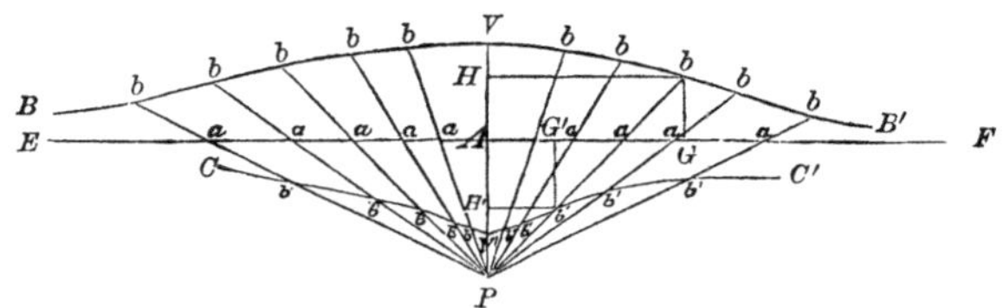

Construction.—Draw PAV, and EF at right angles to it at A. In PAV take *any* point P *below* EAF, and V *above* EAF. Draw *any number* of lines, as Pb, Pb, etc., on each side of PV, and make extensions *beyond* EF from P, as ab, ab, etc., *each equal* to AV, and the curve passing through the point V, and the several points b on each side of V, will be the *superior conchoid.*

Now, on the same lines drawn from P, lay the same *distance equal to* AV *below* EAF, as AV', ab', ab', etc., on each side of AP, and the curve passing through the point V', and the several points b' on each side of V', will be the *inferior conchoid.*

The point P is called the *pole;* the line EAF, the *directrix;* V, the *vertex* of the *superior* conchoid; and V', of the *inferior.*

Property 1.—The *directrix* EAF is an *asymptote to both curves.* For, let fall the perpendiculars bG, $b'G'$ on the directrix, and bH, $b'H'$ on PAV. Then the triangles baG, PaA, taking Pab the line from the remote extremity of which the perpendiculars bG and bH are let fall, are similar, and we have $Pa : PA :: ab\ (AV) : bG$. Hence

* For a treatise on the interesting subject of curves, see Prof. Leslie's *Geometry of Curve Lines*, or, the "Application of Algebra to the Doctrine of Curves," at the close of the second volume of Bonnycastle's *Algebra.*

$Pa \times bG = PA \times AV$, a *constant quantity.* Wherefore $Pa \times bG$ *must also be constant;* and, consequently, since the factor Pa *increases* as the point of the curve gets further from PAV, on either side, so the other factor bG *must decrease,* and the curve approach *nearer* and *nearer* to EF *without the possibility of ever meeting it.*

In like manner, using the similar triangles $b'aG'$, and the same triangle PaA, it may be shown that EAF is an *asymptote to the inferior conchoid.*

Property 2.—The equation of the *superior conchoid* BVB' is $x^4 + 2bx^3 + (b^2 - a^2 + y^2)\,x^2 - 2a^2bx = a^2b^2$.

The equation of the *inferior conchoid* $CV'C'$ is $x^4 - 2bx^3 + (b^2 - a^2 + y^2)\,x^2 + 2a^2bx = a^2b^2$, where x represents AH or bG, and y represents bH or AG in the superior conchoid, or their correspondents in the inferior.

Scholium.—When AV is *greater* than AP, V' will fall below P, and the curve passing through C, P, V', C', and the several points b', as thus formed, is called the *nodated conchoid,* and the part P, V', b', b', etc., is called a *node.* The student would find it interesting to construct the figure under this condition.

THE CISSOID OF DIOCLES.*

To construct the cissoid.

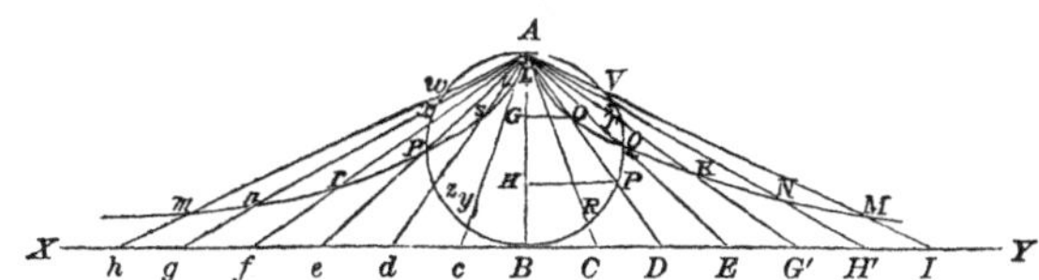

Construction.—Draw XY perpendicular to AB, the diameter of the given circle $AVPBp$, and from A to the line XY draw any number of lines AC, AD, AE, AG', etc., Ac, Ad, Ae, etc., on each side of AB, cutting the circumference of the circle in the points R, P, Q, etc., y, z, p, etc. Then lay the *chord* AR on CA from C to L; the *chord* AP on DA from D to O; the *chord* AT on $G'A$ from G' to K, and so proceed to lay the *intercepted chord* of every line *on that line* from its intersection with XY towards A, and the curve passing through A; and these several points L, O, Q, K, etc.

* See Bonnycastle's *Algebra*, vol. ii. p. 401, London edition, 1820.

thus given successively from A, both ways, will be the *cissoid of Diocles.*

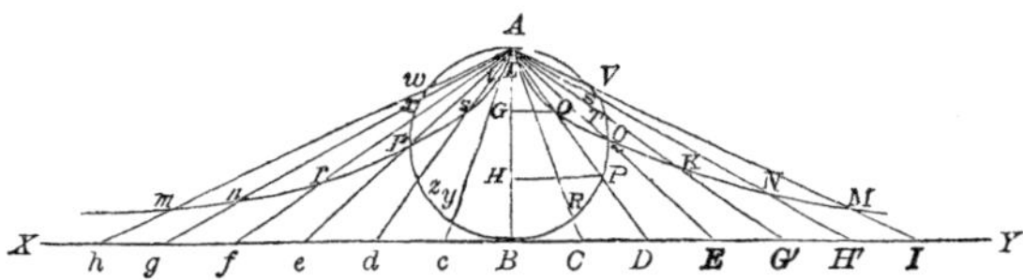

The circle $AVBp$ is called the *generating circle; AB* the *axis of the two branches,* which meet in a *cusp* at A, and pass through the *middle points* Q and p of the two semicircles, drawing continually *nearer* and *nearer* to the *directrix* XY as it extends farther from AB; and the directrix XY is hence *their common asymptote.*

Draw PH and OG parallel to XY; put $AB = a$, $AG = x$, $GO = y$; then the equation of the curve is $x^3 = (a - x)y^2$.

THE QUADRATRIX OF DINOSTRATUS.

Construction.—Let AVB be a semicircle of which the centre is C. At C, perpendicular to AB, draw a line, on which lay $CM =$ the diameter AB, and draw NML parallel to AB. Divide the quadrant BV, and the radius CV, into *any* but the *same* number of *equal* parts, say 9, in the points 1, 2, 3, etc., numbering from B on the quadrant and C on the radius, and continue the equal divisions, at pleasure, *beyond* V on the semicircle, and *above* V, on CM. Then from C to *any number* on the radius CV will be the *same part* of the radius CV, that from B to the *same number* on the quadrant BV is of the quadrant BV; that is, $C5 : B5 :: \text{rad. } CV : \text{quad. } BV ::$ diameter

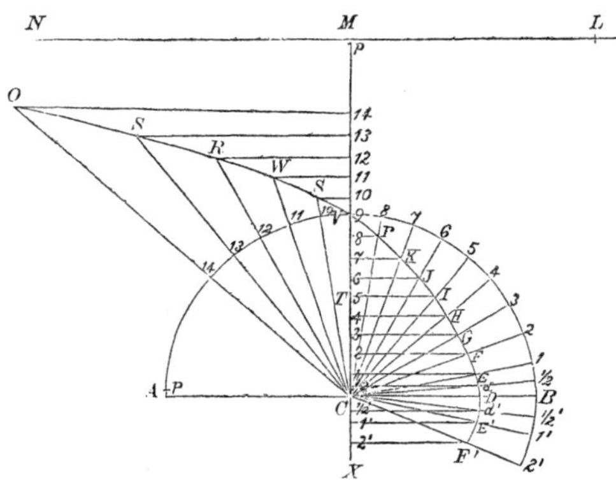

CM: semicircle BVA. Also, $C7 : B7$ :: rad. CV: quad. BV, and $C14 : B14 :: CM : BVA$. Join C with each of the points 1, 2, 3, etc. on the semicircle, and parallel to AB, through the point 1 on the radius, draw a line to meet the line $C1$ in E. Through the point 2 on the radius draw a line parallel to AB to meet the line $C2$ in F. So, through *each point* on CM, successively, parallel to AB, draw a line to meet the line drawn from C to the point of the *same number* on the semicircle, in the points G, H, I, J, V, R, S, O, etc., and the curve passing through these several points E, F, G, H, etc., taken in order, is called the *quadratrix*.

The circle AVB completed is called the *generating circle;* the line NML is called the *directrix*, which is also, as will be readily seen from the mode of construction, the *asymptote* to the quadratrix. For, suppose there be taken on MC a distance Mp a thousandth or a ten-thousandth part of the distance $C1$, and on the semicircle Ap a *like part* of $B1$, join Cp, and produce it to meet a line parallel to MLN, through p, on MC, which will give a point in the curve, and this point will become more and more remote from M as the parts Mp and Ap are made smaller, and the limit is the extension of the two parallels CA and MN, which will never meet. Therefore LMN, produced, is the asymptote of the curve.

Putting the radius $CB = a$, CD, the base of the quadratrix, $= b$, the arc $B5 = z$, and $CT = y$, the equation of the quadratrix is $ay = bz$.

NOTE 1.—This curve obtained much notoriety from the fact that if it were possible to form it accurately by a simple geometrical operation, it would enable mathematicians to determine the rectification and quadrature of the circle. It was from this property that the curve was called the *quadratrix*. It would also afford a means of dividing a given angle, or given arc, into any number of equal parts or in any given ratio.*

NOTE 2.—One principal difficulty in constructing the quadratrix is in finding the point D in CB so as to determine the *base* CD of the quadratrix. The most practical method I have been able to devise is to continue the semicircle *below* B, and also the line VC below AB. Then take $B\frac{1}{2}$ a half, a fourth, or as small a part as may be employed of $B1$, and lay it from B to $\frac{1}{2}'$ *below* B. On VC, produced to X, lay $C\frac{1}{2}$, a *like part* of $C1$, from C to $\frac{1}{2}'$. Join C and the points $\frac{1}{2}$ and $\frac{1}{2}'$ on the arc, and through the points $\frac{1}{2}$ and $\frac{1}{2}'$ on VX, parallel to CB, draw the lines from $\frac{1}{2}$ to d and from $\frac{1}{2}'$ to d', meeting the lines from C to $\frac{1}{2}$ and $\frac{1}{2}'$ in d and d', which will be points in the curve, and in the curve continued *below* CB, and hence the curve must pass through the point D, and thus give CD, the base of the quadratrix, as accurately as can be obtained by mechanical means.

* See Bonnycastle's *Algebra*, vol. ii. p. 403, London edition, 1820.

THE LOGARITHMIC CURVE.

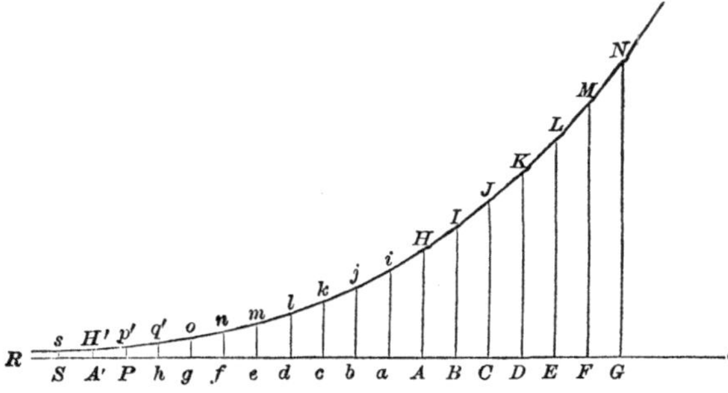

Construction. — Draw the line RQ, on which erect a perpendicular AH of *any length*. In RQ lay off from A, *both ways*, continued at pleasure, the abscissas AB, AC, AD, etc., Aa, Ab, Ac, etc., increasing in an *arithmetical* progression, having the intervening spaces AB, BC, CD, Aa, ab, bc, etc. all equal; and then, parallel to AH, through the points B, C, D, etc., a, b, c, d, etc., draw lines BI, CJ, DK, etc., ai, bj, ck, etc., whose lengths shall be in *geometrical progression,* increasing from AH towards Q by a *common multiplier* or ratio, and decreasing from A towards R by a *common divisor* of *the same value.* In the present figure AH was taken $=12$, the common difference $=5$, and the ratio $1\frac{1}{4}$. Then the curve passing through the upper extremities of all these lines N, M, L, K, etc. will be the *logarithmic curve.*

$BI = 12 \times 1\frac{1}{4} =$	$15,$	ai	$= AH \div 1\frac{1}{4} = 12 \div 1\frac{1}{4} =$	$9.6,$
$CJ = 15 \times 1\frac{1}{4} =$	$18.75,$	bj	$= 9.6 \div 1\frac{1}{4} =$	$7.68,$
$DK = 18.75 \times 1\frac{1}{4} =$	$23.4375,$	ck	$=$	$6.144,$
$EL =$	$29.2969,$	dl	$=$	$4.9152,$
$FM =$	$36.621,$	em	$=$	$3.93216,$
$GN =$	$47.776,$	fn	$=$	$3.1457.$

By laying these respective distances on the lines indicated, the points I, J, K, L, etc. in the curve are obtained, and the curve can readily be drawn through the upper extremities of all these lines, beginning at N, and taking M, L, K, J, I, H, i, j, k, etc. in succession.

Scholium 1.—It is evident that the curve approaches nearer and nearer to RP *without* the possibility of ever arriving at it; for, *however many times* we may divide the value of AH by $1\frac{1}{4}$, or, which is the same thing, multiply it by $\frac{4}{5}$, it *must still give some value to lay on a perpendicular above* RQ; hence the curve, though approaching nearer and nearer, *can never arrive at* RQ, and RQ is the *asymptote to the curve.*

Scholium 2.—Since the ordinates AH, BI, CJ, etc. are in geometrical progression, the square of any one is equal to the product of the two adjacent ones, or of any two lines equally distant from it. Thus, $AH^2 = BI \times ai = CJ \times bj = DK \times ck = EL \times dl$, etc.

Scholium 3.—Putting $r =$ ratio (in this case $1\frac{1}{4}$), and $AH = a$, we have $BI = a \cdot r$; $CJ = a \cdot r^2$; $DK = a \cdot r^3$; $EL = a \cdot r^4$, etc.; and $ai = \frac{a}{r}$; $bj = \frac{a}{r^2}$; $ck = \frac{a}{r^3}$; $dl = \frac{a}{r^4}$, etc. If now we commence the curve at H', where $A'H'$ will be *just equal to* r,* and then lay off the arithmetical ratios, or common differences, $A'P$, Ph, hg, gf, fe, etc., and erect perpendiculars at these successive points, we have $A'H' = r$; $Pp' = A'H' \times r = r^2$; $hq' = Pp' \times r = r^3$; $go = r^4$; $fn = r^5$; $em = r^6$, etc. Putting x to *represent any number* of these arithmetical differences from A' towards Q, calling A' one, P two, h three, etc., and putting y to represent the corresponding ordinate, we have $y = r^x$, which is the *equation of the logarithmic curve;* and from the form of its equation, the logarithmic curve is sometimes called the *exponential curve.*

Now, if we lay $A'S = A'P$, then the ordinate $Ss = \frac{A'H'}{r} = \frac{r}{r} = r^0 = 1$; and beginning at S, the ordinates will be $1, r, r^2, r^3, r^4, r^5$, etc. $= 1, 1\frac{1}{4}, (1\frac{1}{4})^2, (1\frac{1}{4})^3, (1\frac{1}{4})^4, (1\frac{1}{4})^5$, etc.; that is, $Ss = 1$, $A'H' = 1\frac{1}{4}$, $Pp' = (1\frac{1}{4})^2$, $hq' = (1\frac{1}{4})^3$, $go = (1\frac{1}{4})^4$, $fn = (1\frac{1}{4})^5$, etc.

* To find the distance AA' at which the ordinate $A'H'$ of the curve will be equal to the ratio r, having AH and the arithmetical and geometrical ratios given, let $x =$ the number of arithmetical differences in AA', then, by the hypothesis, we have $A'H' = \frac{a}{r^x} = r$. Hence $a = r^{(x+1)}$. Whence, taking the logarithm of each side of the equation, we have $\log. a = (x+1) \times \log. r$, or $x + 1 = \frac{\log. a}{\log. r}$. Taking, as in the example, $a = 12$, and $r = 1\frac{1}{4}$, we have $x + 1 = \frac{\log. 12}{\log. 1.25} = \frac{1.079181}{0.096910} = 11.136$. Hence $x = 11.136 - 1 = 10.136 =$ the number of arithmetical spaces between A and A'; and as each space in the figure is 5, we have $AA' = 10.136 \times 5 = 50.68$.

THE SPIRAL OF ARCHIMEDES, OR EQUABLE SPIRAL.

To construct the equable spiral.

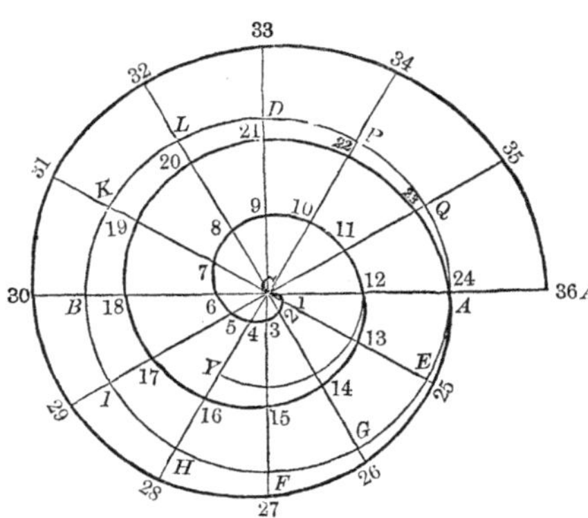

Construction. — Let the distance that the spiral gets from the centre C in one complete revolution be denoted by r; then with radius $CA =$ *any multiple* of r by a *whole number*, describe the circle $AFBD$, and draw the diameters ACB and DCF at right angles. Divide each quadrant into three equal parts, at the points E, G, H, I, K, L, P, Q. Through these points draw diameters, and produce them both ways from the centre. On CE, from C, lay $\frac{1}{12}$ of r to 1; on CG lay $\frac{2}{12}$ of r to 2; on CF lay $\frac{3}{12}$ of r to 3, etc., in order, around to 12 on CA, which will be $\frac{12}{12}r, = r$, and the curve passing through the several points 1, 2, 3, etc. will be *one revolution* of the *spiral of Archimedes.*

Example.—Let $r = 12$, and $CA = 2r = 24$. From C, on CE, lay 1; on CG, 2; on CF, 3, etc., when 12 will come on CA. Then, as before remarked, the curve passing through the points 1, 2, 3, etc. will be *one revolution* of the spiral. From the points 1, 2, 3, 4, 5, etc. lay on each of the radii a distance equal to r, around to A; then $CA = 2r = 24$. Through these points successively, beginning at 12, pass a curve through 13, 14, 15, etc., and it will give the *second revolution of the spiral.*

In like manner, from these several points, beginning at 13, lay on the several radii, produced if necessary, distances each equal to r, to 25, 26, 27, etc., and through these several points pass a curve, beginning at A, and it will give the *third revolution* of the *equable spiral.*

In the same way these successive revolutions may be continued at pleasure; the whole, whatever the number of revolutions, being called the spiral of Archimedes.

With the centre C, and radius $C12 = r$, describe any arc $12Y$, which put $= z$. Put $\pi =$ the whole circumference of that circle, and the corresponding ordinate $C4 = y$; then it is evident that $C4 : CY$

or $r::$ arc $12Y:$ the whole circumference. That is, $y:r::z:\pi$, whence $y=\frac{rz}{\pi}=$ the *equation of the spiral*

THE LEMNISCATE.

To construct the lemniscate from the equation $a^2y^2=a^2x^2-x^4$.

Analysis.—When $x=0, y=0$. Hence the curve passes through the origin O. If $y=0$, we have $x^2=a^2$ or $x=\pm a$, which shows that the curve cuts the axis of x in two points at the distance of a from the origin, one to the *right*, the other to the *left* of O. We have $y=\pm\frac{x}{a}\sqrt{a^2-x^2}$.

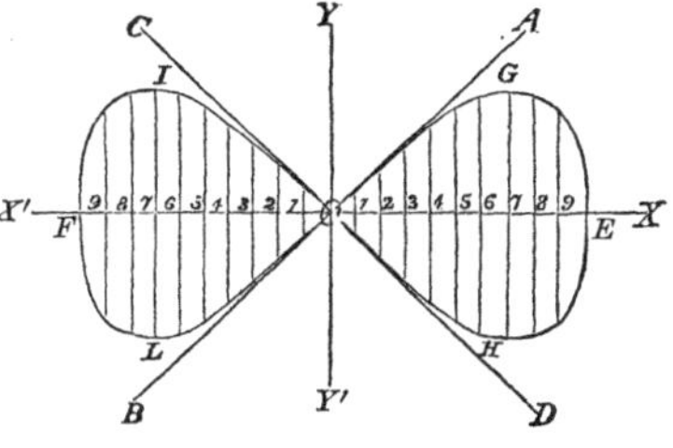

Taking $a=10$,

If $x=\pm1$, $y=\frac{1}{10}\sqrt{99}=\pm0.995$.
" $x=\pm2$, $y=\frac{2}{10}\sqrt{96}=\pm1.96$.
" $x=\pm3$, $y=\frac{3}{10}\sqrt{91}=\pm2.86$.
" $x=\pm4$, $y=$ ±3.666.
" $x=\pm5$, $y=$ ±4.33.
" $x=\pm6$, $y=$ ±4.8.
" $x=\pm7$, $y=$ ±4.999.
" $x=\pm8$, $y=$ ±4.8.
" $x=\pm9$, $y=$ ±3.92.
" $x=\pm10$, $y=$ 0.

Construction.—Draw the axes XX', YY' at right angles, intersecting each other at O, the origin. Lay OE, OF, each $=a=10$, and E and F will be the points in which the curve cuts the axis of x besides the origin.

On OX, OX' lay the successive values of x from O, both ways, to the points 1, 2, 3, 4, 5, etc., through which points, parallel to YY', draw lines, and on these parallel lines from where they intersect the axis XX', lay the corresponding values of y, respectively, both ways, *up* and *down*, and the curve passing through F, O, and E, and the remote extremities of these parallel lines in the order of their succession, will be the *lemniscate*, as in the figure. Commencing at O, the curve extends through O, G, E, H, O, I, F, L, and back again to O. AOB and COD are tangents to the curve.

NOTE.—A straight line, parallel to the axis of x, will evidently cut the curve in *four points*, x being in the fourth power in the equation. y being in only the second power, a line parallel to the axis of y cuts the curve in only two points.

In the ellipse and hyperbola, where both the co-ordinates x and y are in the second power in the equation, a straight line parallel to either axis will cut the curve in two points, regarding the complete hyperbola as consisting of the opposite hyperbolas.

In the parabola, whose equation is $y^2=ax$, a straight line parallel to the axis of y will cut the curve in two points, but parallel to the axis of x in but one, x being in the first power only. See these three curves in the first part of this section.

A TABLE OF SQUARE ROOTS OF NUMBERS FROM 1 TO 200, TO FACILITATE THE CONSTRUCTION OF CURVES BY POINTS FROM THE EQUATIONS.

NOS.	ROOTS.	NOS.	ROOTS.	NOS.	ROOTS.	NOS.	ROOTS.	NOS.	ROOTS.	NOS.	ROOTS.	NOS.	ROOTS.	NOS.	ROOTS.
1	1.	26	5.099	51	7.141	76	8.718	101	10.05	126	11.22	151	12.29	176	13.27
2	1.414	27	5.196	52	7.211	77	8.775	102	10.10	127	11.27	152	12.33	177	13.30
3	1.732	28	5.292	53	7.280	78	8.832	103	10.15	128	11 31	153	12.37	178	13.34
4	2.	29	5.385	54	7.348	79	8.888	104	10.20	129	11.36	154	12.41	179	13.38
5	2.236	30	5.477	55	7.416	80	8.944	105	10.25	130	11.40	155	12.45	180	13.42
6	2.449	31	5.568	56	7.483	81	9.	106	10.30	131	11.45	156	12.49	181	13.45
7	2.646	32	5.657	57	7.550	82	9.055	107	10.34	132	11.49	157	12.53	182	13.49
8	2.828	33	5.745	58	7.616	83	9.110	108	10.39	133	11.53	158	12.57	183	13.53
9	3.	34	5.831	59	7.681	84	9.165	109	10.44	134	11.58	159	12.61	184	13.56
10	3.162	35	5.916	60	7.746	85	9.220	110	10.49	135	11.62	160	12.65	185	13.60
11	3 317	36	6.	61	7.810	86	9.274	111	10.54	136	11.66	161	12.69	186	13.64
12	3.464	37	6.083	62	7.874	87	9.327	112	10.58	137	11.70	162	12.73	187	13.67
13	3.606	38	6.164	63	7.937	88	9.381	113	10.63	138	11.75	163	12.77	188	13.71
14	3.742	39	6.245	64	8.	89	9.434	114	10.68	139	11.79	164	12.81	189	13.75
15	3.873	40	6.325	65	8.062	90	9.487	115	10.72	140	11.83	165	12.85	190	13.78
16	4.	41	6.403	66	8.124	91	9.539	116	10.77	141	11.87	166	12.88	191	13.82
17	4.123	42	6.481	67	8.185	92	9.592	117	10.82	142	11.92	167	12.92	192	13.86
18	4.243	43	6.557	68	8.246	93	9.644	118	10.86	143	11.96	168	12.96	193	13.89
19	4.359	44	6.633	69	8.307	94	9.695	119	10.91	144	12.	169	13.	194	13.93
20	4.472	45	6.708	70	8.367	95	9.747	120	10.95	145	12.04	170	13.04	195	13.96
21	4.583	46	6.782	71	8.426	96	9.798	121	11.	146	12.08	171	13.08	196	14.
22	4.690	47	6.856	72	8.485	97	9.849	122	11.05	147	12.12	172	13.11	197	14.04
23	4.796	48	6.928	73	8.544	98	9.899	123	11.09	148	12.17	173	13.15	198	14.07
24	4.899	49	7.	74	8.602	99	9.950	124	11.14	149	12.21	174	13.19	199	14.11
25	5.	50	7.071	75	8.660	100	10.	125	11.18	150	12.25	175	13.23	200	14.14

THE END.

www.ingramcontent.com/pod-product-compliance
Lightning Source LLC
LaVergne TN
LVHW010229110826
845151LV00004B/1230